PROBLEMS IN THE GENERAL THEORY OF RELATIVITY AND THEORY OF GROUP REPRESENTATIONS

PROBLEMY OBSHCHEI TEORII OTNOSITEL'NOSTI I TEORIYA PREDSTAVLENII GRUPP

ПРОБЛЕМЫ ОБЩЕЙ ТЕОРИИ ОТНОСИТЕЛЬНОСТИ И ТЕОРИЯ ПРЕДСТАВЛЕНИЙ ГРУПП

The Lebedev Physics Institute Series

Editors: Academicians D. V. Skobel'tsyn and N. G. Basov

P. N. Lebedev Physics Institute, Academy of Sciences of the USSR

Recent Volumes in this Series

Volume 57	Theory of Interaction of Elementary Particles at High Energies
Volume 58	Investigations in Nonlinear Optics and Hyperacoustics
Volume 59	Luminescence and Nonlinear Optics
Volume 60	Spectroscopy of Laser Crystals with Ionic Structure
Volume 61	Theory of Plasmas
Volume 62	Methods in Stellar Atmosphere and Interplanetary Plasma Research
Volume 63	Nuclear Reactions and Interaction of Neutrons and Matter
Volume 64	Primary Cosmic Radiation
Volume 65	Stellarators
Volume 66	Theory of Collective Particle Acceleration and Relativistic Electron Beam Emission
Volume 67	Physical Investigations in Strong Magnetic Fields
Volume 68	Radiative Recombination in Semiconducting Crystals
Volume 69	Nuclear Reactions and Charged-Particle Accelerators
Volume 70	Group-Theoretical Methods in Physics
Volume 71	Photonuclear and Photomesic Processes
Volume 72	Physical Acoustics and Optics: Molecular Scattering of Light; Propagation of Hypersound; Metal Optics
Volume 73	Microwave–Plasma Interactions
Volume 74	Neutral Current Sheets in Plasmas
Volume 75	Optical Properties of Semiconductors
Volume 76	Lasers and Their Applications
Volume 77	Radio, Submillimeter, and X-Ray Telescopes
Volume 78	Research in Molecular Laser Plasmas
Volume 79	Luminescence Centers in Crystals
Volume 80	Synchrotron Radiation
Volume 81	Pulse Gas-Discharge Atomic and Molecular Lasers
Volume 82	Electronic Characteristics and Electron–Phonon Interaction in Superconducting Metals and Alloys
Volume 83	Theoretical Problems in the Spectroscopy and Gas Dynamics of Lasers
Volume 84	Temporal Characteristics of Laser Pulses and Interaction of Laser Radiation with Matter
Volume 85	High-Power Lasers and Laser Plasmas
Volume 86	Superconductivity
Volume 87	Coherent Cooperative Phenomena
Volume 88	Cosmic Rays in the Stratosphere and in Near Space
Volume 89	Electrical and Optical Properties of III–IV Semiconductors
Volume 90	The Kinetics of Simple Models in the Theory of Oscillations
Volume 91	Lasers and Their Applications in Physical Research
Volume 93	Techniques and Methods of Radio-Astronomic Reception
Volume 94	Pulsed Neutron Research
Volume 96	Problems in the General Theory of Relativity and Theory of Group Representations
Volume 97	Excitation and Domain Luminescence of Semiconductors
Volume 100	Research on the Condensation of Excitons in Germanium by the Microwave Method

In preparation:

Volume 95	Strong and Electromagnetic Interactions of Elementary Particles and Nuclei
Volume 98	Materials and Apparatus in Quantum Radiophysics
Volume 99	Stimulated Raman Scattering

Proceedings (Trudy) of the P. N. Lebedev Physics Institute

Volume 96

Problems in the General Theory of Relativity and Theory of Group Representations

Edited by
N. G. Basov
P. N. Lebedev Physics Institute
Academy of Sciences of the USSR
Moscow, USSR

Translated from Russian by
Alan Mason

CONSULTANTS BUREAU
NEW YORK AND LONDON

Library of Congress Cataloging in Publication Data

Main entry under title:

Problems in the general theory of relativity and theory of group representations.

(Proceedings (Trudy) of the P. N. Lebedev Physics Institute; v. 96)
Translation of Problemy obshcheĭ teorii otnositel'nosti i teoriiâ predstavleniĭ grupp.

1 General relativity (Physics) – Addresses, essays, lectures. 2. Representation of groups – Addresses, essays, lectures. I. Basov, Nikolaĭ Gennadievich, 1922- II. Series: Akademiiâ nauk SSSR. Fizicheskiĭ institut. Proceedings; v. 96.
QC1.A4114 vol. 96 [QC173.6] 530'.08s [530.1'1]
78-12612

ISBN 978-1-4684-0678-8 ISBN 978-1-4684-0676-4 (eBook)
DOI 10.1007/978-1-4684-0676-4

The original Russian text was published by Nauka Press in Moscow in 1977 for the Academy of Sciences of the USSR as Volume 96 of the Proceedings of the P.N. Lebedev Physics Institute. This translation is published under an agreement with the Copyright Agency of the USSR (VAAP).

Softcover reprint of the hardcover 1st edition 1979
A Division of Plenum Publishing Corporation
227 West 17th Street, New York, N.Y. 10011

PREFACE

This collection contains survey articles dealing with the following topics: The Mach principle and its role in the general theory of relativity, the modern conception of the vacuum, new methods in the theory of Lie group representations, the coherent state method and its application to physical problems, and the Newman–Penrose method and its application to problems in general relativity theory.

CONTENTS

ON THE PROBLEM OF SINGULARITY IN THE DE SITTER MODEL

V. K. Mal'tsev and M. A. Markov

. . . The de Sitter solution in no way corresponds to the case of a world without matter, but corresponds rather to a world in which all matter is concentrated on the surface $r = \pi/2$ which could probably be proved by taking the limit from a volume distribution of matter to a surface distribution [1]. — Einstein

If following Einstein the de Sitter metric is viewed as a limiting metric of a metric with a continuous matter distribution, then the components of the energy-momentum tensor of matter T_i^k become infinite on the "world boundary" as $r \to \pi R/2$. In a stationary reference frame $T_1^1 = T_2^2 = T_3^3 \to \infty$, $T_0^0 \to 0$; in falling frames, on the other hand, $T_0^0 \to \infty$ also.

1. At the present time it is customary to assume that the de Sitter model describes an empty space, i.e., that the condition

$$T_i^k = 0, \tag{1}$$

is satisfied at all points of space in this model, where T_i^k is the energy-momentum tensor of matter.

There is a well-known remark by Einstein concerning the de Sitter solution, however, which we have taken as epigraph to this article. According to Einstein's remark, the de Sitter model is not everywhere regular, since the matter in it must be contained on the horizon $\chi_0 = \pi/2\,(g_{00}(\chi_0) = 0)$. In order to prove this, Einstein proposed considering a "limit transition from a volume distribution of matter to a surface distribution," which apparently has not been done up to the present.

In this note we consider a realization of the limiting procedure proposed by Einstein.

2. By the de Sitter model, we understand a space described by the metric

$$\begin{gathered} ds^2 = \cos^2 \chi dt^2 - R^2 d\chi^2 - R^2 \sin^2 \chi\,(d\theta^2 + \sin^2 \theta d\beta^2), \\ 0 \leqslant \chi \leqslant \pi/2, \quad 0 \leqslant \theta \leqslant \pi, \quad 0 \leqslant \beta \leqslant 2\pi. \end{gathered} \tag{2}$$

A straightforward calculation of the components of the Einstein tensor

$$G_i^k = R_i^k - {}^1/_2\,\delta_i^k R_m^m$$

gives

$$G_i^k = \frac{3}{R^2}\,\delta_i^k, \tag{3}$$

which by the Einstein equations with Λ-term

$$G_i^k + \Lambda \delta_i^k = \varkappa T_i^k \tag{4}$$

corresponds to the absence of matter (1) for a suitable choice of Λ.

However, it is easy to see that on the horizon $\chi_0 = \pi/2$ itself a direct calculation is not possible (since expressions of the form 0/0 arise in the expression for G_i^k).

Thus Eqs. (3) and (1) can only be established for $\chi < \chi_0$, but not for $\chi = \chi_0$. In order to study the horizon itself, we proceed as follows.

Consider a model with a "shifted" horizon

$$ds^2 = \left(\frac{a + R\cos\chi}{a+R}\right)^2 dt^2 - R^2 d\chi^2 - R^2 \sin^2\chi\,(d\theta^2 + \sin^2\theta\, d\beta^2);$$

$$0 \leqslant \chi \leqslant \begin{cases} \pi, & a > R, \\ \arccos(-a/R), & a < R; \end{cases} \tag{5}$$

$$0 \leqslant \theta \leqslant \pi, \quad 0 \leqslant \beta \leqslant 2\pi; \quad a,\ R = \text{const},$$

so that now $\chi_0 = \arccos(-a/R) \neq \pi/2$.

The nonzero components of the Einstein tensor are here equal to

$$G_0^0 = 3/R^2, \quad G_1^1 = G_2^2 = G_3^3 = \frac{3}{R^2} - \frac{2}{R^2}\frac{a}{a + R\cos\chi}. \tag{6}$$

It is easy to see that the right-hand side of the Einstein equations (4) is different from zero in this case for any value of Λ (with the natural assumption that $|\Lambda| = \text{const} < \infty$), i.e., we are dealing with a certain volume distribution of matter in space, and, moreover, for $a < R$ there is a singularity on the horizon $\chi = \chi_0$ which does not depend on the value of a,

$$\frac{a}{a + R\cos\chi} \xrightarrow[\chi \to \chi_0]{} \frac{a}{a - a} = \infty \tag{7}$$

as is easily verified. We now fix R and let $a \to 0$. Then at all points of space Eq. (5) goes over into the de Sitter solution (2) and, for $\chi \neq \chi_0$, the components G_i^k in (6) go over into the corresponding de Sitter components G_i^k (3), as expected.

However, on the horizon $\chi = \chi_0$, passage to the limit as $a \to 0$ does not remove the divergence in (7), which corresponds [by Eqs. (4)] to the fact that the same singularity is also preserved in the components T_i^k, i.e., in the matter distribution.

Thus even if it is possible to eliminate T_i^k at all points $\chi \neq \chi_0$ by a suitable choice of Λ, for $\chi = \chi_0 = \pi/2$ the components $T_i^k \neq 0$ for every $|\Lambda| < \infty$. We have thus explicitly obtained a matter distribution on the surface $\chi = \pi/2$ in the de Sitter model.

This singularity is a real one which cannot be removed by coordinate transformations, for the singularity is possessed by the invariants $R_m^m = g_{ik}R^{ik}$ and $C = R_{iklm}R^{iklm}$:

$$R_m^m = -\frac{12}{R^2} + \frac{6}{R^2}\frac{a}{a + R\cos\chi}, \tag{8}$$

$$C = \frac{12}{R^4}\left[1 + \left(\frac{R\cos\chi}{a + R\cos\chi}\right)^2\right]. \tag{9}$$

At the same time, it should be emphasized that the components T_1^1, T_2^2, and T_3^3 do not vanish for $\Lambda = 3/R^2$, while T_0^0 becomes zero.

Thus passage to the limit to obtain the de Sitter metric does not cause the tensor T_i^k to vanish on the world boundary, in accordance with Einstein's proposal. Can a direct interpretation of this result be given, as Einstein would have wished, to the effect that the de Sitter metric is determined by matter distributed with infinite density on the boundary of this universe?

Speaking more carefully, it can now in any case be asserted that a localization of matter on the boundary of the de Sitter universe in the form $T_1^1 = T_2^2 = T_3^3 = \infty$, $T_0^0 = 0$ is consistent with the de Sitter metric.

To be sure, it is not clear if any physical meaning can be given to such a matter distribution when $T_1^1 = T_2^2 = T_3^3 = \infty$, and $T_0^0 = 0$.*

3. However, in the well-known interpretation it is also possible to speak of the total mass of a de Sitter universe.

In order to determine the value of the total mass of our de Sitter universe, it is advisable to study the continuation of this universe beyond the singular point χ_0 into empty space. We take the metric of this space in the form

$$ds^2 = (1 + \alpha/r + \beta r^2)\, dt^2 + g_{11} dr^2 - r^2 d\Omega^2; \tag{10}$$

here $r = R \sin\chi$, $d\Omega^2 = d\theta^2 + \sin^2\theta d\beta^2$.

This metric is patched together with metric (5) at each moment of time at the point χ_1:

$$\chi_1 = \chi_0 + \varepsilon, \quad \varepsilon \ll 1. \tag{11}$$

The condition that $\overset{\mathrm{I}}{g}_{00} = [(a + R\cos\chi)/(a + R)]^2$ and $\overset{\mathrm{II}}{g}_{00} = 1 + \alpha/r + \beta r^2$ and their first derivatives be continuous at the point $\chi = \chi_1$ $(r = r_1(\chi_1))$ gives

$$\alpha = -{}^2/_3 R \sin\chi_1 \left[1 - \left(\frac{a + R\cos\chi_1}{a + R}\right)^2 - \frac{(a + R\cos\chi_1)}{(a + R)^2}\, \frac{R\sin^2\chi_1}{\cos\chi_1}\right], \tag{12}$$

$$\beta = -\frac{1}{3R^2\sin^2\chi_1}\left[1 - \left(\frac{a + R\cos\chi_1}{a + R}\right)^2 + 2\,\frac{(a + R\cos\chi_1)}{(a + R)^2}\, \frac{R\sin^2\chi_1}{\cos\chi_1}\right]. \tag{13}$$

Letting $\varepsilon \to 0$, we now make the transition to the singular point χ_0, and then

$$\alpha \to -{}^2/_3 R \sin\chi_0, \tag{14}$$

$$\beta \to -(3R^2 \sin^2\chi_0)^{-1}. \tag{15}$$

Passing now to de Sitter space $(a \to 0)$, we obtain

$$\alpha = -{}^2/_3 R, \tag{16}$$

$$\beta = -1/(3R^2). \tag{17}$$

In other words, to an external observer the Λ-term is equal to $-1/R^2$, and the total mass of the de Sitter universe is equal to

$$M = c^2 R/(3k) \tag{18}$$

(k is Newton's gravitational constant).

---

* However, this case apparently does not lead, for example, to an infinite speed of propagation of signals in matter, regardless of the form of the equation of state of matter, which is unknown to us: for $\chi = \chi_0$ the linear element becomes spacelike and the concept of the speed of propagation of a signal on the horizon in general becomes meaningless.

In this model, the Λ-term turns out to be related to the mass M by the expression

$$-\Lambda = c^4/(3k^2M^2). \tag{19}$$

4. The following natural question arises: To what extent does the nonuniqueness related to the possible ways of "shifting" the horizon affect the physical results?

It turns out that under rather natural assumptions concerning the nature of the "shift" there is in general no physical indeterminacy. Consider a model in which the line element has the form

$$ds^2 = \varphi^2 dt^2 - R^2 d\chi^2 - R^2 \sin^2\chi\,(d\theta^2 + \sin^2\theta d\beta^2), \tag{20}$$

where $\varphi = \varphi(\chi, \alpha)$, so that for some value of the parameter α (e.g., for $\alpha = 0$) we have a transition to the de Sitter model, i.e.,

$$\varphi(\chi, \alpha) \xrightarrow[\alpha\to 0]{} \cos\chi, \tag{21}$$

and moreover

$$\varphi(\chi_0, \alpha) = 0; \qquad \chi_0(\alpha \neq 0) \neq \pi/2; \qquad \chi_0 \xrightarrow[\alpha\to 0]{} \pi/2. \tag{22}$$

A direct calculation in terms of the given metric shows easily that if the expansion of $\varphi(\chi, \alpha)$ for small α near χ_0 has the form

$$\varphi(\chi) = \cos\chi + a_0 + \sum_{n=3}^{\infty} a_n(\chi - \chi_0)^n, \tag{23}$$

then the components G_i^k as $\alpha \to 0$ do not depend on a_n and coincide with the components G_i^k in (6).

The fact that terms of the order $(\chi - \chi_0)$ and $(\chi - \chi_0)^2$ are not present in the sum over n simply indicates the requirement that regardless of the concrete form of the function $\varphi(\chi, \alpha)$, the function itself and its first and second derivatives on the horizon $\chi = \chi_0$ should coincide with the corresponding values of the unperturbed de Sitter universe. But since the curvature is defined by precisely these quantities, such a requirement is physically justified.

5. The result obtained above is determined by the fact that the de Sitter metric in the form (2) has a singularity at $\chi = \pi/2$. This singularity, in contrast to the singularity in the Schwarzschild metric, cannot be removed by a coordinate transformation, as was already remarked by Einstein himself [1].

Indeed, in the metric (2) the determinant

$$-g = R^6 \cos^2\chi \sin^4\chi \sin^2\theta \tag{24}$$

has a nonremovable singularity [1] for $\chi = \pi/2$.

Solutions of the homogeneous Einstein equations with Λ-terms are known, however, which do not have the property that g_{00} vanishes in some region [2]. Nevertheless, it is easy to see first of all that these models are not equivalent to the de Sitter model we consider, and second, an independent analysis of them as limiting metrics leads to the same results.

Indeed, as is well known, we have

$$\sqrt{-g(x')} = \frac{\partial(x)}{\partial(x')}\sqrt{-g(x)}, \tag{25}$$

under a transformation of coordinates, where $\partial(x)/\partial(x')$ is the Jacobian of the transformation; hence if $(-g(x_0))^{1/2} = 0$ while $(-g[x'(x_0)])^{1/2} \neq 0$, this implies that the Jacobian is singular (in the corresponding domain), so that it is not possible in general to assert that two spaces whose points are related by such a transformation are equivalent.

However, such spaces can of course be studied independently (some of these spaces are cited in [2]). In doing this, it turns out that even though the boundaries of these spaces are not equivalent to the boundary of the de Sitter model (2), it is not possible in any of the spaces cited in [2] to avoid a singularity on some part of the boundary.

We first consider one of these spaces:

$$ds^2 = a^2dt^2 - a^2e^{2t}\,[dr^2 + r^2\,(d\theta^2 + \sin^2\theta d\beta^2)]. \tag{26}$$

We remark that the points $t = \pm\infty$ cannot be removed from consideration: Indeed, one point or the other of these points, say P, must be taken to belong to the space if the invariant distance from it to any other point P_0 of the space (say, to the coordinate origin) is to be finite [1]:

$$\int_{P_0}^{P} ds < \infty. \tag{27}$$

This integral is finite for the entire boundary of the de Sitter model and is an invariant.

At the same time it is easy to see that when $t = \pm\infty$, the determinant $-g$ for the model (26) becomes singular and we run up against the same situation encountered in the case of the de Sitter model (2).

We therefore proceed in analogous fashion. Put

$$ds^2 = a^2dt^2 - a^2\,(e^t - \lambda)^2\,[dr^2 + r^2\,(d\theta^2 + \sin^2\theta d\beta^2], \quad \ln\lambda \leqslant t < \infty. \tag{28}$$

The nonvanishing components of the Einstein tensor are here equal to

$$G_0^0 = \frac{3}{a^2}\frac{e^{2t}}{(e^t - \lambda)^2}, \qquad G_1^1 = G_2^2 = G_3^3 = \frac{3}{a^2}\frac{e^{2t}}{(e^t - \lambda)^2} - \frac{2}{a^2}\frac{\lambda e^{2t}}{(e^t - \lambda)^2}, \tag{29}$$

so that as $\lambda \to 0$ (28) goes everywhere into (26), and for $t \neq t_0 = \ln\lambda$ (29) goes over into the de Sitter G_i^k, while on the horizon $t_0 = \ln\lambda$ we have the characteristic nonvanishing singularity

$$[\lambda/(\lambda - \lambda)]^2. \tag{30}$$

We remark that there is no creation or disappearance of matter in time here: The situation is simply that in the given coordinates, a signal from the boundary where matter is located progagates for an infinitely long time, and hence at finite times t an observer cannot describe the boundary in his own coordinates. Everywhere except on the horizon, the metric (26) is obtained from the de Sitter metric by going to a moving reference frame, while at the same time motion on the boundary is generally impossible, although it can be said formally that the observer began to fall from the boundary in the infinitely distant past and will also return to the boundary in the infinitely distant future (in contrast, for example to the coordinate Schwarzschild singularity, which is reached after a finite proper time). It is this motion which is associated with the circumstance that T_0^0 and the spatial components T_α^β are mixed up together, so that the singularity appears also in T_0^0.

We next consider the metric defined by the element

$$ds^2 = a^2dt^2 - a^2\mathrm{ch}^2t\,[d\chi^2 + \sin^2\chi\,(d\theta^2 + \sin^2\theta d\beta^2)]. \quad (31)^*$$

The horizon $\chi = \pi/2$ of the ordinary de Sitter metric corresponds here to the surface $\mathrm{ch}\,t\cdot\sin\chi = 1$ on which the line element (31) degenerates in exactly the same way as the usual line element for $\chi = \pi/2$:

$$ds^2 = -a^2\,(d\theta^2 + \sin^2\theta d\beta^2). \quad (32)$$

At the same time, when $t \to \pm\infty$ the line element not only becomes degenerate, but individual components of the metric (g_{11}) vanish; the surface $t = \pm\infty$ intersects the singular surface $\mathrm{ch}\,t\cdot\sin\chi = 1$ for $\chi = \pm\pi$.

We consider the metric (31) in more detail. To do this we put

$$ds^2 = a^2\,dt^2 - \frac{a^2}{\lambda - \mathrm{th}^2\,t}[d\chi^2 + \sin^2\chi\,(d\theta^2 + \sin^2\theta\,d\beta^2)], \quad (33)$$

or

$$ds^2 = a^2\,dt^2 - \frac{a^2\,\mathrm{ch}^2\,t}{\lambda\,\mathrm{ch}^2\,t - \mathrm{sh}^2\,t}[d\chi^2 + \sin^2\chi\,(d\theta^2 + \sin^2\theta\,d\beta^2)], \quad (34)$$

so that for $\mathrm{th}^2\,t < \lambda$ we have

$$ds^2 \xrightarrow[\lambda\to 1]{} a^2\,dt^2 - a^2\,\mathrm{ch}^2\,t\,[d\chi^2 + \sin^2\chi\,(d\theta^2 + \sin^2\theta d\beta^2)], \quad (35)$$

while the spatial metric becomes degenerate for

$$t_0 = \mathrm{th}^{-1}\sqrt{\lambda} < \infty, \quad \text{and} \quad t_0 \xrightarrow[\lambda\to 1]{} \infty. \quad (36)$$

The components G_i^k of the Einstein tensor are here equal to

$$G_0^0 = \frac{3}{a^2}\left[\lambda - \mathrm{th}^2\,t + \frac{\mathrm{th}^2\,t}{(\lambda\,\mathrm{ch}^2\,t - \mathrm{sh}^2\,t)^2}\right], \quad (37)$$

$$G_1^1 = G_2^2 = G_3^3 = \frac{1}{a^2}\left[\lambda - \mathrm{th}^2\,t + \frac{2\lambda + \mathrm{th}^2\,t + 4\,(1-\lambda)\,\mathrm{sh}^2\,t}{(\lambda\,\mathrm{ch}^2\,t - \mathrm{sh}^2\,t)^2}\right], \quad (38)$$

so that for $t < t_0$

$$G_0^0 \xrightarrow[\lambda\to 1]{} 3/a^2, \quad (39)$$

$$G_1^1 = G_2^2 = G_3^3 \xrightarrow[\lambda\to 1]{} 3/a^2; \quad (40)$$

while at the same time when $t = t_0$

$$G_0^0 = \frac{3}{a^2}\,\frac{\lambda\,(1-\lambda)^2}{(\lambda-\lambda)^2} = \infty, \quad (41)$$

$$G_1^1 = G_2^2 = G_3^3 = \frac{7}{a^2}\,\frac{\lambda\,(1-\lambda)^2}{(\lambda-\lambda)^2} = \infty \quad (42)$$

---

* Here and elsewhere ch stands for cosh, sh for sinh, and th for tanh.

independently of λ, so that this singularity is also preserved as $\lambda \to 1$. The same singularity also appears in the scalar curvature R_m^m:

$$-R_m^m = \frac{6}{a^2}\left[\lambda - \mathrm{th}^2 t + \frac{\lambda + \mathrm{th}^2 t + 2(1-\lambda)\,\mathrm{sh}^2 t}{(\lambda\,\mathrm{ch}^2 t - \mathrm{sh}^2 t)^2}\right] \xrightarrow[t \to t_0]{} \frac{24}{a^2}\frac{\lambda(1-\lambda)^2}{(\lambda-\lambda)^2} = \infty. \quad (43)$$

As for the condition that the integral (27) over all of four-dimensional space be finite, this expression for the form (2) is given in the same paper of Einstein [1]: "The points of the surface $r = (\pi/2)R$ must evidently be considered as points lying in a bounded domain if we choose P_0 to be the point $r = t = 0$, since the integral $\int_0^{\pi R/2} dr$, taken for constant ψ, θ, and t is finite."

Consider the corresponding integral for the metrics (26) and (31). In the case of (31) the path of integration corresponding to the Einstein path is [2]

$$y \cos\chi = \mathrm{th}\, t, \quad (44)$$

where $y = \mathrm{th}\,\tau$ (τ is the original time (2), τ = const); moreover, the line element is

$$ds^2 = a^2 \frac{y^2 - 1}{(1 - y^2\cos^2\chi)^2} d\chi^2 \leqslant 0, \quad (45)$$

$$\int \sqrt{-ds^2} = a\sqrt{1-y^2} \int \frac{d\chi}{1 - y^2\cos^2\chi} = \begin{cases} 0, & |\tau| = \infty,\ |y| = 1, \\ a\pi/2, & \tau = 0,\ y = 0 \end{cases} \quad (46)$$

(the upper limit is defined by $(1 - y^2)\cos^2\chi = 0$, the lower limit by $\mathrm{ch}\,t \cdot \sin\chi = 0$).

In the case of (26) the path of integration is [2]

$$e^{-2t} = r^2 + (1 - y)/(1 + y), \quad (47)$$

and the line element is

$$ds^2 = a^2 \frac{y-1}{y+1} \frac{dr^2}{[r^2 + (1-y)/(1+y)]^2} \leqslant 0, \quad (48)$$

$$\int \sqrt{-ds^2} = \begin{cases} 0, & |\tau| = \infty,\ |y| = 1, \\ a\dfrac{\pi}{2}, & |\tau| < \infty,\ |y| < 1 \end{cases} \quad (49)$$

(the upper limit is taken from $r^2 + (1 - y)/(1 + y) = r^2$, the lower limit from $e^t r = 0$).

6. The preceding considerations bring to mind the fact that the equation

$$R_{ik} = \Lambda g_{ik} \quad (50)$$

cannot in general have regular nontrivial solutions. This conclusion could have been supported by the following argument.

As is well known, the Einstein equations are obtained from the following variational principle:

$$\delta \int (R - 2\Lambda)\sqrt{-g}\, d\Omega + \delta S_m = 0, \quad (51)$$

where S_m is the action of matter.

Assume that (50) has a nontrivial solution; then by the fact that

$$\delta S_m \sim \int T_{ik}\delta g^{ik}\sqrt{-g}\, d\Omega, \quad (52)$$

we have from (51) that

$$\delta \int (R - 2\Lambda) \sqrt{-g} d\Omega = 0. \tag{53}$$

Regarding (50) as a condition imposed on (53) (we recall that the curvature is introduced independently of the metric; relation (50) is holonomic), we obtain

$$\int [g^{ik}\delta R_{ik} + (R_{ik} - {}^1/_2 g_{ik}R + \Lambda g_{ik})\delta g^{ik}] \sqrt{-g}\, d\Omega = \int [\Lambda g^{ik}\delta g_{ik} + \Lambda g_{ik}\delta g^{ik} + (R_{ik} - {}^1/_2 g_{ik}R)\,\delta g^{ik}] \sqrt{-g}\, d\Omega = 0. \tag{54}$$

The first two terms in the integral cancel one another, and we have finally

$$\int (R_{ik} - {}^1/_2 g_{ik}R)\,\delta g^{ik} \sqrt{-g}\, d\Omega = 0. \tag{55}$$

If we assume that the Einstein tensor in the integrand of (55) is continuous, we have

$$R_{ik} - {}^1/_2 g_{ik}R = 0, \tag{56}$$

which is consistent with (50) only in the case

$$g_{ik} = 0, \tag{57}$$

i.e., in the case of a trivial solution.

If on the other hand we do not assume continuity of the Einstein tensor, (56) is incorrect and there are two possibilities: 1) either the metric g_{ik} is discontinuous and it is necessary to introduce a term on the right-hand side of the Einstein equations compensating for the discontinuities of R_{ik}, which means that matter is localized in certain regions of the model; or else 2) the homogeneous Einstein equations (50) are valid in a nontrivial way, but then one must assume corresponding discontinuities of the metric g_{ik}, which is scarcely admissible from a physical point of view.

7. Thus the assertion that the de Sitter metric describes a space everywhere devoid of matter ($T_i^k = 0$) can be disputed if one adopts the Einstein point of view, according to which the de Sitter metric must be regarded as the limiting metric of some metric with a distribution of matter.

If for some reason such a limiting distribution of matter is not physically admissible, then the de Sitter solution is thereby physically inadmissible also, and must from this point of view be excluded from discussion in the general theory of relativity.

The authors express their gratitude to I. D. Novikov for a discussion of this problem and in particular for pointing out the paper [2] and the metric in the form (31), in which the anti-Einstein point of view is expressed most clearly, since in the metric (31) the spatial universe is closed and the T_i^k vanish everywhere for finite t and in the passage to the limit.

LITERATURE CITED

1. A. Einstein, Sitzungsber. Kgl. Preuss. Akad. Wiss., 1:270 (1918); H. Weyl, Phys. Z., 20:31 (1919); F. Klein, Nachr. Acad. Wiss., Göttingen, Math.-Phys. Kl., 394 (1918).
2. A. L. Zel'manov, Dokl. Akad. Nauk SSSR, 124:1030 (1959).

ON THE INTEGRAL FORMULATION OF THE MACH PRINCIPLE IN A CONFORMALLY FLAT SPACE

V. K. Mal'tsev and M. A. Markov

The well-known integral formulation of Mach's principle, $g_{ik}(x) = \frac{8\pi k}{c^4} \int G_{ik}^{\alpha\beta}(x, y) T_{\alpha\beta}(y) \sqrt{-g(y)}\, d^4y + \Lambda_{ik}(x)$ with the requirement $\Lambda_{ik}(x) = 0$ constitutes a rather complicated mathematical formalism in which many aspects of the physical content of the theory are not evident. In this article an attempt is made to consider the integral formulation for the simplest case of conformally flat spaces: $ds^2 = \varphi^2(x)(c^2dt^2 - dx^2 - dy^2 - dz^2)$. The presence of only one scalar function $\varphi(x)$ in this formalism makes it possible to analyze in a more detailed way the highly unusual physical features of this formulation of the Mach principle: the absence of asymptotically flat spaces, the compatibility of the condition $\Lambda_{ik} = 0$ with the existence of local free radiation, problems of inertia and gravitation, restrictions on the equations of state of matter, etc.

One of the possible means for formally realizing the Mach principle is by the method of integral equations in the general theory of relativity (GTR). This method is based on a well-known assertion of Einstein to the effect that a gravitational field must be completely determined by the energy-momentum tensor of matter [1], and it consists in the requirement that the metric satisfy the integral equations [2-6]

$$g_{ik}(x) = \frac{8\pi k}{c^4} \int G_{ik}^{\alpha\beta}(x, y)\, T_{\alpha\beta}(y) \sqrt{-g(y)}\, d^4y, \tag{1}$$

where $G_{ik}^{\alpha\beta}(x, y)$ is some bitensor (the indices α, β refer to the point y, and i, k to the point x) [7]; the tensor $T_{\alpha\beta}$ describes matter and is constructed from the energy-momentum tensor of matter. The tensor $G_{ik}^{\alpha\beta}(x, y)$ is an analog of the Green functions for well-known equations.

That condition (1) is nontrivial follows from the fact that the Einstein equations are in general equivalent to the following integral equations:

$$g_{ik}(x) = \frac{8\pi k}{c^4} \int G_{ik}^{\alpha\beta}(x, y)\, T_{\alpha\beta}(y) \sqrt{-g(y)}\, d^4y + \Lambda_{ik}(x), \tag{2}$$

where the term Λ_{ik} describes the metric in the absence of matter ($T_{\alpha\beta} \equiv 0$), and as follows from the Einstein equations, it is nonzero in the general case.

It should be stressed that in view of the nonlinearity of the Einstein equations, the quantities $G_{ik}^{\alpha\beta}(x, y)$ and $\Lambda_{ik}(x)$ can be extremely complicated functionals of the metric g_{ik}:

$$\begin{aligned} G_{ik}^{\alpha\beta}(x, y) &= G_{ik}^{\alpha\beta}(x, y \mid g_{mn}, g_{\mu\nu}), \\ \Lambda_{ik}(x) &= \Lambda_{ik}(x \mid g_{mn}, g_{\mu\nu}), \end{aligned} \tag{3}$$

so that (2) is a nonlinear integral equation.

As will be shown below, $\Lambda_{ik}(x)$ can be represented as an integral over a surface containing all of a given cosmological model. Thus, the condition

$$\Lambda_{ik}(x) = 0 \tag{4}$$

is a boundary condition imposed on the solutions of the Einstein equations, and in view of the tensor nature of $\Lambda_{ik}(x)$, it is a general covariant.

The main difficulty associated with the analysis of the theory in the form (1) consists in the nonuniqueness of choosing the differential operator of the equation for the Green function $G^{\alpha\beta}_{ik}$:

$$D^{mn}_{ik}(x)\, G^{\alpha\beta}_{mn}(x, y) = \delta^{(\alpha}_{(i}\delta^{\beta)}_{k)} \frac{\delta^4(x-y)}{\sqrt{-g(x)}}, \tag{5}$$

and hence in the nonuniqueness of the choice of the Green function itself. Unfortunately, the variants of this theory which have been proposed [2-5] reflect only isolated aspects of the problem.

This nonuniqueness is associated with the fact that the only requirement on the operator D known at present which is universally accepted,

$$D^{mn}_{ik} g_{mn} = R_{ik} - {}^1/_2 g_{ik} R \tag{6}$$

or

$$D^{mn}_{ik} g_{mn} = R_{ik}, \tag{6'}$$

is too weak.

Naturally, this nonuniqueness also carries over to the free term Λ_{ik}, with the result that the same cosmological model treated by various authors turns out sometimes to be consistent with the Mach principle and sometimes not [2, 6].

Because Eqs. (1), (5) are very complex (apparently even more complicated than the Einstein equations themselves), the analysis of their physical consequences is made extremely difficult. It is not clear to what degree this formalism is adequate for what is understood by the Mach principle.

It is clear only that in this formulation, if matter is absent ($T_{ik} \equiv 0$), then space is also absent ($g_{ik} \equiv 0$) (strictly speaking, this formalism is constructed so that just this requirement should hold). It is also clear that the asymptotically flat solutions (of the type of the Schwarzschild solution) do not satisfy the requirements of this formalism [2, 6]. But in that setup it is difficult to obtain and analyze other consequences of the Mach principle which are associated with the formulation given here.

It is therefore of interest to apply the procedure discussed above to spaces of some simpler, special type, e.g., conformally flat spaces, and to analyze the various consequences of our formulation of Mach's principle.

1. As was observed above, the main reason for the difficulties obstructing the realization and analysis of the Mach principle in the form of boundary conditions (4) is the extremely complicated structure of the Einstein equations for the ten independent components of the gravitational potential g_{ik}. However, the study of spaces of the conformally flat type is made substan-

tially easier from a formal point of view, since in the metric of these spaces

$$g_{ik}(x) = \varphi^2(x)\eta_{ik}(x) \tag{7}$$

the quantities $\eta_{ik}(x)$ (the metric of flat space) are defined up to a choice of reference frame and may be assumed to be known. Thus, all the information concerning the gravitational field lies in the single "potential" $\varphi(x)$. At the same time, in spite of this formal simplification, studies show that the physical aspect of the problem is not significantly distorted, while at the same time it becomes more transparent.

2. Since only one "potential" $\varphi(x)$ is to be determined, the Einstein equations [8]

$$R_{ik} - {}^1/_2\, g_{ik}R = \frac{8\pi k}{c^4} T_{ik}, \tag{8}$$

viewed as equations for $\varphi(x)$, are not all independent. In order to avoid overdeterminacy and to give the equations a more symmetric form, we take the contracted Einstein equations

$$R = -\frac{8\pi k}{c^4} T, \qquad T = T_{ik}g^{ik}. \tag{9}$$

as the equation for $\varphi(x)$. For the metric (7) the Christoffel symbols are equal to [9]

$$\Gamma^l_{ik} = \gamma^l_{ik} + \delta^l_i \frac{\partial}{\partial x^k} \ln\varphi + \delta^l_k \frac{\partial}{\partial x^i} \ln\varphi - \eta_{ik}\eta^{lm} \frac{\partial}{\partial x^m} \ln\varphi, \tag{10}$$

where the γ^l_{ik} are the Christoffel symbols in the space with metric η_{ik}. Using these Christoffel symbols we find the scalar curvature

$$R = -6\varphi^{-3}\Box\varphi, \tag{11}$$

where $\Box\varphi$ is the covariant D'Alembertian of φ with respect to the metric η_{ik}. From this we can write the explicit form of (9) as

$$\Box\varphi = \frac{4\pi k}{3c^4} T\varphi^3. \tag{12}$$

Of course, the same equation can be obtained also from the Lagrangian formalism.

3. Equation (12) is nonlinear in φ, although the special form of the nonlinearity permits obtaining an integral equation by analogy with the linear theory [10].

Indeed, (12) can be rewritten as the identity

$$\Box\varphi(x) = \frac{4\pi k}{3c^4} \int T(y)\,\varphi^3(y)\delta^3(x-y)\sqrt{-\eta(y)}\,d^4y, \tag{13}$$

after which, "dividing" by $\Box$, we get

$$\varphi(x) = \frac{4\pi k}{3c^4} \int T(y)\,\varphi^3(y)\,G(x, y)\sqrt{-\eta(y)}\,d^4y + \Lambda(x), \tag{14}$$

where $G(x, y)$ and $\Lambda(x)$ satisfy the equations

$$\Box G(x, y) = \delta^4(x-y),$$
$$\Box\Lambda(x) = 0. \tag{15}$$

The explicit form of $\Lambda(x)$ can be determined by the standard method (cf., e.g., [11]):

$$\Lambda(x) = -\oint_S \left(G \frac{\partial}{\partial y^i} \varphi - \varphi \frac{\partial}{\partial y^i} G\right) \sqrt{-\eta}\, dS^i; \tag{16}$$

here S is the boundary of the domain over which the integration in the first term of (14) is carried out. Finally the equation for $\varphi(x)$ takes the form

$$\varphi(x) = \frac{4\pi k}{3c^4} \int T\varphi^3 G(x, y) \sqrt{-\eta}\, d^4y - \oint_S \left(G \frac{\partial}{\partial y^i} \varphi - \varphi \frac{\partial}{\partial y^i} G\right) \sqrt{-\eta}\, dS^i. \tag{17}$$

The two integrals on the right-hand side of Eq. (17) have the following meaning: the volume integral takes into account the contribution to $\varphi(x)$ of all sources localized in the given volume. The surface integral takes into account the contribution from both the sources located outside the given volume and the free radiation incident from infinity which does not arise from any sources.

Since all the observable masses of the universe play a role in the Mach principle, the domain of integration in (17) must be understood in the sense of passing to the limit over all the space of a given cosmological model [6].

4. In accordance with the program discussed above, the Mach principle is satisfied by models for which

$$\oint_S \left(G \frac{\partial}{\partial y^i} \varphi - \varphi \frac{\partial}{\partial y^i} G\right) \sqrt{-\eta}\, dS^i = 0 \tag{18}$$

or, what is the same,

$$\varphi(x) = \frac{4\pi k}{3c^4} \int T(y) \varphi^3(y) G(x, y) \sqrt{-\eta(y)}\, d^4y. \tag{19}$$

Thus we are associating the formal expression of Mach's principle with a certain boundary condition.

We consider this boundary condition in somewhat more detail. Because the integration in (17) is performed over all space of one or another cosmological model, all the sources T are accounted for by the first terms on the right-hand side of (17). Thus the free term Λ is not associated with any material source of the given cosmological model [6]; it describes a wave of the field $\varphi(x)$ for which no material source is present at any time or place throughout the entire evolution of the given model. In other words, after passing to the limit the free term must satisfy the homogeneous equation

$$\Box \Lambda = 0.$$

at every point of the given cosmological model (i.e., everywhere and at every time) throughout the whole evolution of the model.

Boundary condition (18) thus excludes a "global" free wave in the sense indicated above. In the special case when all the space of the given model is devoid of matter throughout its entire history, this simply means that the concept of space and time becomes degenerate, in full agreement with a qualitative study of the Mach principle [(19) gives $\varphi \equiv 0$ for $T \equiv 0$].

On the other hand, it cannot be asserted that condition (19) excludes any free wave whatever. Indeed, in the case where a source T is present in some bounded space-time domain, the integral in (19) is also different from zero (generally speaking) outside the domain occupied by the source. At the same time, the field $\varphi(x)$ satisfies a homogeneous field equation in the domain exterior to the source and can be considered to be a locally free wave by an observer in this domain. Thus in our formulation the Mach principle permits waves for which at some place and some time there exists a source that can be described as T (cf. Paragraph 6, for example).

This circumstance is closely related to the fact that the Mach principle in the form of Eq. (19) or boundary condition (18) imposes certain restrictions on the sources T also, as we shall see below. Of course, it can turn out that no physically acceptable source satisfies this condition. Then the entire theory will be inconsistent in this sense with the Mach principle. Examples will be presented below showing that in this respect the theory given has a physical content and realizes a concrete version of the Mach principle.

5. For the concrete calculation of G(x, y) we pass to a system of coordinates in which the metric η_{ik} has the form

$$\eta_{ik}=\begin{Vmatrix}1 & 0 & 0 & 0\\ 0 & -1 & 0 & 0\\ 0 & 0 & -1 & 0\\ 0 & 0 & 0 & -1\end{Vmatrix}. \tag{20}$$

The covariant derivatives with respect to η_{ik} reduce to partial derivatives, and choosing retarded potentials we obtain from (15) [10]

$$G(x,y)=\frac{\delta(t'-t+|\mathbf{r}-\mathbf{r}'|/c)}{4\pi|\mathbf{r}-\mathbf{r}'|} \tag{21}$$

and for static problems

$$G(x,y)=\frac{1}{4\pi|\mathbf{r}-\mathbf{r}'|}\qquad (x=\{ct,\mathbf{r}\},\ y=\{ct',\mathbf{r}'\}). \tag{21'}$$

Equation (19) now has the form

$$\varphi(\mathbf{r},t)=\frac{k}{3c^4}\int T(\mathbf{r}',\tau)\,\varphi^3(\mathbf{r}',\tau)\,\frac{d\mathbf{r}'}{|\mathbf{r}-\mathbf{r}'|}\qquad\left(\tau=t-\frac{|\mathbf{r}-\mathbf{r}'|}{c}\right) \tag{22}$$

and we can turn to a consideration of the concrete physical consequences of this equation.

6. What was said in Paragraph 4 concerning locally free fields can be illustrated by the following example. Assume given a source

$$T(r,t)=T_0\frac{\delta(r-a)}{r^2}\delta(t) \tag{23}$$

Then (22) gives

$$\varphi(r,t)=a^2\sqrt{\frac{3c^3}{2\pi kT_0}}\,\frac{1}{r}\,\theta(ct-|r-a|)\,\theta(a+r-ct)$$
$$\left(\theta(X)=\begin{cases}1, & X>0\\ 0, & X<0\end{cases}\right). \tag{24}$$

This solution has the nature of a traveling wave and does not vanish in the domain exterior to the source (i.e., for $r\neq a$, $t>0$), while at the same time it satisfies the homogeneous wave equation

$$\Box\varphi=0 \tag{25}$$

in this domain. Condition (18) prevents us from adding arbitrary solutions of the homogeneous equation (25) to (24), i.e., we cannot add globally free waves, which would have been possible on the basis of the differential equation (12) alone.

The following example illustrates what was also said in Paragraph 4 concerning the restrictions on the source imposed by the Mach principle.

Assume given the following matter distribution

$$T(r) = T_0 (a^2 - r^2)\theta(R - r), \qquad T_0,\ a,\ R = \text{const}. \tag{26}$$

Differential equation (12) has the solution

$$\varphi(r) = \begin{cases} c^2 \sqrt{\dfrac{3}{2\pi k T_0}} \dfrac{3}{a^2 + 3r^2}, & r < R, \\ \text{const}_1/r + \text{const}_2, & r > R, \end{cases} \tag{27}$$

for arbitrary R and constants const_1 and const_2 defined by certain additional considerations (behavior at the boundary of a body, at infinity, etc.).

On the other hand, we obtain from integral equation (22)

$$\varphi(r) = \begin{cases} c^2 \sqrt{\dfrac{3}{2\pi k T_0}} \dfrac{3}{a^2 + 3r^2}, & r < R, \\ \dfrac{3}{2} c^2 \sqrt{\dfrac{1}{2\pi k T_0}} \dfrac{1}{ar}, & r > R, \end{cases} \tag{28}$$

where moreover we have the condition $R = a/\sqrt{3}$.

Cases can exist, of course, where the Mach principle does not impose any restrictions on the source. For example, because of the special form of the nonlinearity of the field equation in our theory, delta-function sources linearize (12). One such example was given at the beginning of this paragraph.

Another example: take a source of the form

$$T(r) = T_0 \frac{\delta(r - a)}{4\pi r^2} \tag{29}$$

Equation (12) gives

$$\varphi(r) = \frac{c_1}{r}(c_2 r + c_3 - |r - a|), \qquad c_1, c_2, c_3 = \text{const}, \tag{30}$$

and from (22) we have

$$\varphi(r) = c^2 \sqrt{\frac{3a}{kT_0}} \frac{a + r - |r - a|}{2r} \tag{31}$$

for arbitrary a and T_0. (However, $a = 0$ is excluded, i.e., there are no point particles; for $a = 0$ we have $\varphi = 0$, as is clear from (31) and which can of course be verified by a direct calculation).

The peculiar dependence $\varphi \sim T_0^{-1/2}$ is striking. Since, however, this dependence is not continuous at zero, the limit of $T_0 \to 0$ must be determined directly from the cubic equation for the constant factor of φ, which gives $\varphi = 0$ for $T_0 = 0$, in complete accord with the Mach principle.

It should also be emphasized that because of its interpretation $\varphi(x)$ is defined only up to a constant factor, the variation of which is tantamount simply to a change of scale of the coordinates and does not lead to any physical consequences. This arbitrariness corresponds to the choice in selecting an additive constant in the potential of the Newtonian theory (thus, $\ln \varphi$ appears in the equations of motion in the Newtonian approximation, and the scale factor turns into an additive constant).

7. We now consider the problem of the inertia of a body. Since the Mach principle does not refer the accelerated motion of a body to absolute space, but to all the other bodies in the universe, the inertia of a body manifested in this motion is determined by the presence of the other bodies, so that for an isolated body in the universe the concept of mass loses its meaning [1, 12-14].

A precise notion of the mass of a body is closely related to the existence of flat asymptotic behavior of the space surrounding it. Such behavior is absent in the examples given above.

This is a result of extreme generality.

Consider an arbitrary island distribution of matter. At sufficiently large distances from this distribution the problem becomes spherically symmetric and Eq. (12) gives

$$\varphi(r) = a/r + b, \quad a, \quad b = \text{const}, \tag{32}$$

i.e., at infinity we have, generally speaking, a flat space

$$ds^2 = b^2 \{c^2 dt^2 - dx^2 - dy^2 - dz^2\}. \tag{33}$$

If we require that condition (18) hold,

$$\lim_{r' \to \infty} \left\{ \frac{1}{4\pi |\mathbf{r} - \mathbf{r}'|} \frac{\partial}{\partial r'} \left(\frac{a}{r'} + b \right) - \left(\frac{a}{r'} + b \right) \frac{\partial}{\partial r'} \frac{1}{4\pi |\mathbf{r} - \mathbf{r}'|} \right\} 4\pi r'^2 = 0, \tag{34}$$

we obtain

$$b = 0. \tag{35}$$

Thus what we have called the realization of the Mach principle excludes not only solutions with $T \equiv 0$ but also asymptotically flat solutions, in complete accord with a qualitative analysis of Mach's principle.*

But let us consider the dynamics in bounded domains of a given universe, where the quadratic form ds^2 is meaningful and $g_{ik} \neq 0$. A characteristic feature of the dynamics is the fact that the effect of a test body on the metric cannot be ignored since otherwise the concepts of inertia and gravitating mass do not arise.

Indeed, the field has the form

$$\varphi(r) = a/r. \tag{36}$$

We introduce a test particle into this space without perturbing the metric, the metric being

---

* The Petrov classification hardly makes sense for the peculiar space obtained in this way ($g_{ik} \to 0$ as $r \to \infty$) since the Weyl tensor (conformal curvature) is identically equal to zero.

determined by the field φ. Then the force acting on the particle is equal to [8]

$$\mathbf{f} = -\frac{mc^2}{\sqrt{1-v^2/c^2}}\operatorname{grad}\ln\sqrt{g_{00}} = \frac{mc^2}{\sqrt{1-v^2/c^2}}\frac{\mathbf{r}}{r^2}, \tag{37}$$

where m is the mass of the test particle and v its velocity; for $v \ll c$ we have the equations of motion

$$\ddot{\mathbf{r}} = \frac{c^2}{r}\frac{\mathbf{r}}{r}. \tag{38}$$

Thus the equations of motion in general do not contain any parameters of a central body; thus whatever central gravitational body we take, we cannot distinguish one body from another by observing the motion of the test particle: There is no quantity which can be attributed to a central body which would determine the motion of the test particle (e.g., its acceleration). Since on the other hand such a parameter of a gravitating body is given by definition by the gravitational mass of the body, this result means by the equivalence principle that an isolated body (not interacting with and isolated from other bodies) has no inertia, in complete agreement with the spirit of Mach's principle.

If use is made of [8], it can be shown that the mass of a central body is equal to zero, Indeed, the inertial mass of a body at rest can be defined in terms of the zeroth component of its four-momentum by

$$M = \frac{c^2}{16\pi k}\oint\frac{\partial}{\partial x^m}[(-g)(g^{00}g^{\alpha m} - g^{0\alpha}g^{0m})]\,df_\alpha, \tag{39}$$

which gives for the case of a spherically symmetric field

$$M = \lim_{r\to\infty}\frac{c^2}{k}\varphi^3\frac{\partial\varphi}{\partial r}r^2, \tag{40}$$

and for our field* (36)

$$M = \lim_{r\to\infty}\left(-\frac{c^2}{k}\frac{a^4}{r^3}\right) = 0. \tag{41}$$

The situation changes essentially if we allow for the interaction of the test particle on the metric.

We remark as a preliminary that in defining the field of two particles (central and test particles) in this theory, one must not represent them in the form of mathematical points, i.e., define a δ-function matter distribution

$$T(\mathbf{r}) = T_1\delta(\mathbf{r}) + T_2\delta(\mathbf{r}-\mathbf{R}), \tag{42}$$

since the field of a particle in our theory (as is seen from Paragraph 6) tends to zero if the particle dimensions tend to zero.

We therefore assume that both particles have a radius ρ (with $\rho \ll$ R, where R is the distance between the particles), so that

$$T(\mathbf{r}) = T_1\frac{\delta(r-\rho)}{4\pi r^2} + T_2\frac{\delta(|\mathbf{r}-\mathbf{R}|-\rho)}{4\pi(\mathbf{r}-\mathbf{R})^2}. \tag{43}$$

---

* This holds as in the case of a closed universe, although this model cannot be considered to be three-dimensionally closed.

The corresponding field has the form

$$\varphi(\mathbf{r}) = \frac{kT_1}{3c^4}\int\frac{\delta(r'-\rho)}{4\pi r'^2}\varphi^3(\mathbf{r}')\frac{d\mathbf{r}'}{|\mathbf{r}-\mathbf{r}'|} + \frac{kT_2}{3c^4}\int\frac{\delta(|\mathbf{r}'-\mathbf{R}|-\rho)}{4\pi(\mathbf{r}'-\mathbf{R})^2}\varphi^3(\mathbf{r}')\frac{d\mathbf{r}'}{|\mathbf{r}-\mathbf{r}'|} =$$

$$= \frac{kT_1}{3c^4}\varphi^3(\mathbf{r}^*)\int\frac{\delta(r'-\rho)}{4\pi r'^2}\frac{d\mathbf{r}'}{|\mathbf{r}-\mathbf{r}'|} + \frac{kT_2}{3c^4}\varphi^3(\mathbf{r}^{**}+\mathbf{R})\int\frac{\delta(|\mathbf{r}'-\mathbf{R}|-\rho)}{4\pi(\mathbf{r}'-\mathbf{R})^2}\frac{d\mathbf{r}'}{|\mathbf{r}-\mathbf{r}'|}, \tag{44}$$

where

$$|\mathbf{r}^*| = |\mathbf{r}^{**}| = \rho \ll R, \tag{45}$$

and between the particles

$$\varphi(\mathbf{r}) = \frac{kT_1}{3c^4}\varphi^3(\mathbf{r}^*)\frac{1}{r} + \frac{kT_2}{3c^4}\varphi^3(\mathbf{R}+\mathbf{r}^{**})\frac{1}{|\mathbf{r}-\mathbf{R}|}. \tag{46}$$

Ignoring now the dimensions of the particles in comparison with the distance between them and putting $|\mathbf{r}-\mathbf{R}| = \rho$, we obtain

$$\varphi(R) = \frac{kT_1}{3c^4}\varphi^3(0)\frac{1}{R} + \frac{kT_2}{3c^4}\varphi^3(R)\frac{1}{\rho}. \tag{47}$$

Assume now that R is so large that the first term on the right-hand side of (47) represents a small correction to the second term. If we ignore the second and higher powers of this correction, the solution of the cubic equation (47) will have the form

$$\varphi(R) = \pm\sqrt{\frac{3\rho}{kT_2}c^2} - \frac{kT_1}{6c^4}\varphi^3(0)\frac{1}{R}. \tag{48}$$

Choosing the plus sign in the first term on the right, making the coordinate transformation

$$x'^i = \sqrt{\frac{3\rho}{kT_2}}c^2x^i \tag{49}$$

and recalling the definition of $\varphi(\mathbf{r})$, we obtain

$$\varphi(R') = 1 - \frac{kT_1}{6c^4}\varphi^3(0)\frac{1}{R'}. \tag{50}$$

From the equation for a geodesic in the nonrelativistic approximation, and bearing in mind that the term $\sim 1/R'$ is small, we have the equation of motion

$$\ddot{\mathbf{R}}' = -\frac{kM}{(R')^2}\frac{\mathbf{R}'}{R'} \tag{51}$$

for the second particle, where $M = (T_1/6c^2)\varphi^3(0)$ denotes the mass of the first particle.

We have thus obtained Newton's law, the allowance made for the field of both particles being essential.

This same result can be illustrated as follows. Assume two concentric massive spheres are given,

$$T(r) = T_1\frac{\delta(r-a)}{4\pi r^2} + T_2\frac{\delta(r-b)}{4\pi r^2}, \quad a, b = \text{const}, \quad a < b. \tag{52}$$

A solution of the integral equation is

$$\varphi(r)=\frac{kT_1}{3c^4}\varphi^3(a)\frac{r+a-|r-a|}{2ar}+\frac{kT_2}{3c^4}\varphi^3(b)\frac{r+b-|r-b|}{2br}. \tag{53}$$

In the space between the spheres (i.e., for $a \le r \le b$)

$$\varphi(r)=\frac{kT_1}{3c^4}\varphi^3(a)\frac{1}{r}+\frac{kT_2}{3c^4}\varphi^3(b)\frac{1}{b}. \tag{54}$$

The system

$$\begin{aligned}\varphi(a)&=\frac{kT_1}{3c^4a}\varphi^3(a)+\frac{kT_2}{3c^4b}\varphi^3(b),\\ \varphi(b)&=\frac{kT_1}{3c^4b}\varphi^3(a)+\frac{kT_2}{3c^4b}\varphi^3(b).\end{aligned} \tag{55}$$

for determining the constants $\varphi(a)$ and $\varphi(b)$ is obtained in an obvious way from (54). It is easy to see that this system has a solution with

$$\varphi(a)<0, \qquad \varphi(b)>0; \tag{56}$$

indeed, writing $\varphi(a) \equiv -\alpha < 0$ and $\varphi(b) \equiv \beta > 0$ and considering (55) to be identities, it is easy to obtain from (56) the inequalities

$$T_2\beta^3<\frac{b}{a}T_1\alpha^3, \qquad T_2\beta^3>T_1\alpha^3 \tag{57}$$

which do not contradict each other (since $a < b$). For a suitable renormalization of the coordinate, we obtain from (54) the Schwarzschild field

$$\varphi(r')=1-\frac{kM}{c^2r'} \qquad \left(M\equiv-\frac{T_1}{3c^2}\varphi^3(a)>0\right), \tag{58}$$

which leads to the Newtonian equations of motion

$$\ddot{\mathbf{r}}'=-\frac{kM}{r'^2}\frac{\mathbf{r}'}{r'}. \tag{59}$$

At the same time, outside a large sphere (i.e., for $r \ge b$) the field has a form which does not lead to a Newtonian law:

$$\varphi(r)=\left[\frac{kT_1}{3c^4}\varphi^3(a)+\frac{kT_2}{3c^4}\varphi^3(b)\right]\frac{1}{r}. \tag{60}$$

It is clear from these examples that in the theory of gravitation based on integral equations, one can speak of Newton's law (and hence also of mass) only to the extent that there exists an interaction of several bodies, so that it is impossible to introduce the mass of an isolated body.

From a formal viewpoint this result is quite transparent. Indeed, in the ordinary theory Newton's law is obtained in an approximation based on the existence of an asymptotically flat space around an island distribution of matter. On the other hand, the integral formalism discards a solution possessing flat asymptotics. It is nevertheless clear that flat asymptotic behavior is not necessary for the phenomenon of Newtonian force. It is sufficient just that the field have the Schwazschild form

$$\varphi(r) = \text{const}_1 + \text{const}_2/r, \tag{61}$$

in the domain of localization of a test particle, and this condition is achieved by allowing for the field of the second body participating in the interaction* (while at the same time in the usual theory in which a flat asymptotic behavior is given at the start, the self-field of the test particle only leads to insignificant corrections).

Thus in our theory the inertia of a body is indeed determined completely by its interaction with other bodies in accordance with Mach's principle.

8. The problem of whether the Mach principle is satisfied in the real universe is also of great interest.

If we assume that the real universe is described with sufficient accuracy by the Friedmann model, then it must be explained under which conditions the Friedmann model is compatible with boundary conditions of the type (18). As was remarked in the introduction, the results of the known studies [2, 3, 6] are in mutual contradiction.

If we assume that the equation of state of matter in the Friedmann model has the form

$$p = \gamma\varepsilon, \tag{62}$$

where γ = const and p and ε are the pressure and energy density of matter, then [2, 3] gives

$$p = -\varepsilon \quad (\gamma = -1), \tag{63}$$

while [6] gives

$${}^1/_3\varepsilon \leqslant p \leqslant \varepsilon \quad ({}^1/_3 \leqslant \gamma \leqslant 1). \tag{64}$$

Since the Friedmann model is conformally flat, the problem posed above turns out to be solvable within the framework of our theory also.

Consider the simplest case of a flat model

$$ds^2 = \varphi^2(\tau)\,[d\tau^2 - d\chi^2 - \chi^2\,(d\theta^2 + \sin^2\theta d\beta^2)], \tag{65}$$

where [6]

$$\begin{aligned} \varphi(\tau) &= A_0^{m/2}\left(\frac{\tau}{m}\right)^m, \qquad m = \frac{2}{3\gamma+1}, \\ A_0 &= \frac{8\pi k\varepsilon_0}{3c^4}, \qquad \varepsilon_0 = \varepsilon(\tau)\,\varphi^{3\gamma+3}(\tau) = \text{const}. \end{aligned} \tag{66}$$

The field has a singularity at $\tau = 0$ so that the integration in (22) must be performed for $\tau \geq \alpha > 0$ after which $a \to 0$.

After an elementary integration (22) gives

$$A_0^{m/2}\left(\frac{\tau}{m}\right)^m = \lim_{\alpha\to 0}\left\{A_0^{m/2}\left[\left(\frac{\tau}{m}\right)^m - \left(\frac{\alpha}{m}\right)^{m-1}\tau + \frac{1-3\gamma}{2}\left(\frac{1}{m}\right)^{m-1}\alpha^m\right]\right\}, \tag{67}$$

* To a certain approximation the self-field of a particle can be assumed independent of the mutual location of the particles.

with $\gamma \neq {}^1/_3$; it follows from this that

$$\lim_{\alpha\to 0}\left[\left(\frac{\alpha}{m}\right)^{m-1}\tau - \frac{1-3\gamma}{2}\left(\frac{1}{m}\right)^{m-1}\alpha^m\right] = 0, \tag{68}$$

which is possible only for $m > 1$, from which we obtain

$$-{}^1/_3 < \gamma < {}^1/_3. \tag{69}$$

For an open model

$$ds^2 = A^2(\tau)\,[d\tau^2 - d\chi^2 - \mathrm{sh}^2\,\chi\,(d\theta^2 + \sin^2\theta d\beta^2)], \tag{70}$$

where [6]

$$A(\tau) = A_0^{m/2}\mathrm{sh}^m(\tau/m), \tag{71}$$

the well-known substitution [8]

$$r = \mathrm{const}\cdot e^{\tau}\mathrm{sh}\,\chi, \qquad ct = \mathrm{const}\cdot e^{\tau}\mathrm{ch}\,\chi, \tag{72}$$

reduces it to the conformally flat form

$$ds^2 = \varphi^2(r, t)\,[c^2dt^2 - dr^2 - r^2(d\theta^2 + \sin^2\theta d\beta^2)], \tag{73}$$

where taking the constant in (72) equal to $({}^1\!/_2 A_0^{1/2})^m$ we have

$$\varphi(r, t) = \left[1 - \frac{A_0}{4}(c^2t^2 - r^2)^{-1/m}\right]^m. \tag{74}$$

In this model the field has a singularity at $c^2t^2 - r^2 = (A_0/4)^m$, so that the integration in (22) is performed over the domain

$$c^2t^2 - r^2 \geqslant (A_0/4)^m\,(1 + \alpha), \tag{75}$$

and we then let $\alpha \to 0$. Moreover, Eq. (22) gives ($\gamma \neq 1/3$)

$$\left[1 - \frac{A_0}{4}(c^2t^2 - r^2)^{-1/m}\right]^m = \lim_{\alpha\to 0}\left\{\left[1 - \frac{A_0}{4}(c^2t^2 - r^2)^{-1/m}\right]^m - \right.$$
$$\left. -\left(\frac{\alpha}{m}\right)^{m-1}\left[1 - \left(\frac{A_0}{4}\right)^m\left(1 - \frac{1-m}{m}\alpha\right)\frac{1}{c^2t^2 - r^2}\right]\right\}, \tag{76}$$

from which

$$\lim_{\alpha\to 0}\left(\frac{\alpha}{m}\right)^{m-1}\left[1 - \left(\frac{A_0}{4}\right)^m\left(1 - \frac{1-m}{m}\alpha\right)\frac{1}{c^2t^2 - r^2}\right] = 0, \tag{77}$$

which is possible only for $m > 1$, i.e., $-1/3 < \gamma < 1/3$ as in the case of the flat model.

Taking into account the general relation [16]

$$0 \leqslant \gamma \leqslant 1, \tag{78}$$

we obtain that the Friedmann model is a Mach model for

$$0 \leqslant p < {}^1/_3\varepsilon, \tag{79}$$

at least for the open and flat cases.

If we also take into account that by (67) and (76) the agreement between the model and the Mach principle is mainly determined by the beginning of the evolution, when the difference among all three types is slight, we may evidently assume that the result obtained is also valid for a closed model.

The fact that Eq. (22) forbids the value $\gamma = 1/3$ is probably connected with the "scalar" nature of our theory ($\gamma = 1/3 \Rightarrow T = 0$), although since only universes consisting solely of photons and neutrinos are thus excluded from consideration, this can hardly be considered significant for the real universe.

Mention must also be made of the following fact, viz. the rather widespread point of view that the Mach principle necessarily requires three-dimensional closedness of cosmological models. This requirement is usually derived from the fact that the Mach principle forbids asymptotically flat spaces. However, this is not so in the study of the version of Mach's principle proposed by us. It seems to us that a convincing example of this is given by the flat and the open Friedmann models which are not three-dimensionally closed.

9. The expression of the gravitation equations in integral form in order to express the Mach principle mathematically (which strictly speaking goes back to the well-known work of Einstein [15]), applied to GTR in the general case in [2-6], is not transparent physically because the formalism is too cumbersome. However, the study of conformally flat spaces does indeed simplify the understanding of the formalism of this version of Mach's principle in an essential way.

The version of Mach's principle considered is characterized by its unique geometric interpretation and by the highly specific nature of its physical content. The fact that subjecting the Einstein equations to certain boundary conditions leads to the same consequences which we are accustomed to obtain from a qualititative analysis of Mach's principle gives a basis for believing that the integral equation formalism (although requiring further development and improvement) is indeed sufficient for Mach's principle, and that the general covariant boundary conditions imposed on the Einstein equations in this formalism indeed make GTR into a Mach theory, as was envisaged by A. Einstein.

APPENDIX TO PARAGRAPH 3

As was remarked in the introduction, condition (6) imposed on the operator of the equation for the Green function is too weak.

Thus in the case considered by us of conformally flat spaces, an operator of the form

$$D = \alpha \Box + (1 - \alpha)\, \varphi^{-1} \Box \varphi \tag{81}$$

can be taken as an operator satisfying only a condition analogous to (6) [so that $D\varphi = \Box\varphi$ gives the left-hand side of Eq. (12)] with arbitrary (nonzero) α.

The formalism considered by us corresponds to $\alpha = 1$, whereas in the cases in [2] and [4] $\alpha = -1/3$, as is easily seen, while in [3] and [5] $\alpha = 1/3$.

The equation for the Green function can now be written in the form

$$\alpha \Box G + (1 - \alpha)\, \varphi^{-1} G \Box \varphi = \delta^4 (x - y). \tag{82}$$

Equation (17) remains formally unchanged, although since G(x, y) depends in some way or another on α, the meaning of the expressions appearing in (17) [and hence also in (18), (19)] changes.

All the formalisms coincide only in the case where $\Box\varphi = 0$ (e.g., empty space and the Schwarzschild solution, cf. the introduction), whereas in the case where $\Box\varphi \neq 0$ the results in general are different (e.g., in the case of the Friedmann model, cf. Paragraph 8).

APPENDIX TO PARAGRAPH 7

We underline the difference of our formalism from the well-known formalism of Brans and Dicke (cf. the article by Dicke in [17]), which consists in imposing certain boundary conditions on the metric instead of introducing a scalar field Φ supplementary to the metric.

As far as the scalar interaction is concerned, the fact remarked by Einstein in [18]) that this interaction leads to violation of the law of conservation of energy is treated in the Brans–Dicke formalism as making it necessary to introduce a dependence of the inertia on a scalar potential. However, as was shown by Einstein [19], this violation is removed if the scalar field Φ changes the metric g_{ik} in such a way that $g_{ik} = \text{const}\cdot\Phi^2\eta_{ik}$ is the metric of flat space, i.e., if a metric field of type (7) is considered. Indeed, for a metric field one can take the equations of motion to be the equations of geodesics (which is what we have done in this article), which do not lead to violation of conservation laws. On the other hand, the imposition of boundary conditions on the metric defining the geodesics can as in any theory alter the nature of the motion; for our type of boundary conditions, this leads, as was shown in Paragraph 7, to the appearance of a relation between inertia and gravitation which is characteristic of Mach's principle.

APPENDIX TO PARAGRAPH 8

The Friedmann model for all three types of spaces is conformally flat. The open and flat types are considered in this article. As for the closed type, we remark that since the closed Friedmann model is obviously conformal to the Einstein model, it suffices to consider the latter. Calculating the Weyl tensor for the Einstein model, we see that it is equal to zero, i.e., the Einstein model along with the Friedmann model is conformally flat. A coordinate transformation taking the metric of the Einstein model to the form (7) is given in [20].

LITERATURE CITED

1. A. Einstein, Collected Works, Vol. 1 [Russian translation], Nauka, Moscow (1965), p. 613.
2. B. L. Al'tshuler, Zh. Éksp. Teor. Fiz., 51:1143 (1966).
3. B. L. Al'tshuler, Zh. Éksp. Teor. Fiz., 55:1311 (1968).
4. D. Linden-Bell, Mon. Not. Roy. Astron. Soc., 135:413 (1967).
5. D. W. Sciama, P. C. Waylen, and R. C. Gilman, Phys. Rev., 187:1762 (1969).
6. R. C. Gilman, Phys. Rev. D, 2:1400 (1970).
7. B. S. DeWitt and R. W. Brehme, Ann. Phys., 9:220 (1960).
8. L. D. Landau and E. M. Lifshitz, Field Theory [in Russian], Nauka, Moscow (1967).
9. L. P. Eisenhart, Riemannian Geometry, Princeton University Press (1960).
10. D. Ivanenko and A. Sokolov, The Classical Theory of Fields [in Russian], Gostekhizdat, Moscow-Leningrad (1951).
11. G. Barton, Introduction to Dispersion Techniques in Field Theory, Benjamin, New York (1965).
12. E. Mach, The Science of Mechanics, Open Court Publ. Co., Lasalle, Illinois (1960).
13. A. Einstein, Collected Works, Vol. 1 [Russian translation], Nauka, Moscow (1965), p. 223.
14. W. Pauli, The Theory of Relativity, Pergamon Press, New York (1958).
15. A. Einstein, Collected Works, Vol. 2 [Russian translation], Nauka, Moscow (1966), p. 5.
16. B. Harrison, K. Thorne, M. Wakano, and J. Wheeler, Gravitational Theory and Gravitational Collapse, University Chicago Press, Chicago (1965).

17. Gravitation and Relativity, Benjamin, New York (1964).
18. A. Einstein, Collected Works, Vol. 1 [Russian translation], Nauka, Moscow (1965), p. 227.
19. A. Einstein, Collected Works, Vol. 1 [Russian translation], Nauka, Moscow (1965), p. 273.
20. M. Gürsey and Y. Gürsey, Nuovo Cimento, 25B:786 (1975).

CANONICAL TRANSFORMATIONS AND THE THEORY OF REPRESENTATIONS OF LIE GROUPS

A. N. Leznov, I. A. Malkin, and V. I. Man'ko

A new approach is introduced for constructing representations of Lie groups based on canonical transformations in group phase space of the Lie group which reduce the regular representation of the Lie group to canonical form. Examples are considered of the groups (semisimple, nilpotent, solvable) having application to problems in physics, and for these groups formulas are given for the generators, finite rotation operators, characters, and Plancherel measure. The Lie group is viewed as a dynamical system described by a Hamiltonian which is a linear form in the momenta with coefficients depending on the coordinates. This permits the use of the methods of quantum mechanics, in particular integrals of motion, coherent states, and path integrals, for constructing representations of Lie groups.

INTRODUCTION

Representations of Lie groups in recent years have found increasing application to the most diverse questions in theoretical physics. While at the dawn of the development of quantum mechanics compact groups were mainly used (the group of rotations of three-dimensional space, the isotopic invariance group), at the present time extensive application is made of the theory of infinite dimensional representations of Lie groups of a rather broad class. Numerous examples of the application of the representation theory of semisimple Lie groups to problems in elementary particle physics can be found in [1] as well as in the survey [2]. The appearance of dynamical symmetries of a number of classical quantum problems [3-5] has led to the necessity of using solvable Lie groups (cf. [6]) as well as arbitrary Lie groups. It suffices to mention the Poincaré group and the inhomogeneous Poincaré group, a group lying at the foundation of all contemporary high-energy physics (cf. [7]). The infinite dimensional representations of the Poincaré group were constructed in a paper of Wigner [8] and applied to the needs of quantum field theory in [9]. In the study of any physical problem for which an equation describing the situation has been formulated, an important role is played by the symmetry conditions and properties of both the physical phenomenon itself and the equation adequate to describe it. As is well known, according to Noether's theorem the group of physical observables (the integrals of motion) is directly related to the group of symmetries of the equations describing the physical phenomenon under discussion.

The role of symmetries in various branches of physics such as the general theory of relativity, quantum field theory, elementary particle theory, and quantum mechanics is extremely great, and the mathematical apparatus for describing symmetries is the theory of groups. Thus all the possible Einstein spaces suitable for describing four-dimensional space-time can be classified on the basis of a study of precisely the groups of motion of these spaces (cf. [10]). In field theory and the modern theory of elementary particles, the very concept of the state of an elementary particle is rigorously formulated only in the language of the theory of group representations (representations of the Poincaré group). Physical quantities such as the mass of a particle (gravitational and inertial) make their appearance here as purely group-theoretic characteristics (the eigenvalue of one of the Casimir operators of the Poincaré group defining an irreducible representation of the group). The wave function of an elementary particle (plane wave in the free case) is also a group object, viz., a basis vector of an irreducible representation of the Poincaré group characterized by the momentum, mass, spin, and spin projection on the direction of motion. The symmetry groups SU_2, SU_3, SU_6, SU_{12}, ISL_6, G_2, $O(4, 2)$ are involved in the description of the internal structure of elementary particles and of quantum numbers such as charge, hypercharge, strangeness, etc. [2-5]. The application of various groups has allowed the formulation of an algebraic approach [4] to the description of quantum systems. The advantage of this approach lies in the fact that it is possible, without resorting to the equation describing the quantum system, to immediately obtain such physical quantities as the energy levels, their degeneracy, transition amplitudes, Green functions, relations between the cross sections of various physical processes, the particle masses and other characteristics of the particles, etc. This is possible thanks to the formulation of the problem in purely group-theoretic terms, which permits the application of the subsequently developed methods of the theory of representations of the Lie groups or Lie algebras adequate for the description of the problem being considered. Thus in the application to physics of the methods of group theory, it is necessary to have a well-developed and convenient mathematical apparatus, viz., the machinery of group representations, most frequently Lie group representations. It should be remarked that this machinery has been developed by mathematicians, but often in such a general form that it is rather hard to apply it to concrete physical problems. In physics, it is the machinery of infinitesimal (differential) operators of Lie group representations that has customarily been successfully applied. The infinitesimal operators describe the behavior of the system under infinitesimally small transformations, the behavior being dictated directly by the physical formulation of the problem, in which the eigenvalues of the generators (or of operators constructed from them) determine the physically observable quantities (of the type of moment, momentum, energy, etc.). At the same time, in the mathematical literature the machinery of finite (noninfinitesimal) representations is used most often, and formulas are constructed for global transformations. Although these formulas do contain the formulas for the infinitesimal transformations (which must still be extracted), they are sometimes less convenient. The mathematical literature devoted to the representation theory of general Lie groups is extremely vast. Lie group representations are usually constructed by the method of induced representations introduced and studied in [11] (cf. also [12, 13]). We refer also, for example, to the book of Kirillov [14] where numerous references can be found on this subject (cf. [15]), although the goals and methods of the physicist and the mathematician differ essentially. The mathematician as a rule is interested in proving existence theorems for a solution of the problem which are as general as possible, while the physicist is primarily interested in concrete computational formulas. The purpose of this paper is to give a constructive method for constructing representations of an arbitrary Lie group. In this survey we will emphasize calculational formulas (without always giving sufficient attention to their rigorous justification) which are convenient in the applications, and, in particular, we emphasize the infinitesimal formulation for the irreducible representations of the Lie algebras which we discuss. The representations of semisimple Lie groups along these lines were studied in detail in the papers [16-20]. Representations of nonsemisimple Lie groups are constructed in [21]. It was announced in that paper that it is possible to construct the irreducible representations of arbitrary Lie groups by the methods of canonical transformations and integration over trajectories.

In this survey what we want to do is demonstrate the very close relation of the theory of Lie group representations with canonical transformations in group phase space and the Feynman path integral. We also mention an approach, recently undergoing intensive development, to the construction of Lie group representations which used the ideas of classical and quantum mechanics (cf. Kirillov [14] on the orbit method, Kostant [22], the book of Arnol'd [23], as well as [24-28]). The use of the ideas of classical and quantum physics in the construction of Lie group representations leads in turn to a deeper understanding of the nature of the dynamical symmetries of physical systems (cf. [4] and the survey [29]).

In this paper the central task is to elucidate a very simple geometric picture which can be associated with the regular representation and the irreducible representations of Lie groups. To accomplish this it is necessary to use the language of group phase space (symplectic manifolds), on which an action of the Lie group is given. It turns out that the problem of constructing irreducible Lie group representations can be related to finding a certain canonical transformation in phase space to new variables in which the generators of the regular representation have a form which is maximally convenient for further study. In the new variables it becomes obvious that the irreducible representations of the Lie group are in correspondence with the motions in phase space along special surfaces which are fixed by the values of the set of canonical momenta (in essence, the eigenvalues of the Casimir operators) conjugate to the cyclic variables. The cyclic variables, on which the generators of the regular representation do not depend, parametrize the conjugacy classes of elements of the Lie group, and their change under the action of the group carries important information on the structure of the irreducible representations.

The general scheme for constructing Lie group representations with the aid of classical canonical transformations is presented in Section 1. In Section 4, a detailed illustration of the method is given based on the very simple nilpotent Heisenberg group, which is important in physical applications. In Section 2 an exposition is given of the method of induced representations of Lie groups, but in infinitesimal form, in contrast to the usual treatment. In Section 3 the construction of representations of nonsemisimple Lie groups with ideal (cf. [21]) is considered. Section 5 briefly discusses the construction of the irreducible representations of the noncompact semisimple Lie group SL(2, R) in the context of the method of canonical transformations. Section 6 is devoted to the construction using the asymptotic method [16] of the irreducible representations of semisimple Lie groups and the Casimir operators in terms of root-space techniques (cf. [19]). Section 7 considers examples of compact and solvable groups as well as Mautner's example of a group which is not of type I. In Section 8 we state some concluding remarks concerning the new method and discuss problems still requiring solution.

1. CANONICAL TRANSFORMATIONS AND REPRESENTATIONS OF LIE GROUPS

We now introduce the necessary notation and definitions. Assume given an n-dimensional Lie group G parametrized by n real coordinates [i.e., an element of the group is given by $g = q(q_1, q_2, \ldots, q_n)$] which we represent as an n-dimensional vector $\mathbf{q} = (q_1, \ldots, q_n)$. Naturally, it is necessary here to take into account the domain of variation of the parameters q_i. The parameters can all vary in a noncompact domain, or else all in a compact domain (compact groups), or else, finally, some of the parameters may be compact and some noncompact. Corresponding periodicity conditions arise here for functions in the compact parameters. We introduce the phase space in the group by adding n variable momenta $\mathbf{p} = (p_1, \ldots, p_n)$ which are canonically conjugate to $\mathbf{q}$. The Poisson brackets are given here by

$$\{p_i, q_k\} = \delta_{ik}; \quad i, k = 1, 2, \ldots, n. \tag{1.1}$$

Consider now the space of square-integrable functions $\psi(\mathbf{q})$ on the group such that

$$\int \psi^*(\mathbf{q})\psi(\mathbf{q})\,d\mathbf{q} < \infty, \tag{1.2}$$

where $d\mathbf{q}$ is an invariant measure on the group, for which we obtain an expression below. Now consider the right and left regular representations of G,

$$T_r^r(g)\,\psi[g_0(\mathbf{q})] = \psi(g_0 g),$$
$$T_l^r(g)\,\psi[g_0(\mathbf{q})] = \psi(g^{-1}g_0). \qquad (1.3)$$

The group G acts on the point $(\mathbf{q}, \mathbf{p})$ by right translations in the phase space of the group by the following rule: the $\mathbf{q}_g$ are the coordinates defining the group element $g_0 g$, where the $\mathbf{q}$ are the coordinates of the group element g_0; the $\mathbf{p}_g$ are by definition associated to the numbers $\mathbf{p}$ via the following procedure: Calculate the derivatives $\partial/\partial\mathbf{q}_g$ in terms of the derivatives $\partial/\partial\mathbf{q}$, and then replace $\partial/\partial\mathbf{q}$ by $\mathbf{p}$ and $\partial/\partial\mathbf{q}_g$ by $\mathbf{p}_g$ in the expressions obtained. The action of the group G in the group phase space by left translations is introduced in exactly the same way. Here the point $(\mathbf{q}, \mathbf{p})$ goes into the point $(\mathbf{q}_g, \mathbf{p}_g)$ whose coordinates are obtained in analogy to the coordinates $(\underset{\sim}{\mathbf{q}}_g, \underset{\sim}{\mathbf{p}}_g)$. The Poisson brackets are preserved under the action of right or left translations on a point $(\mathbf{q}, \mathbf{p})$, i.e., we define canonical transformations

$$\{p_{g_k}, q_{g_i}\} = \{p_k, q_i\} = \{\underset{\sim}{p}_{g_k}, \underset{\sim}{q}_{g_i}\} = \delta_{ki}. \qquad (1.4)$$

Consequently, we have invariants (Poincaré integral invariants), in particular, we have that the phase volume is invariant under left and right translations

$$d\mathbf{q}d\mathbf{p} = d\mathbf{q}_g d\mathbf{p}_g = d\underset{\sim}{\mathbf{q}}_g d\underset{\sim}{\mathbf{p}}_g. \qquad (1.5)$$

The group G acts nontransitively in phase space. The right and left translations determine canonical point transformations here (the new coordinates are expressed in terms of the old coordinates only, and the new momenta in terms of the old momenta only), and the infinitesimal transformations of the right L_i and left $\underset{\sim}{L}_i$ regular representations are given by linear forms in the operators $\partial/\partial q_i$:

$$L_i = \Psi_{ik}(\mathbf{q})\,\partial/\partial q_k, \quad \underset{\sim}{L}_i = -\,\underset{\sim}{\Psi}_{ik}(\mathbf{q})\,\partial/\partial q_k. \qquad (1.6)$$

Here the matrices Ψ and $\underset{\sim}{\Psi}$ are functions such that any left generator commutes with any right generator (the matrix $\underset{\sim}{\Psi}^{-1}\Psi$ is the matrix of the adjoint representation). This brings to mind the fact that they indeed depend on different variables and on general quantities which commute with one another (and with the remaining variables); cf., for example, the quantity $\mathcal{P}_z$ in the Heisenberg group in Section 4 below. We assume that the generators (1.6) can be written in "classical" form, and we introduce the n-dimensional vector $\mathbf{L}(\underset{\sim}{\mathbf{L}})$ such that $\mathbf{L} = (L_1, L_2, \ldots, L_n)$.

We have

$$\mathbf{L} = \Psi\mathbf{p}, \quad \underset{\sim}{\mathbf{L}} = -\,\underset{\sim}{\Psi}\mathbf{p}. \qquad (1.7)$$

The Poisson brackets of the generators $\mathbf{L}$ and $\underset{\sim}{\mathbf{L}}$ reproduce the commutation relations for the generators (1.6). The matrices Ψ and $\underset{\sim}{\Psi}$ are nonsingular, and hence the right and left generators are related by

$$\underset{\sim}{\mathbf{L}} = -\,[\underset{\sim}{\Psi}(\mathbf{q})\,\Psi^{-1}(\mathbf{q})]\,\mathbf{L}, \quad \mathbf{L} = -\,[\Psi(\mathbf{q})\,\underset{\sim}{\Psi}^{-1}(\mathbf{q})]\,\underset{\sim}{\mathbf{L}}. \qquad (1.8)$$

There exist invariant operators C_μ (μ = 1, 2, ..., r) (the Casimir operators) such that the Poisson brackets with the right and left generators of the Lie group are equal to zero. They satisfy the conditions

$$c_{ik}^j L_j(\mathbf{q}, \mathbf{p})\,\partial C_\mu/\partial L_k = 0 \qquad (1.9)$$

(summation is understood over indices which are repeated twice).

It turns out that there exists a nonpoint canonical transformation from the variables (q, p) to the variables $\mathbf{X} = (X_1, \ldots, X_p)$, $\mathbf{Y} = (Y_1, \ldots, Y_p)$, $\mathbf{Z} = (Z_1, \ldots, Z_r)$, $\mathcal{P}_X = (\mathcal{P}_{X_1}, \ldots, \mathcal{P}_{X_p})$, $\mathcal{P}_Y = (\mathcal{P}_{Y_1}, \ldots, \mathcal{P}_{Y_p})$, $\mathcal{P}_Z = (\mathcal{P}_{Z_1}, \ldots, \mathcal{P}_{Z_r})$, where $p = (n - r)/2$, such that the operators $\mathbf{L}$ and $\underset{\sim}{\mathbf{L}}$ have the form

$$L_i = \sum_{s=1}^{p} \varphi_{is}(\mathbf{X}) \mathcal{P}_{X_s} + \sum_{\alpha=1}^{r} \varphi_{i\alpha}(\mathbf{X}) \mathcal{P}_{Z_\alpha},$$
$$\underset{\sim}{L}_i = \sum_{s=1}^{p} \underset{\sim}{\varphi}_{is}(\mathbf{Y}) \mathcal{P}_{Y_s} + \sum_{\alpha=1}^{r} \underset{\sim}{\varphi}_{i\alpha}(\mathbf{Y}) \mathcal{P}_{Z_\alpha}. \tag{1.10}$$

There do not exist canonical point transformations reducing (1.6) to the form (1.10). Thus the right generators depend only on the right variables $\mathbf{X}$ and $\mathcal{P}_X$ and on the invariant operators $\mathcal{P}_Z$, while the left generators depend only on the left variables $\mathbf{Y}$ and $\mathcal{P}_Y$ and on the invariants $\mathcal{P}_Z$. The right and left generators of the regular representation are linear forms in the canonical momenta. The coordinates $\mathbf{X}(\mathbf{q}, \mathbf{p})$, $\mathcal{P}_X(q, p)$, $\mathbf{Y}(\mathbf{q}, \mathbf{p})$, $\mathcal{P}_Y(q, p)$, $\mathbf{Z}(\mathbf{q}, \mathbf{p})$, $\mathcal{P}_Z(q, p)$ are very convenient variables since with respect to them it is clear explicitly that the right and left generators, which commute with each other, depend on different variables. Naturally, canonical transformations which mix up $\mathbf{X}$ with $\mathcal{P}_X$ and $\mathbf{Y}$ with $\mathcal{P}_Y$, give new left and right variables which are just as well suited for separating the variables in the right and left generators as the original variables $\mathbf{X}$, $\mathcal{P}_X$, $\mathbf{Y}$, $\mathcal{P}_Y$. This nonuniqueness in the choice of "good" right and left variables can be exploited so as to realize the irreducible representations in various forms. This will be demonstrated below using various groups as examples. The left variables $\mathbf{Y}$ and $\mathcal{P}_Y$ remain unchanged under right translations, while the variables $\mathbf{X}$, $\mathcal{P}_X$ and $\mathbf{Z}$ become the variables

$$\begin{aligned} &\mathbf{X}(g) = \mathbf{X}_g(\mathbf{X}), \\ &\mathcal{P}_X(g) = \mathcal{P}_{X_g}(\mathcal{P}_X, \mathcal{P}_Z), \\ &\mathbf{Z}(g) = \mathbf{Z} + \mathfrak{f}_g(\mathbf{X}), \\ &\mathfrak{f}_{g_1 g_2}(\mathbf{X}) = \mathfrak{f}_{g_1}(\mathbf{X}(g_2)) + \mathfrak{f}_{g_2}(\mathbf{X}). \end{aligned} \tag{1.11}$$

Under the action of left translation by a point with coordinates (q, p), a point with coordinates $(\mathbf{X}, \mathbf{Y}, \mathbf{Z}, \mathcal{P}_X, \mathcal{P}_Y, \mathcal{P}_Z)$ goes into the point with coordinates $\underset{\sim}{\mathbf{Z}}(g)$, $\mathcal{P}_X$, $\mathcal{P}_Y(g)$, $\mathcal{P}_Z)$, where

$$\begin{aligned} &\mathbf{Y}(g) = \mathbf{Y}_g(\mathbf{Y}), \\ &\mathcal{P}_Y(g) = \mathcal{P}_{Y_g}(\mathcal{P}_Y, \mathcal{P}_Z), \\ &\underset{\sim}{\mathbf{Z}}(g) = \mathbf{Z} + \underset{\sim}{\mathfrak{f}}_g(\mathbf{Y}), \\ &\underset{\sim}{\mathfrak{f}}_{g_1 g_2}(\mathbf{Y}) = \underset{\sim}{\mathfrak{f}}_{g_1}(\mathbf{Y}(g_2)) + \underset{\sim}{\mathfrak{f}}_{g_2}(\mathbf{Y}). \end{aligned} \tag{1.12}$$

Thus the coordinates $\mathcal{P}_Z$ are unchanged under left and right translations, the right variables are unchanged under left translations, and the left variables are unchanged under right translations.

It is important that the cyclic variable $\mathbf{Z}$ is only translated under the action of right translations, and moreover, by a vector $\mathbf{f}_g(\mathbf{X})$ depending only on the right coordinates $\mathbf{X}$, while under left translation it is changed by the vector $\underset{\sim}{\mathbf{f}}_g(\mathbf{Y})$ depending only on the left coordinates $\mathbf{Y}$ [cf. (1.11) and (1.12)].

Equations (1.11), (1.12) play a very important role in our considerations. They show what happens to points of group phase space under the group action corresponding to either the right or the left regular representation. The geometric picture of this action naturally does not depend on which variables are chosen for studying the action, i.e., on the choice of coordinates in phase space. But whereas in the old coordinates q and p the picture is extremely complicated, in the new variables it is clear by inspection that most of the coordinates of a point in phase space, viz., all the left (right) coordinates and momenta, are unchanged under right (left) translation. The momentum invariants corresponding to the cyclic variables are unchanged under all translations. The cyclic coordinates themselves undergo pure translations depending only on their coordinates. The nature of the new canonical variables is especially important, i.e., whether they are noncompact or compact or defined by univalent or multivalued functions of the old variables $\mathbf{q}$ and $\mathbf{p}$, or the kinds of properties the functions $f_g(\mathbf{X})$ and $\underset{\sim}{f}_g(\mathbf{Y})$ have in Eqs. (1.11) and (1.12). Knowledge of these properties allows us to give a precise meaning to the equations for the kernels of the right and left regular representations, viz., the terms with δ-function in the cyclic variables.

In order to obtain irreducible representations in Eqs. (1.10) it is necessary to replace the canonical momenta in accordance with the quantization rule $\mathcal{P}_X \to \partial/\partial X$, $\mathcal{P}_Y \to \partial/\partial Y$, $\mathcal{P}_Z \to \partial/\partial Z = \rho$. Since the coordinate $\mathbf{Z}$ is cyclic, the quantities $\partial/\partial Y$ may be assumed to be numbers. The procedure described in classical language is equivalent in the quantum language to the fact that there exists an operator $\hat{G}$ with kernel $G(\mathbf{X}, \mathbf{Y}, \mathbf{Z}, \mathbf{q})$, and functions in the regular representation are replaced as follows:

$$\psi' = \hat{G}\psi. \tag{1.13}$$

In the new representation the kernels of the right $\hat{G}\hat{T}_r^{\mathrm{r}}\hat{G}^{-1}$ and left $\hat{G}\hat{T}_l^{\mathrm{r}}\hat{G}^{-1}$ regular representations factorize and have the form

$$\begin{aligned} T_r^r(\mathbf{X}, \mathbf{X}', \mathbf{Y}, \mathbf{Y}', \mathbf{Z}, \mathbf{Z}', g) &= T_r^r(\mathbf{X}, \mathbf{X}', g)\,\delta(\mathbf{Y}-\mathbf{Y}')\,\delta(\mathbf{Z}-\mathbf{Z}'-\mathbf{f}_g(\mathbf{X})),\\ T_l^r(\mathbf{X}, \mathbf{X}', \mathbf{Y}, \mathbf{Y}', \mathbf{Z}, \mathbf{Z}', g) &= T_l^r(\mathbf{Y}, \mathbf{Y}', g)\,\delta(\mathbf{X}-\mathbf{X}')\,\delta(\mathbf{Z}-\mathbf{Z}'-\underset{\sim}{\mathbf{f}}_g(\mathbf{Y})). \end{aligned} \tag{1.14}$$

Since the coordinate Z is cyclic and the momentum $\mathcal{P}_Z$ appears linearly in the generators of the regular representation (1.14) (the kernels $T_r^r(\mathbf{X}, \mathbf{X}', \mathbf{Z}, \mathbf{Z}', g)$, $T_l^r(\mathbf{Y}, \mathbf{Y}', \mathbf{Z}, \mathbf{Z}', g)$ define reducible representation equivalent to the regular representation), the irreducible components are chosen by expanding the function $\psi(\mathbf{X}, \mathbf{Z})$ or $\psi(\mathbf{Y}, \mathbf{Z})$ into a Fourier integral with respect to the variable $\mathbf{Z}$. In the Fourier components a representation is realized which is already irreducible, which is determined by a vector $\underset{\sim}{\rho} = \rho(\rho_1, \ldots, \rho_r)$, where $\rho = \mathcal{P}_Z$. The infinitesimal operators of the left and right regular representations are obtained from the generators (1.10) by the quantization rule described above and have the form

$$\begin{aligned} L_i &= \sum_{s=1}^{p} \varphi_{is}(\mathbf{X}) \frac{\partial}{\partial X_s} + \sum_{\alpha=1}^{r} \varphi_{i\alpha}(\mathbf{X}) \frac{\partial}{\partial Z_\alpha},\\ \underset{\sim}{L}_i &= \sum_{s=1}^{p} \underset{\sim}{\varphi}_{is}(\mathbf{Y}) \frac{\partial}{\partial Y_s} + \sum_{\alpha=1}^{r} \underset{\sim}{\varphi}_{i\alpha}(\mathbf{Y}) \frac{\partial}{\partial Z_\alpha}. \end{aligned} \tag{1.15}$$

From the point of view of the motion of points in group phase space, it is very easy to determine the irreducible components of the left or right regular representation. Assume that we are considering the right regular representation. Then if we take an initial point $\mathbf{X}_0$, $\mathcal{P}_{X_0}$, $\mathbf{Z}_0$ in phase space with fixed coordinates $\mathcal{P}_Z$ and act on this point by all right translations (the left coordinates Y and $\mathcal{P}_Y$ being fixed and unchanged in this process), the initial point sweeps out some surface in phase space. This surface corresponds precisely to an irreducible representation of the Lie group. The set of such surfaces determined by all possible values of the momenta $\mathcal{P}_Z$, corresponds to the set of irreducible representations of the Lie group. Thus the construction of irreducible representations can be viewed from the point of view of geometric constructions in phase space. This same problem can also be formulated as the problem of ordinary quantization of a simple classical motion along a surface in group phase space. We remark that the functions $\varphi_{i\alpha}(\mathbf{X})$ and $\underset{\sim}{\varphi}_{i\alpha}(\mathbf{Y})$ in Eq. (1.15) for the generators are related to the vectors $\mathbf{f}_g(\mathbf{X})$ and $\underset{\sim}{\mathbf{f}}_g(\mathbf{Y})$, by which the cyclic variable $\mathbf{Z}$ is translated, by an obvious equation since the functions $\varphi_{i\alpha}(X)$ and $\underset{\sim}{\varphi}_{i\alpha}(Y)$ are obtained from the functions $\mathbf{f}_g(X)$ and $\mathbf{f}_g(Y)$ by taking the Taylor series expansion if the group element g is taken in the infinitesimal form $g = 1 + F$. Consequently, if the generators of the regular representation in the simplest form (1.15) are known, the infinitesimal (and hence the integral) translations of the cyclic momenta can be found from the coefficients of the cyclic momenta $\mathcal{P}_Z$. And conversely, if the translations of the cyclic variable **Z** are known, we know the important terms in the generators which are proportional to the cyclic momenta.

A well-known result from the theory of functional groups (cf. [30]) can serve to suggest that the conjecture stated above is correct. The commutation relations (in the language of the classical Poisson brackets) of the generators of a Lie group represent a special case of functional groups. It is proved for such groups that there exist canonically conjugate coordinates $X_s(L)$ and momenta $\mathcal{P}_{X_s}(L)$ (i.e., there exist functions X_s and $\mathcal{P}_{X_m}$, for which the Poisson brackets satisfy $\{\mathcal{P}_{X_m}, X_s\} = \delta_{ms}$). In addition, there exist cyclic momenta $\mathcal{P}_Z(L)$, for which $\{\mathcal{P}_{Z_\alpha}, L_i\} = 0$. The system of relations defining X_s, $\mathcal{P}_{X_m}$, $\mathcal{P}_{Z_\alpha}$, can be solved in the opposite direction and the group generators turn out to be expressed in terms of a system of canonically conjugate coordinates and momenta plus cyclic variables. Thus,

$$L_i = L_i\,(X_s, \mathcal{P}_{X_m}, \mathcal{P}_{Z_\alpha}); \quad s, m = 1, \ldots, p, \ \alpha = 1, \ldots, r. \tag{1.16}$$

The only fact which has not been proved is the assertion that there exists a transformation reducing L_i in (1.16) to a form linear in the momenta ($\mathcal{P}_X$ and $\mathcal{P}_Z$), which is essential for the application of a simple quantization rule. This point of view also clarifies the commutativity of the right and left Lie group generators expressed as functions of the right and left variables X, $\mathcal{P}_X$, Y, $\mathcal{P}_Y$. Indeed, the left coordinates and momenta depend only on the generators of the left regular representation, while the right variables depend only on the generators of the right regular representation.

We remark a simple but extremely important fact connected with the passage from certain coordinates to others. If the functions expressing one set of coordinates in terms of another, viz., the coordinates L in terms of **X**, $\mathcal{P}_X$, $\mathcal{P}_Z$ [cf. (1.16)] and conversely, possess singularities, then it is easy to see by studying the Jacobian of the transformation composed of the quantities $\partial L/\partial X$, $\partial L/\partial \mathcal{P}_X$, $\partial L/\partial \mathcal{P}_Z$, that the transformation from the old variables to the new ones becomes difficult in the equations where, for example, integration is performed over the variables **L**. Moreover, if there is a δ-type function in the variables **L**, one cannot pass from it to δ-type functions of the variables **X**, $\mathcal{P}_X$, $\mathcal{P}_Z$ in the obvious way. In this sense, equations with integration over the variables **L** are not equivalent to equations with integration over phase space. Of course, it is possible to develop a procedure for determining the nonuniqueness and give a meaning to the singularities in the Jacobian of the transformation, but this is a separate problem. The equations involving integration over group phase space are in this sense more general than those involving integration with respect to the variables **L**. This fact is essential in the study of the "wild" Lie groups, examples of which are given in [14].

In principle, the form of the generators (1.15) is sufficient for the construction of the representations of Lie groups and their group algebras, and in fact all known methods for solving these problems either explicitly or implicitly reduce to finding the generators (1.15) (or their exponentials) in the required number of parameters. The usual standard procedure for constructing representations from (1.15) reduces to integration of system (1.15), which is equivalent to solving the first-order linear equation of the form

$$i\,\frac{\partial S}{\partial t} = \left(\sum_{i=1}^{n} iL_i u_i\right) S, \tag{1.17}$$

where t is a continuous variable in the direction of the corresponding group translation u_i.

It is easy to see that the Green function $S(\mathbf{X}, \mathbf{X}', \mathbf{Z}, \mathbf{Z}', g)$ satisfies the system of 2n equations

$$\begin{aligned} &\mathbf{X}_0(\mathbf{X}, \mathbf{Z}, g) S = \mathbf{X}' S, \\ &\mathbf{Z}_0(\mathbf{X}, \mathbf{Z}, g) S = \mathbf{Z}' S, \\ &\mathcal{P}_{0\mathbf{X}}(\mathbf{X}, \mathbf{Z}, g) S = -\frac{\partial}{\partial \mathbf{X}'} S, \quad \frac{\partial S}{\partial \mathbf{Z}} = -\frac{\partial S}{\partial \mathbf{Z}'}. \end{aligned} \tag{1.18}$$

The operators $\mathbf{X}_0$, $\mathbf{Z}_0$, $\mathcal{P}_{0\mathbf{X}}$ in Eqs. (1.18) coincide with $\mathbf{X}(g^{-1})$, $\mathbf{Z}(g^{-1})$, $\mathcal{P}_{\mathbf{X}}(g^{-1})$ in Eqs. (1.11) and correspond to the 2n integrals of motion which describe the initial points of a trajectory of the Lie group (dynamical system) in group phase space (cf. the discussion of this point in Section 8 below). If these operators are known, it is sometimes technically easier to solve the system of equations (1.18) rather than (1.17).

We observe that the canonical transformation from the old variables **q**, **p** in phase space to the new variables $(\mathbf{X}, \mathbf{Y}, \mathbf{Z}, \mathcal{P}_{\mathbf{X}}, \mathcal{P}_{\mathbf{Y}}, \mathcal{P}_{\mathbf{Z}})$ is not a point transformation. Its generating function $S(\mathbf{q}, \mathbf{X}, \mathbf{Y}, \mathbf{Z})$ (the action) satisfies the equations

$$\begin{aligned} &\sum_{k=1}^{n} \Psi_{ik}(\mathbf{q}) \frac{\partial S}{\partial q_k} = -\sum_{s=1}^{p} \varphi_{is}(\mathbf{X}) \frac{\partial S}{\partial X_s} + \sum_{\alpha=1}^{r} \varphi_{i\alpha}(\mathbf{X}) \left(-\frac{\partial S}{\partial Z_\alpha}\right), \\ &\sum_{k=1}^{n} \underset{\sim}{\Psi}_{ik}(\mathbf{q}) \frac{\partial S}{\partial q_k} = -\sum_{s=1}^{p} \underset{\sim}{\varphi}_{is}(\mathbf{Y}) \frac{\partial S}{\partial Y_s} - \sum_{\alpha=1}^{r} \underset{\sim}{\varphi}_{i\alpha}(\mathbf{Y}) \frac{\partial S}{\partial Z_\alpha}. \end{aligned} \tag{1.19}$$

The consistency of system (1.19) determines the conditions under which there exists a generating function in the variables chosen. These conditions are equivalent to the requirement that the corresponding Jacobian be nonzero and are worked out, e.g., in [23]. In the case where system (1.19) is inconsistent, it is necessary to choose generating functions of other variables, as in the usual problems of mechanics and thermodynamics. (We remark that in the quantum mechanical case, the corresponding action always exists.)

The Green function $G(\mathbf{q}, \mathbf{X}, \mathbf{Y}, \mathbf{Z})$ [the kernel of the operator $\hat{G}$ in (1.13)] satisfies equations which follow from comparing Eqs. (1.15) and (1.6), e.g.,

$$\sum_{k=1}^{n} \hat{G}^{-1} \left(\Psi_{ik} \frac{\partial}{\partial q_k}\right) \hat{G} = \sum_{s=1}^{p} \varphi_{is}(\mathbf{X}) \frac{\partial}{\partial X_s} + \sum_{\alpha=1}^{r} \varphi_{i\alpha}(\mathbf{X}) \frac{\partial}{\partial Z_\alpha}, \tag{1.20}$$

which gives the equations

$$\begin{aligned} &\sum_{k=1}^{n} \Psi_{ik}(\mathbf{q}) \frac{\partial G}{\partial q_k} = -\sum_{s=1}^{p} \varphi_{is}(\mathbf{X}) \frac{\partial G}{\partial X_s} + \sum_{\alpha=1}^{r} \varphi_{i\alpha}(\mathbf{X}) \left(-\frac{\partial G}{\partial Z_\alpha}\right) + \sum_{s=1}^{p} \frac{\partial \varphi_{is}}{\partial X_s} G, \\ &\sum_{k=1}^{n} \underset{\sim}{\Psi}_{ik}(\mathbf{q}) \frac{\partial G}{\partial q_k} = -\sum_{s=1}^{p} \underset{\sim}{\varphi}_{is}(\mathbf{Y}) \frac{\partial G}{\partial Y_s} + \sum_{s=1}^{p} \frac{\partial \underset{\sim}{\varphi}_{is}}{\partial Y_s} G - \sum_{\alpha=1}^{r} \underset{\sim}{\varphi}_{i\alpha}(\mathbf{Y}) \frac{\partial G}{\partial Z_\alpha}. \end{aligned} \tag{1.21}$$

We replace the function in (1.21) by D, where

$$G = \exp(D). \tag{1.22}$$

For the function D we have the system of inhomogeneous equations

$$\begin{aligned} &\sum_{k=1}^{n} \Psi_{ik}(\mathbf{q}) \frac{\partial D}{\partial q_k} = -\sum_{s=1}^{p} \varphi_{is}(\mathbf{X}) \frac{\partial D}{\partial X_s} - \sum_{\alpha=1}^{r} \varphi_{i\alpha}(\mathbf{X}) \frac{\partial D}{\partial Z_\alpha} - \sum_{s=1}^{p} \frac{\partial \varphi_{is}}{\partial X_s}, \\ &\sum_{k=1}^{n} \underset{\sim}{\Psi}_{ik}(\mathbf{q}) \frac{\partial D}{\partial q_k} = -\sum_{s=1}^{p} \underset{\sim}{\varphi}_{is}(\mathbf{Y}) \frac{\partial D}{\partial Y_s} - \sum_{\alpha=1}^{r} \underset{\sim}{\varphi}_{i\alpha}(\mathbf{Y}) \frac{\partial D}{\partial Z_\alpha} - \sum_{s=1}^{p} \frac{\partial \underset{\sim}{\varphi}_{is}}{\partial Y_s}. \end{aligned} \tag{1.23}$$

If the sum $\sum_{s=1}^{p} \partial\varphi_{is}/\partial X_s = 0$, system (1.23) coincides with the system of equations for the action S and S coincides with D (cf. Section 4, the example of the Heisenberg group). The case

$$\sum_{s=1}^{p} \frac{\partial \varphi_{is}(\mathbf{X})}{\partial X_s} = \sum_{\alpha=1}^{r} \varphi_{i\alpha}(\mathbf{X}) \mu_\alpha \tag{1.24}$$

can occur (here the μ_α are numbers). Then system (1.23) for the Green function D coincides with system (1.19) for the action S if the substitution $\partial/\partial Z_\alpha \to \partial/\partial Z_\alpha + \mu_\alpha$ is made. It is obvious from system (1.19) that the solution S depends on the cyclic variable Z in the following standard way:

$$S = -\rho \mathbf{Z} + S'(\mathbf{X}, \mathbf{Y}, \mathbf{q}) \tag{1.25}$$

(here $\rho = (\rho_1, \ldots, \rho_r)$ is an r-tuple of arbitrary numbers). The truncated action $S'(\mathbf{X}, \mathbf{Y}, \mathbf{q})$ satisfies the inhomogeneous system of equations

$$\begin{aligned} \sum_{k=1}^{n} \Psi_{ik} \frac{\partial S'}{\partial q_k} &= -\sum_{s=1}^{p} \varphi_{is}(\mathbf{X}) \frac{\partial S'}{\partial X_s} + \sum_{\alpha=1}^{r} \rho_\alpha \varphi_{i\alpha}(\mathbf{X}), \\ \sum_{k=1}^{n} \underset{\sim}{\Psi}_{ik} \frac{\partial S'}{\partial q_k} &= -\sum_{s=1}^{p} \underset{\sim}{\varphi}_{is}(\mathbf{Y}) \frac{\partial S'}{\partial Y_s} + \sum_{\alpha=1}^{r} \rho_\alpha \underset{\sim}{\varphi}_{i\alpha}(\mathbf{Y}). \end{aligned} \tag{1.26}$$

The function D also breaks up into a sum

$$D = -\rho \mathbf{Z} + D'(\mathbf{X}, \mathbf{Y}, \mathbf{q}). \tag{1.27}$$

The function $D'(\mathbf{X}, \mathbf{Y}, \mathbf{q})$ satisfies an inhomogeneous system of equations which coincides with (1.26) except for an altered right-hand side. The term $\sum_{\alpha=1}^{r} \rho_\alpha \varphi_{i\alpha}$ is replaced by

$$\sum_{\alpha=1}^{r} \rho'_\alpha \varphi_{i\alpha} = \sum_{\alpha=1}^{r} \rho_\alpha \varphi_{i\alpha} + \sum_{s=1}^{p} \frac{\partial \varphi_{is}}{\partial X_s}. \tag{1.28}$$

However, solving a system of equations with right-hand side (1.28) is no more difficult than solving system (1.26). If the explicit form of the generators of the regular representation in the "good" variables is unknown, we can decide if the function D coincides with S by considering the matrices $\Psi_{ik}(\mathbf{q})$, $\underset{\sim}{\Psi}_{ik}(\mathbf{q})$ in the usual group variables. To do this it is necessary to consider the equation for the generating function of the inverse canonical transformation $S'(\mathbf{X}, \mathbf{Y}, \mathbf{Z}, \mathbf{q})$ and the kernel G' of the inverse transformation $\hat{G}^{-1}$. The kernel G' is equal to exp (D') in the case where

$$\sum_{k=1}^{n} \frac{\partial \Psi_{ik}(\mathbf{q})}{\partial q_k} = 0, \qquad \sum_{k=1}^{n} \frac{\partial \underset{\sim}{\Psi}_{ik}(\mathbf{q})}{\partial q_k} = 0. \tag{1.29}$$

These conditions are necessary and are obtain by writing down the equations analogous to Eqs. (1.19)-(1.23). It is of interest that in the case where Eq. (1.24) holds for the equation for the function G, we have the analogous equality for the function G' also, except that its left-hand side is expressed in terms of the variables **q**:

$$\begin{aligned} \sum_{k=1}^{n} \frac{\partial \Psi_{ik}}{\partial q_k}(\mathbf{q}) &= \sum_{\alpha=1}^{r} \varphi_{i\alpha}(\mathbf{X}) \mu'_\alpha, \\ \sum_{k=1}^{n} \frac{\partial \underset{\sim}{\Psi}_{ik}(\mathbf{q})}{\partial q_k} &= \sum_{\alpha=1}^{r} \underset{\sim}{\varphi}_{i\alpha}(\mathbf{Y}) \mu''_\alpha, \end{aligned} \tag{1.30}$$

where the μ'_α, μ''_α are numbers.

If we now make use of the fact that the transformations of (q, p) under right and left translations are canonical, we can rederive the well-known expressions for the invariant group measure. For both the right and left translations, the fact that the phase volumes are equal implies the Jacobians are equal to one:

$$d\mathbf{q}\,d\mathbf{p} = d\mathbf{q}_g\,d\mathbf{p}_g,$$
$$\begin{Vmatrix} \partial\mathbf{q}_g/\partial\mathbf{q} & \partial\mathbf{q}_g/\partial\mathbf{p} \\ \partial\mathbf{p}_g/\partial\mathbf{q} & \partial\mathbf{p}_g/\partial\mathbf{p} \end{Vmatrix} = 1. \tag{1.31}$$

Because the transformations are point transformations,

$$\partial\mathbf{q}_g/\partial\mathbf{p} = \partial\mathbf{p}_g/\partial\mathbf{q} = 0. \tag{1.32}$$

Therefore,

$$\|\partial\mathbf{q}_g/\partial\mathbf{q}\| = \|\partial\mathbf{p}/\partial\mathbf{p}_g\|. \tag{1.33}$$

But under right translations the left generators are unchanged,

$$\underset{\sim}{\mathbf{L}}(\mathbf{q}_g, \mathbf{p}_g) = \underset{\sim}{\mathbf{L}}(\mathbf{q}, \mathbf{p}), \quad \underset{\sim}{\Psi}(\mathbf{q}_g)\,\mathbf{p}_g = \underset{\sim}{\Psi}(\mathbf{q})\,\mathbf{p},$$
$$[\Psi(g_1)\,\underset{\sim}{\Psi}^{-1}(g_2)]\,[\Psi(g_2)\,\underset{\sim}{\Psi}^{-1}(g_2)] = [\Psi(g_1g_2)\,\underset{\sim}{\Psi}^{-1}(g_1g_2)]. \tag{1.34}$$

Thus

$$\|\partial\mathbf{p}/\partial\mathbf{p}_g\| = \|\underset{\sim}{\Psi}^{-1}(\mathbf{q})\,\underset{\sim}{\Psi}(\mathbf{q}_g)\| = \det\,\underset{\sim}{\Psi}(\mathbf{q}_g)/\det\,\underset{\sim}{\Psi}(\mathbf{q}). \tag{1.35}$$

This and (1.33) imply the equality

$$[\det\,\underset{\sim}{\Psi}(\mathbf{q})]^{-1}\,d\mathbf{q} = [\det\,\underset{\sim}{\Psi}(\mathbf{q}_g)]^{-1}\,d\mathbf{q}_g. \tag{1.36}$$

Consequently, the right-invariant measure $\mu(g) = \det\underset{\sim}{\Psi}^{-1}(\mathbf{q})$ on the group is given by the determinant of the matrix $\underset{\sim}{\Psi}$ of left generators, and the left-invariant measure by the matrix Ψ of right generators. For these measures to be equal, it is necessary that

$$|\det\Psi| = |\det\underset{\sim}{\Psi}|, \quad |\det\Psi\underset{\sim}{\Psi}^{-1}| = 1. \tag{1.37}$$

The central point in the program described above is the determination of the explicit form of the generators (1.15). To this end we translate the construction of induced representations, widely used by mathematicians, into the infinitesimal language.

2. INDUCED REPRESENTATIONS (INFINITESIMAL FORM)

The construction of induced representations [11] mentioned in the introduction is an effective method for constructing representations of groups, including Lie groups. In order to make the discussion complete, we present this construction; in certain cases it also permits us to obtain irreducible group representations. Let the group G have a subgroup R. Then every element of the group G can be written in the form

$$g = kr. \tag{2.1}$$

The set of elements g such that for given k the element r varies over all of R is called the left coset of the group G by the subgroup R (written G/R) defined by the element $k \subset G$. It is easy to show that all the elements of G lie in coset classes in such a way that no element can simultaneously belong to two distinct coset classes defined by different elements k of G. The set of coset classes forms a homogeneous space on which the group G acts transitively. This means that any two coset classes can be taken into each other by an element of G. Let functions be given on the group G on which we impose an additional restriction as compared to the definition of the regular representation, viz., assume that f(g) is such that (g = kr)

$$f(gr_0) = \exp\,[i\chi(r_0)]\cdot f(g). \tag{2.2}$$

The function $\chi(r_0)$ is the character of a one-dimensional representation of the group R.

We then define a representation by the same formula as that for the left regular representation

$$T(g_0)\,f(g) = f(g_0^{-1}g). \tag{2.3}$$

It is clear that the transformed functions have the same property (2.2), and we have the equation of a representation since $T(g_0)T(g_1) = T(g_0g_1)$, as in the case of the regular representation. This representation is called an induced representation. We remark that one can construct a representation analogously by starting from the decomposition g = rk, where r is an element of the subgroup R. We can then consider the right coset classes and construct the induced representation of G using the formula analogous to that for the right regular representation. Now consider a Lie group G with Lie subgroup R. It is then possible to write the generators of the left regular representation $\hat{\underset{\sim}{F}}^{\alpha}$ of G in the parameterization (2.1). They have the form

$$\hat{\underset{\sim}{F}}^{\alpha} = \hat{Y}^{\alpha}(k) + \sum_{\mu=1}^{m} a_{\mu}^{\alpha}(k)\,\hat{\underset{\sim}{X}}^{\mu}(r) \qquad (\alpha = 1, \ldots, n). \tag{2.4}$$

Here the first-order differential operators $\hat{Y}^{\alpha}(k)$ and $\hat{\underset{\sim}{X}}^{\mu}(r)$ possess the following properties:

$$\hat{Y}^{\alpha}r = 0,\ \hat{\underset{\sim}{X}}^{\mu}r = -X^{\mu}r,\ \hat{\underset{\sim}{X}}^{\mu}k = 0 \quad (\mu = 1, 2, \ldots, m;\ m = \dim R). \tag{2.5}$$

The operators $\hat{\underset{\sim}{X}}^{\mu}(r)$ are the generators of the left regular representation on the subgroup R (the index μ varies over all the generators of R). The differential operators $\hat{Y}^{\alpha}(k)$ are the generators of the left quasiregular representation of G on the homogeneous manifold G/R, i.e., we have

$$[\hat{Y}^{\alpha}, \hat{Y}^{\beta}] = c_{\alpha\beta}^{\gamma}\hat{Y}^{\gamma}. \tag{2.6}$$

The indices α, β, γ run over all the values of the generators of G. In Eq. (2.5) X^{μ} is the matrix in the matrix representation chosen of Eq. (2.1) corresponding to a generator of the subgroup R. The functions $a_{\mu}^{\alpha}(k)$ on the homogeneous manifold G/R are easily calculated from the following equation:

$$a_{\mu}^{\alpha}(k) = \mathrm{Tr}\,[X_{\mu}(k^{-1}\hat{\underset{\sim}{F}}^{\alpha}k + k^{-1}F^{\alpha}k)]. \tag{2.7}$$

Here the matrices X_{μ} satisfy the equation

$$\mathrm{Tr}\,(X_{\mu}X^{\rho}) = \delta_{\mu}^{\rho}, \tag{2.8}$$

while the operators $\hat{\underset{\sim}{F}}^{\alpha}$ are the generators of the left regular representation of G and act on the matrix elements of the matrix k in Eq. (2.7); F^{α} is the matrix of a generator of G in the matrix representation chosen. The equation

$$\hat{\underset{\sim}{F}}^{\alpha}g = -F^{\alpha}g, \tag{2.9}$$

holds which together with (2.1) and (2.5) easily implies Eq. (2.7). Since it has been shown that we have constructed representations of G, the functions $a^{\alpha}_{\mu}(k)$ satisfy the following relations which follow from the commutation relations for the generators of the group G:

$$\underset{\sim}{\hat{F}}^{\alpha} a^{\beta}_{\sigma} - \underset{\sim}{\hat{F}}^{\beta} a^{\alpha}_{\sigma} = c^{\gamma}_{\alpha\beta} a^{\gamma}_{\sigma} - a^{\alpha}_{\mu} \underset{\sim}{c}^{\sigma}_{\mu\rho} a^{\beta}_{\rho}. \tag{2.10}$$

Here the $\underset{\sim}{c}^{\sigma}_{\mu\rho}$ are the structure constants of the subgroup R:

$$[X^{\mu}, X^{\rho}] = \sum_{\sigma=1}^{m} \underset{\sim}{c}^{\sigma}_{\mu\rho} X^{\sigma}. \tag{2.11}$$

The expressions in (2.7) are a solution of the nonlinear equations (2.10).

Let us see how the generators of the left regular representation of G (2.4) look when applied to the functions (2.2). Acting on a function of the form (2.2), we have the following generators for the induced representation:

$$\underline{\hat{F}}^{\alpha} = \hat{Y}^{\alpha}(k) + i \sum_{\mu=1}^{m} a^{\alpha}_{\mu}(k)\, \underset{\sim}{\hat{X}}^{\mu} \chi(r). \tag{2.12}$$

The function $\underset{\sim}{\hat{X}}^{\mu}\chi(r)$ is a number (independent of r). In the case where a more general induced representation is given, when f(g) is a vector function on which a finite-dimensional representation of the subgroup R is realized, the generators of the induced representation are given by the formula

$$\underline{\hat{F}}^{\alpha} = \hat{Y}^{\alpha}(k) + \sum_{\mu=1}^{m} a^{\alpha}_{\mu}(k)\, X^{\mu}. \tag{2.13}$$

In Eq. (2.13) the X^{μ} are the generators of a finite-dimensional representation of R. Equation (2.12) is a particular case of Eq. (2.13) when X^{μ} is the generator of a one-dimensional representation of the subgroup R ($i\underset{\sim}{X}^{\mu}\chi(r) = X^{\mu}$).

In the case where the matrices k in Eq. (2.1) form a subgroup K of G, it is possible to write the generators $\hat{Y}^{\alpha}(k)$ of the quasiregular representation of G in terms of the generators of translations on the group K($\hat{X}^{\nu}(k)$), viz.,

$$Y^{\alpha}(k) = \sum_{\nu=1}^{\dim K} b^{\alpha}_{\nu}(k)\, \hat{X}^{\nu}(k). \tag{2.14}$$

We now consider in more detail the algebraic structure of the infinitesimal operators of the regular representation for a number of typical cases.

Let G be an arbitrary nonsemisimple Lie group. Then by the Levi–Mal'tsev theorem there exists a decomposition of the form

$$G = KZ, \tag{2.15}$$

where K is a semisimple subgroup of G and Z is the radical. Assume further that K and Z are nontrivial. In the Lie algebra of G, choose a basis of generators $\{K_{\alpha}, Z_i, \alpha = 1, 2, \ldots, A, i = 1, 2, \ldots, N\}$. The generators K_{α} and Z_i are basis generators in the Lie algebras of the groups K and Z respectively, and they satisfy the commutation relations

$$[K_{\alpha}, K_{\beta}] = A^{\gamma}_{\alpha\beta} K_{\gamma}; \quad [K_{\alpha}, Z_i] = D^{l}_{\alpha i} Z_l; \quad [Z_i, Z_k] = C^{l}_{ik} Z_l.$$

The structure constants $A^{\gamma}_{\alpha\beta}$, $D^{l}_{\alpha i}$, C^{l}_{ik} satisfy a number of relations which are a consequence of the Jacobi identity. Every element X can be expanded in the basis $\{K_\alpha, Z_i\}$

$$X = \sum_\alpha \mathrm{Tr}\,(XK^\alpha)\cdot K_\alpha + \sum_i \mathrm{Tr}\,(XZ^i)\,Z_i,$$

where $\{K^\alpha, Z^i\}$ is the basis dual to $\{K_\alpha, Z_i\}$, so that

$$\mathrm{Tr}\,(K_\alpha K^\beta) = \delta_{\alpha\beta}, \quad \mathrm{Tr}\,(K_\alpha Z^i) = \mathrm{Tr}\,(Z_i K^\alpha) = 0, \quad \mathrm{Tr}\,(Z_i Z^k) = \delta_{ik}.$$

In this case it is easy to calculate the generators of the left regular representation of G using the techniques developed in Sections 1-2.

The generator $\hat{F}_X$ corresponding to the element X of the Lie algebra of the group G has the form

$$\hat{F}_X = \sum_\alpha \tilde{\Omega}_\alpha \hat{\hat{K}}_\alpha - \sum_i \omega_i \hat{Z}_i, \tag{2.16}$$

where $\hat{Z}_i$ is the generator of the left regular representation of the group Z corresponding to the element Z_i (Z_i is a basis in the Lie algebra of Z); $\hat{\hat{K}}_\alpha$ is the generator of the right regular representation of the group K corresponding to the element K_α (K_α is a basis in K). The operators $\hat{\hat{K}}_\alpha$ and $\hat{Z}_i$ commute. The functions $\tilde{\Omega}_\alpha$ and ω_i are given by the following expressions:

$$\tilde{\Omega}_\alpha = -\mathrm{Tr}\,(K^{-1}XKK^\alpha), \quad \omega_i = -\mathrm{Tr}\,(K^{-1}XKZ^i), \tag{2.17}$$

where $K^{-1}XK$ denotes the element of the Lie algebra of G obtained by applying the adjoint representation $\mathrm{Ad}_{X^{-1}}X$ of the group G to X. We remark that $\tilde{\Omega}_\alpha$ and ω_i do not depend explicitly on the group parameters Z, i.e., $[\hat{Z}_i, \tilde{\Omega}_\alpha] = [\hat{Z}_i, \omega_k] = 0$. It is easy to verify using the identities

$$\begin{gathered}[\hat{\hat{K}}_\alpha, \hat{\hat{K}}_\beta] = A^\gamma_{\alpha\beta}\hat{\hat{K}}_\gamma; \quad [\hat{Z}_i, \hat{Z}_k] = C^l_{ik}\hat{Z}_l; \quad [\hat{\hat{K}}_\alpha, \tilde{\Omega}_\beta] = -A^\beta_{\alpha\gamma}\tilde{\Omega}_\gamma; \\ [\hat{\hat{K}}_\alpha, \omega_l] = D^l_{\alpha k}\omega_k,\end{gathered} \tag{2.18}$$

that the operators $\hat{F}_X$ determine a representation of the Lie algebra of G, i.e., $[\hat{F}_X, \hat{F}_Y] = \hat{F}_{[X,Y]}$.

For the right regular representation of G, we have

$$\hat{\hat{F}}_X = \sum_\alpha \Omega_\alpha \hat{K}_\alpha - \sum_i \tilde{\omega}_i \hat{\hat{Z}}_i, \tag{2.19}$$

$$\Omega_\alpha = \mathrm{Tr}\,(ZXZ^{-1}K^\alpha), \quad \tilde{\omega}_i = \mathrm{Tr}\,(ZXZ^{-1}X^i), \tag{2.20}$$

where ZXZ^{-1} denotes the element of the Lie algebra obtained by applying to X the operator $\mathrm{Ad}_Z X$ from the adjoint representation of the group G. The functions Ω_α, $\tilde{\omega}_i$ do not depend on the parameters of K, i.e., $[\hat{K}_\alpha, \Omega_\beta] = [\hat{K}_\alpha, \tilde{\omega}_i] = 0$. The operators $\hat{F}_X$ determine a representation of the Lie algebra of the group G, i.e.,

$$[\tilde{F}_X, \tilde{F}_Y] = \tilde{F}_{[X,\,Y]}.$$

We now construct new representations of the Lie algebra of G induced by representations of the Lie algebra of K or the Lie algebra of Z, making essential use of the construction (2.16) and (2.19). Let W_i be the operators of the representation of the Lie algebra of the group Z corresponding to the generators Z of Z acting in the vector space V. We define generators $\hat{F}_X^{\mathrm{Ind}(W,\,Z)}$ of the representation of the Lie group G induced by the representation W of the Lie algebra of Z acting in the space of functions on the group K with values in L by the formula

$$\hat{F}_X^{\mathrm{Ind}(W,Z)} = \sum_\alpha \tilde{\Omega}_\alpha \hat{\tilde{K}}_\alpha - \sum_i \omega_i \hat{W}_i, \tag{2.21}$$

where $\tilde{\Omega}_\alpha$ and ω_i are given by (2.17). It is easy to see by a straightforward verification that (2.21) defines a representation of the Lie algebra of G, i.e., $[\hat{F}_X^{\mathrm{Ind}(W,Z)}, \hat{F}_Y^{\mathrm{Ind}(W,Z)}] = \hat{F}_{[X,Y]}^{\mathrm{Ind}(W,Z)}$. In order to show this, it is necessary to use a number of the identities (2.18) stated above for the operator $\hat{\tilde{K}}_\alpha$.

A representation of the Lie algebra of the group G induced by a representation Y of the Lie albebra of K (the generator K_α corresponds to the operator $\hat{Y}_\alpha$) is constructed analogously starting from the right regular representation, and it has the form

$$\hat{F}_X^{\mathrm{Ind}(Y,K)} = \sum_\alpha \Omega_\alpha \hat{Y}_\alpha - \sum_i \tilde{\omega}_i \hat{\tilde{Z}}_i, \tag{2.22}$$

where Ω_α, $\tilde{\omega}_i$ are given by (2.20).

We observe that Eqs. (2.21) and (2.22) for an induced representation of the Lie algebra remain valid for any Lie group G which is a semidirect product of subgroups K and Z, G = KZ, where Z is normal in G.

The construction of induced Lie algebra representations described above can also be carried out for any semisimple group. Let G be a complex semisimple Lie group, u a maximal compact subgroup, τ a maximal noncompact Cartan subgroup, and let z be a maximal nilpotent subgroup. In this case the Iwasawa decomposition

$$g = z\tau u \tag{2.23}$$

is valid. Choose a Cartan–Weyl basis in the Lie algebra consisting of the generators h_i of a Cartan subalgebra, the positive roots X_α, $[h_i, X_\alpha] = \alpha_i X_\alpha$, and the negative roots $X_{-\alpha}$, $[h_i, X_\alpha] = X_{-\alpha}$. We parametrize the noncompact Cartan subgroup τ in the form $\tau = \exp\left(\sum_i \tau_i h_i\right)$, where the h_i are the generators of the Cartan subalgebra and the τ_i are real numbers. It is clear that $\tau X_\alpha \tau^{-1} = e^{\alpha(h)} X^\alpha$, where $\alpha(h) = \sum_i \alpha_i h_i$. The Lie algebra of the maximal nilpotent group z corresponds to the algebra of negative roots $\{X_{-\alpha}\}$.

A generator of the left regular representation of the group G has the form

$$\hat{F}_X = \sum_\alpha (\Omega_\alpha^* \hat{K}_{-\alpha} - \Omega_\alpha \hat{K}_\alpha + \tilde{\omega}_{-\alpha}^{(1)} \hat{\tilde{X}}_{-\alpha}^{(1)} + \tilde{\omega}_{-\alpha}^{(2)} \hat{\tilde{X}}_{-\alpha}^{(2)}) + \sum_i \left(-\dot{\varphi}_i \hat{K}_i + \dot{\tau}_i \frac{\partial}{\partial \tau_i}\right), \tag{2.24}$$

where $\hat{X}_{-\alpha}^{(1)}$, $\hat{X}_{-\alpha}^{(2)}$ are the generators of the right regular representation of the nilpotent group z corresponding to $X_{-\alpha}$ and $iX_{-\alpha}$; $\hat{K}_\alpha$ and $\hat{K}_i$ are the operators of the left regular representation of the compact group u corresponding to the roots X_α and to the generators ih_i of the compact Cartan subalgebra; the $\partial/\partial\tau_i$ are the differential operators with respect to the parameters of the maximal noncompact Cartan subgroup.

The functions Ω_α, $\omega_\alpha^{(i)}$, φ_i, τ_i have the form

$$\begin{aligned} \Omega_\alpha &= e^{-\alpha(h)} \operatorname{Tr}(z^{-1} X z X^\alpha), \\ \omega_\alpha &= \omega_\alpha^{(1)} + i\omega_\alpha^{(2)} = -\operatorname{Tr}(z^{-1} X z X^\alpha) + e^{2\alpha(h)} [\operatorname{Tr}(z^{-1} X z X^{-\alpha})]^*, \\ \dot{\tau}_i + i\dot{\varphi}_i &= -\operatorname{Tr}(z^{-1} X z h^i). \end{aligned} \tag{2.25}$$

Let Y be a representation of a compact Lie algebra of the group G acting in a vector space L. We define the induced representation of the Lie algebra of G induced by the representation Y in the space of vector functions on the group z with values in L according to the formula

$$\hat{F}_X^{\mathrm{Ind}(Y,u)} = \sum_\alpha (\Omega_\alpha^* \hat{Y}_\alpha - \Omega_\alpha \hat{Y}_{-\alpha} + \tilde{\omega}_\alpha^{(1)} \hat{\tilde{X}}_{-\alpha}^{(1)} + \tilde{\omega}_\alpha^{(2)} \hat{\tilde{X}}_{-\alpha}^{(2)}) + \sum_i \left(-\dot{\varphi}_i \hat{Y}_i + \dot{\tau}_i \frac{\partial}{\partial \tau_i} \right), \tag{2.26}$$

where Ω_α, $\hat{\omega}_\alpha^{(i)}$, $\dot{\varphi}_i$, $\dot{\tau}_i$ are given by (2.25) and the generators K_α, ih_i correspond to the operators $\hat{Y}_\alpha$, $\hat{Y}_i$. Equation (2.26) coincides with the infinitesimal operator of the representation of G induced by the representation of the compact group u corresponding to Y.

A generator of the right regular representation of the group G has the form

$$F_X = \sum_\alpha (\Omega_\alpha^* \hat{K}_{-\alpha} - \Omega_\alpha \hat{K}_\alpha + \tilde{\omega}_{-\alpha}^{(1)} \hat{\tilde{X}}_{-\alpha}^{(1)} + \tilde{\omega}_{-\alpha}^{(2)} \hat{\tilde{X}}_{-\alpha}^{(2)}) + \sum_i \left(-\dot{\varphi}_i \hat{K}_i + \dot{\tau}_i \frac{\partial}{\partial \tau_i} \right), \tag{2.27}$$

where $\hat{X}_{-\alpha}^{(1)}$, $\hat{X}_{-\alpha}^{(2)}$ are generators of the right regular representation of z; $\hat{K}_\alpha$ $\hat{K}_i$ are generators of the left regular representation of u. The functions Ω_α, $\omega_{-\alpha}^{(i)}$, $\dot{\varphi}_i$, $\dot{\tau}_i$ are given by the expressions

$$\begin{aligned} \Omega_\alpha &= \mathrm{Tr}\,(uXu^{-1}X^\alpha), \\ \omega_{-\alpha} &= \omega_{-\alpha}^{(1)} + i\omega_{-\alpha}^{(2)} = e^{-\alpha(h)}\,\{\mathrm{Tr}\,(uXu^{-1}X^{-\alpha}) + [\mathrm{Tr}\,(uXu^{-1}X^{-\alpha})]^*\}, \\ \dot{\varphi}_i + i\tau_i &= \mathrm{Tr}\,(uXu^{-1}h^i). \end{aligned} \tag{2.28}$$

Let W be the set of operators given by the representation of the Lie algebra of z acting in a vector space L, where the generators $W_{-\alpha}^{(1)}$, $W_{-\alpha}^{(2)}$ of the representation W of the Lie algebra of z correspond to the elements $X_{-\alpha}$, $iX_{-\alpha}$. We define the representation of the Lie algebra of G induced by W in the space of vector valued functions on G with values in L by the formula

$$F_X^{\mathrm{Ind}(W,z)} = \sum_\alpha (\Omega_\alpha^* \hat{K}_{-\alpha} - \Omega_\alpha \hat{K}_\alpha + \tilde{\omega}_{-\alpha}^{(1)} W_{-\alpha}^{(1)} + \tilde{\omega}_{-\alpha}^{(2)} W_{-\alpha}^{(2)}) + \sum_i \left(-\dot{\varphi}_i \hat{K}_i + \dot{\tau}_i \frac{\partial}{\partial \tau_i} \right) \tag{2.29}$$

where Ω_α, $\tilde{\omega}_{-\alpha}^{(1)}$, $\tilde{\omega}_{-\alpha}^{(2)}$, $\dot{\varphi}_i$, $\dot{\tau}_i$ are given by (2.29)

We remark that we can take the limit in Eq. (2.29) as $\tau \to \infty$, $\alpha(\tau) > 0$ with respect to the parameters of the noncompact Cartan subgroup τ. In this case the limit representation coincides with the limit representation obtained directly from the regular representation [17-20].

3. CONSTRUCTION OF REPRESENTATIONS OF NONSEMISIMPLE GROUPS WITH IDEAL

In the sequel we will keep to the following notation. An arbitrary nonsemisimple group G possesses a closed Abelian normal subgroup N, and we denote the corresponding basis elements of the Lie algebra by X_i. We introduce the notation F = G/N for the quotient group of G by N. Let $\hat{Y}_\alpha$ denote a basis for the Lie algebra of the quotient group, and let $\hat{F}_\alpha(\hat{F}'_\alpha)$ be the infinitesimal operators of the left (right) translations of the regular representation of F. In this notation the commutation relations for the Lie algebra of G have the form

$$[Y_\alpha, Y_\beta] = D_{\alpha\beta}^\gamma Y_\gamma + A_{\alpha\beta}^i X_i, \quad [Y_\alpha, X_i] = C_{\alpha i}^j X_j; \quad [X_i, X_j] = 0, \tag{3.1}$$

where $D_{\alpha\beta}^\gamma$, $A_{\alpha\beta}^\gamma$, $C_{\alpha i}^j$ are the structure constants of the Lie algebra of G and satisfy the usual Jacobi identities

$$\begin{aligned} &D_{\alpha\beta}^\gamma D_{\sigma\gamma}^\delta + D_{\beta\sigma}^\gamma D_{\alpha\gamma}^\delta + D_{\sigma\alpha}^\gamma D_{\beta\gamma}^\delta = 0, \\ &D_{\beta\alpha}^\gamma C_{\gamma i}^j + C_{\beta i}^j C_{\alpha j}^l = C_{\alpha i}^j C_{\beta j}^l, \\ &D_{\alpha\beta}^\gamma A_{\sigma\gamma}^i + A_{\alpha\beta}^i C_{\sigma i}^j + D_{\beta\sigma}^\gamma A_{\alpha\gamma}^j + A_{\beta\sigma}^i C_{\alpha i}^j + D_{\sigma\alpha}^\gamma A_{\beta\gamma}^j + A_{\sigma\alpha}^i C_{\beta i}^j = 0 \end{aligned} \tag{3.2}$$

(here and below, summation between the appropriate limits is understood over indices that are repeated twice).

We define in the space of some representation of the group G an invertible matrix Q ($QQ^{-1} = I$) satisfying the system of equations

$$\hat{F}_\alpha Q = -(Y_\alpha + h^i_\alpha X_i)\, Q,$$
$$\hat{\hat{F}}_\alpha Q = Q\,(Y_\alpha + \tilde{h}^i_\alpha X_i), \tag{3.3}$$

where the h^i_α, $\tilde{h}^i_\alpha$ are scalar functions depending on the parameters of the quotient group F, and Y_α, X_i are the infinitesimal operators of the representation considered corresponding to (3.1). Using the well-known equality

$$\hat{F}_\alpha = -F_{\alpha\beta}\hat{\hat{F}}_\beta, \tag{3.4}$$

where $F_{\alpha\beta}$ is the matrix of the adjoint representation of the group F, together with definition (3.3), it is easy to show that the relations

$$F_{\alpha\gamma} = \mathrm{Tr}\,(Q^{-1}Y_\alpha Q Y^\gamma),$$
$$F_{\alpha\beta}\tilde{h}^r_\beta = \mathrm{Tr}\,[Q^{-1}(Y_\alpha + h^i_\alpha X_i)\, Q X^r] \tag{3.5}$$

are satisfied, where the symbol $\mathrm{Tr}(MY^\gamma)$ denotes the coefficient in the expansion of the operator M in the Lie algebra with respect to the system of tangent matrices,

$$M = [\mathrm{Tr}\,(MY^\gamma)]\, Y_\gamma + [\mathrm{Tr}\,(MX^r)]\, X_r. \tag{3.6}$$

The consistency condition for Eqs. (3.3) leads to a system of equations for the definition of the functions h^i_α:

$$\hat{F}_\beta h^i_\alpha - \hat{F}_\alpha h^i_\beta = D^\gamma_{\beta\alpha} h^i_\gamma + C^i_{\alpha j} h^j_\beta - C^i_{\beta j} h^j_\alpha - A^i_{\alpha\beta}. \tag{3.7}$$

In the case of a non-Abelian normal subgroup all the arguments remain in force, the equations for h^i_α now being nonlinear. An analogous system of equations is obtained for the definition of $\tilde{h}^i_\alpha$. We show how to obtain Eq. (3.7).

Expanding the commutator, we have from (3.2), (3.3), and the definition of $\hat{F}_\alpha$

$$\begin{aligned}[\hat{F}_\beta, F_\alpha]Q &= [\hat{F}_\beta\,(Y_\alpha + h^i_\alpha X_i) - \hat{F}_\alpha\,(Y_\beta + h^i_\beta X_i)]Q = \\ &= -D^\gamma_{\alpha\beta}\,(Y_\gamma + h^i_\gamma X_i)Q = \{[Y_\beta, Y_\alpha] + A^i_{\beta\alpha}X_i + h^i_\beta\,[X_i, Y_\alpha] + \\ &\quad + h^i_\alpha\,[Y_\beta, X_i] + [[\hat{F}_\beta, h^i_\alpha] + [\hat{F}_\alpha, h^i_\beta]]X_i + h^k_\alpha C^i_{\beta k} X_i - C^i_{\alpha k} h^k_\beta X_i\}\,Q.\end{aligned} \tag{3.8}$$

Multiplying both sides of this equality by Q^{-1} on the right, expanding the commutators, and reducing similar terms, we obtain Eqs. (3.7) by comparing the coefficients of the basis matrices X_i. Using the commutation relations $[\hat{F}_\alpha, F_\beta] = D^\gamma_{\alpha\beta}F_\gamma$ and taking into account the Jacobi identity (3.2), it can be shown that system (3.7) is consistent. In this case, when the quotient group F is a subgroup of the group G ($A^i_{\alpha\beta} = 0$) system (3.7) admits the obvious solution: $h^s_\beta = 0$. In particular, this holds for a semisimple quotient group F. The matrix Q can easily be reconstructed in terms of the functions h^i_α found from [3.7]).

We now turn to the determination of the infinitesimal operators of a representation of the group G.

In the above notation we have the following realization for a representation T_g^ρ of G in infinitesimal form:

$$\hat{Y}_\alpha = \hat{F}_\alpha + \varphi_\alpha, \tag{3.9}$$

$$X_i = \mathrm{Tr}\,(Q^{-1}X_iQX^j)\,\rho_j. \tag{3.10}$$

Moreover, the scalar functions φ_α depending on the parameters f_μ satisfy the system of equations

$$\hat{F}_\beta\varphi_\alpha - \hat{F}_\alpha\varphi_\beta = D^\mu_{\beta\alpha}\varphi_\mu + A^i_{\beta\alpha}\hat{X}_i \tag{3.11}$$

which is consistent by virtue of the Jacobi identities (3.2). A straightforward verification using (3.3) and (3.7) shows easily that we have

$$\varphi_\alpha = -\,\mathrm{Tr}\,[(Q^{-1}X_iQX^j)\,h^i_\alpha]\,\rho_j \tag{3.12}$$

up to a "gradient" transformation.

We show that system (3.11) is consistent and verify that (3.12) gives solutions of this system. To do this we let the operator $\hat{F}_\gamma$ act on the left on both sides of Eq. (3.11) and then write two more equalities which follow from cyclic permutation of the indices α, β, γ. Adding the left- and right-hand sides of these three equations, we have

$$[\hat{F}_\alpha, \hat{F}_\beta]\,\varphi_\gamma + [\hat{F}_\gamma, \hat{F}_\alpha]\,\varphi_\beta + [\hat{F}_\beta, \hat{F}_\gamma]\,\varphi_\alpha = D^\delta_{\alpha\beta}\hat{F}_\delta\varphi_\gamma + D^\delta_{\gamma\alpha}\hat{F}_\delta\varphi_\beta + D^\delta_{\beta\gamma}\hat{F}_\delta\varphi_\alpha. \tag{3.13}$$

Expanding the commutators $[\hat{F}_\alpha, \hat{F}_\beta] = D^\gamma_{\alpha\beta}F_\gamma$, reducing similar terms, and using Eq. (3.11), we obtain

$$\begin{aligned}(D^\delta_{\alpha\beta}D^\varepsilon_{\delta\gamma} + D^\delta_{\gamma\alpha}D^\varepsilon_{\delta\beta} + D^\delta_{\beta\gamma}D^\varepsilon_{\delta\alpha})\,\varphi_\varepsilon &= \\ = -\,(D^\delta_{\alpha\beta}A^i_{\delta\gamma} + D^\delta_{\gamma\alpha}A^i_{\delta\beta} + D^\delta_{\beta\gamma}A^i_{\delta\alpha})\,\hat{X}_i &+ A^i_{\beta\alpha}\hat{F}_\gamma\hat{X}_i + A^i_{\beta\gamma}\hat{F}_\alpha\hat{X}_i + A^i_{\gamma\alpha}\hat{F}_\beta\hat{X}_i,\end{aligned} \tag{3.14}$$

which holds identically by virtue of the relation $\hat{F}_\alpha\hat{X}_i = C^j_{\alpha i}\hat{X}_j$, which follows from definitions (3.3) and (3.8) and the Jacobi identities (3.2). We now verify that Eq. (3.12) gives a solution of system (3.11). In order to do this we first multiply (3.7) by $\hat{X}_i$ on the right. If on the left-hand side of the resulting equality we add and subtract terms giving a "total derivative" of the type $\hat{F}_\alpha\,(h^i_\alpha\hat{X}_i) - \hat{F}_\beta\,(h^i_\beta\hat{X}_i)$, we see that (3.9) holds if φ_α is defined by Eq. (3.12).

The number of distinct solutions of system (3.7) is equal to the dimension of the second cohomology group of F with coefficients in the ideal N. It will be shown by examples below how the representation (3.10) is related to the regular representation of the group G. Let us consider in more detail the structure of the infinitesimal operators $\hat{X}_i$ [cf. (3.10)] belonging to the ideal J corresponding to the normal subgroup N. Under arbitrary right translation transformations with infinitesimal operators $\tilde{F}_\alpha$ [cf. (3.3)], the parameters ρ_i transform via a certain matrix b

$$\rho_s = b_{si}\rho_i,$$

satisfying the following system of differential equations:

$$\hat{F}_\alpha b = C_\alpha b, \qquad \hat{\tilde{F}}_\alpha b = -\,bC_\alpha, \qquad (C_\alpha)_{ij} = C^j_{\alpha i}, \tag{3.15}$$

which is consistent by virtue of the Jacobi identities. We remark that the matrices $(-C_\alpha)$

are the infinitesimal operators of a representation of the group F in the ideal N with respect to the basis X_i. This representation is induced by the adjoint representation of G. The matrices $b(f_\mu)$ realize a representation of F. For the vector $R_s = b_{si}\rho_i$ we have from (3.15)

$$\hat{F}_\alpha R_i = (C_\alpha)_{ij} R_j. \tag{3.16}$$

The question of the number of independent parameters ρ_i in (3.10) or the number of invariants of the adjoint representation acting in the ideal N thus reduces to finding the invariants of system (3.16), i.e., to solving the following system of first-order differential equations for the function $\Phi(R_i)$:

$$\frac{\partial \Phi}{\partial R_i}(C_\alpha)_{ij} R_j = 0. \tag{3.17}$$

We remark that the invariants of right and left regular representations coincide by (3.4). The number of distinct functionally independent first integrals of (3.17) coincides with the number of null vectors of the rectangular matrix $C^i_{\alpha j} R_j$. We will denote the functionally independent solutions of (3.17) by ρ_R. The set of invariant parameters ρ_R characterize the representation. Finally,

$$\hat{X}_i = \sum_R \mathrm{Tr}(Q^{-1} X_i Q X^R)\rho_R; \qquad \varphi_\alpha = \sum_{i,R} \mathrm{Tr}(Q^{-1} X_i Q X^R h^i_\alpha)\rho_R. \tag{3.18}$$

In the case where there are no invariants of the system, we have

$$\hat{X}_i = \mathrm{Tr}(Q^{-1} X_i Q X^s)\rho_s, \tag{3.19}$$

where X_s is an arbitrary operator in the ideal N.

The representation defined in infinitesimal form by Eqs. (3.10), (3.12) is, generally speaking, reducible, since in the general case these exists a subgroup S of the group G/N commuting with the infinitesimal operators belonging to the ideal, and hence it commutes with all the infinitesimal operators (3.10). We let $\hat{\hat{F}}^{\underline{\sigma}}$ denote infinitesimal operators of S such that

$$\hat{\hat{F}}^{\underline{\sigma}} \hat{X}_i = 0, \qquad [\hat{\hat{F}}^{\underline{\sigma}}, \hat{\hat{F}}^{\underline{\delta}}] = D^{\underline{\varepsilon}}_{\underline{\sigma\delta}} \hat{\hat{F}}^{\underline{\varepsilon}}, \tag{3.20}$$

where the $\hat{X}_i$ are defined by Eqs. (3.18). The underlined Greek index runs over the values of the invariance subgroup S.

Conditions (3.19) are equivalent by (3.15), (3.16) to the requirements

$$C^R_{\underline{\sigma} j} = 0, \tag{3.21}$$

where j is arbitrary, σ belongs to the invariance subgroup, and R is the index set of the invariant parameter ρ_R.

From (3.9), and taking (3.20) into account, we find the following system of equations for the functions $\hat{\hat{F}}^{\underline{\sigma}}\varphi_\alpha$

$$\hat{F}_\beta(\hat{\hat{F}}^{\underline{\sigma}}\varphi_\alpha) - \hat{F}_\alpha(\hat{\hat{F}}^{\underline{\sigma}}\varphi_\beta) = D^\gamma_{\beta\alpha}(\hat{\hat{F}}^{\underline{\sigma}}\varphi_\gamma), \tag{3.22}$$

for which the conditions of integrability are

$$\hat{\hat{F}}^{\underline{\sigma}}\varphi_\alpha = \hat{F}_\alpha f^{\underline{\sigma}}. \tag{3.23}$$

It follows from (3.23) that the operators $\hat{F}^{\underline{\sigma}} + f^{\underline{\sigma}}$ commute with all the infinitesimal operators of the representation of G [cf. (3.18)]. We clarify the structure of the commutation relations for the operators $\hat{F}^{\underline{\sigma}} + f^{\underline{\sigma}}$. From (3.20) we have

$$[\hat{\tilde{F}}^{\underline{\sigma}} + f^{\underline{\sigma}}, \quad \hat{\tilde{F}}^{\underline{\delta}} + f^{\underline{\delta}}] = D^{\gamma}_{\underline{\delta\sigma}}(\hat{\tilde{F}}^{\underline{\gamma}} + f^{\underline{\gamma}}) + \{\hat{\tilde{F}}^{\underline{\sigma}} f^{\underline{\delta}} - \hat{\tilde{F}}^{\underline{\delta}} f^{\underline{\sigma}} - D^{\underline{\gamma}}_{\underline{\sigma\delta}} f^{\underline{\gamma}}\}. \tag{3.24}$$

We denote the expression in the curly brackets in (3.24) by $\Theta_{\underline{\sigma\delta}}$; it is a c-number and commutes by what has been said above with all the operators $\hat{F}_{\alpha}$. This can be seen by letting the operators $\hat{F}_{\alpha}$ act directly on the bracket and using (3.19)-(3.23),

$$\hat{\tilde{F}}^{\underline{\sigma}} f^{\underline{\delta}} - \hat{\tilde{F}}^{\underline{\delta}} f^{\underline{\sigma}} - D^{\underline{\gamma}}_{\underline{\sigma\delta}} f^{\underline{\gamma}} = \Theta_{\underline{\sigma\delta}}. \tag{3.25}$$

Thus the operators $\hat{\tilde{F}}^{\underline{\sigma}} + f^{\underline{\sigma}}$ are the infinitesimal operators of some projective representation of the group S with projective constants $\Theta_{\underline{\sigma\delta}}$.

In order to calculate the constants $\Theta_{\underline{\sigma\delta}}$ in terms of the group parameters of G and ρ_R, we make use of relation (3.4) between operators with and without tildes together with Eqs. (3.9), (3.23) for φ_{α}. We find the result

$$\Theta_{\underline{\sigma\delta}} = - F^{-1}_{\underline{\sigma}\mu} F^{-1}_{\underline{\delta}\nu} (D^{\underline{\gamma}}_{\nu\mu} \varphi_{\gamma} + A^{i}_{\nu\mu} \hat{X}_i) - D^{\underline{\nu}}_{\underline{\sigma\delta}} f_{\underline{\nu}}. \tag{3.26}$$

Taking the parameters of the quotient groups to be zero in the last equality (the number on the left does not depend on these parameters) and taking into account that $Q(0) = I$, $F_{\alpha\beta}(0) = \delta_{\alpha\beta}$, we obtain

$$\Theta_{\underline{\sigma\delta}} = A^{R}_{\underline{\sigma\delta}} \rho_R + D^{\underline{\gamma}}_{\underline{\sigma\delta}} (\varphi_{\underline{\gamma}}(0) - f_{\underline{\gamma}}(0)). \tag{3.27}$$

Adding the constant term $\varphi_{\underline{\sigma}}(0) - f_{\underline{\sigma}}(0)$ to the operators $F^{\underline{\sigma}} + f^{\underline{\sigma}}$, we arrive at a projective representation of the invariance subgroup with constants

$$\Theta_{\underline{\sigma\delta}} \equiv A^{R}_{\underline{\sigma\delta}} \rho_R. \tag{3.28}$$

From the second Jacobi identity (3.2), bearing in mind (3.21), it follows that the constants $\Theta_{\underline{\sigma\delta}}$ satisfy the conditions for a projective representation

$$D^{\underline{\gamma}}_{\underline{\delta\sigma}} \Theta_{\underline{\alpha\gamma}} + D^{\underline{\gamma}}_{\underline{\sigma\alpha}} \Theta_{\underline{\delta\gamma}} + D^{\underline{\gamma}}_{\underline{\alpha\delta}} \Theta_{\underline{\sigma\gamma}} = 0. \tag{3.29}$$

The set parameters (l_{α}, ρ_R) characterizing the projective representation of the invariance subgroup S forms together with the set of parameters ρ_R a complete set of parameters defining the representation of the group G.

We now turn to the integral form of the realization of Lie group representations.

The explicit form (3.18) of the infinitesimal operators allows us to perform the integration of the system of Lie equations and obtain explicit expressions for the global operator $T^{\rho l}(g)$ realizing the representation of the group denoted by the indices (ρl).

Representation (3.18) is realized on a space of differentiable functions depending on the parameters of the quotient group F. From (3.18) we obtain

$$T^{\rho l}_{g} f(f_{\alpha}) = \prod_{R} |R_{\alpha}|^{\rho_R} f(\tilde{f}_{\alpha}), \tag{3.30}$$

where $\tilde{f}_{\alpha}$ is the transformed value of the quotient group parameters, and $R_R = \exp[\tilde{\tau}_R - \tau_R]$

(where $R_\alpha = \partial/\partial \tau_\alpha$) must be obtained by integrating the system of Lie equations. However, it seems more convenient to do the following. We write an arbitrary transformation as follows:

$$T_g^\rho = T_N^\rho T_F^\rho = \exp\,[J_i\, \mathrm{Sp}\,(Q^{-1}XQX^R)\,\rho_R]\, R_\rho \exp\,[f_\alpha \hat{F}_\alpha + f_\alpha \varphi_\alpha). \tag{3.31}$$

The second factor in (3.31) can easily be written in the form

$$\exp\,(f_\alpha \hat{F}_\alpha + f_\alpha \varphi_\alpha) = \exp V_1 \exp V_2, \qquad V_2 = f_\alpha \hat{F}_\alpha,$$
$$V_1 = f_\alpha \int_0^1 dt\,[\exp\,(f_\beta \hat{F}_\beta t)\,\varphi_\alpha \exp\,(-\,t f_\beta \hat{F}_\beta)]. \tag{3.32}$$

This expansion follows from the fact that the operator is linear with respect to the derivatives in the parameters of the quotient group. From (3.31) we obtain

$$T_g^\rho f\,(F) = \exp\,[J_i\, \mathrm{Sp}\,(Q^{-1}X_iQX^R)\,\rho_R] \exp V_1 f\,(\bar{F}). \tag{3.33}$$

In the case of a nontrivial invariance subgroup S, representation (3.33) is reducible, and to find its irreducible components we expand the function f with respect to the matrix elements of the invariance subgroup $Y^l_{M,M'}(S)$. From (3.33) we obtain

$$T_g^{(\rho l)} f_M(\underline{F}) = \exp\,[J_i\, \mathrm{Sp}\,(Q^{-1}X_iQX^R)\,\rho_R] \exp V_1 \sum_{M'} Y^l_{M,M'}(\tilde{S})\, f_{M'}(\underline{\bar{F}}), \tag{3.34}$$

where $\underline{F}$ is the set of parameters of the quotient group F modulo the invariance subgroup S, and $\tilde{S}$ is the transformed set of parameters of S. Writing $T_g^{\rho l}$ as a kernel operator with δ-function kernel, it is easy to obtain an expression for the character $\chi^{\rho l}$,

$$\chi^{\rho l}(g) = \int d\underline{F} \exp\,[J_i\, \mathrm{Sp}\,(Q^{-1}X_iQX^R)\,\rho_R] \exp\left\{f_\alpha \int_0^1 dt\,[\exp\,(f_\beta \hat{F}_\beta t)]\,\varphi_\alpha\,[\exp\,(-\,f_\gamma \hat{F}_\gamma t)]\right\} \chi^l(\tilde{s})\,\delta\,(\underline{F} - \bar{F}), \tag{3.35}$$

where $\chi^l(s)$ is the character of the projective representation (l, ρ) of the invariance subgroup S and $\underline{F}$ is the set of parameters of the quotient group F/S. Expression (3.35) for the character has been given purely formally, with no explanation of the conditions for the integral (3.35) to exist. In the case where the invariance subgroup reduces to the trivial group, Eq. (3.35) for the character simplifies substantially and takes the form

$$\chi_\rho\,(g) = \delta\,(F - 1) \int dF \exp\,[J_i\, \mathrm{Sp}\,(Q^{-1}X_iQX^R)\,\rho_R]. \tag{3.36}$$

In particular, the expression for the character in the form (3.36) is valid for all nilpotent groups (cf. [14]). To conclude this section, we remark that in the case of semisimple Lie groups all the main concepts such as the parameter set and the notion of invariance subgroup preserve their validity (cf. [20]).

We now discuss the relation of the representation constructed to the regular representation. We write an arbitrary element of the group G in the form

$$g = Q_1 N Q_2, \tag{3.37}$$

where Q_1, Q_2 are defined by Eqs. (3.3) and N is a normal subgroup. In order for expansion (3.37) to be unique, those of the parameters Q_2 associated with the invariance subgroup must be considered to be contained in Q_1. In order to find the infinitesimal operators, we make use of the fundamental group relation in the infinitesimal form

$$\dot{g} = -\,F^s g, \tag{3.38}$$

the dot indicating differentiation with respect to some parameter, and F_s is the corresponding tangential operator. Subsituting (3.37) into (3.38) and carrying out the obvious manipulations, we find a system of equations for the parameters $Q_1^{-1}\dot{Q}_1 = \omega^1$, $\dot{Q}_2 Q_2^{-1} = \omega^2$, $\dot{\tau}$:

$$\dot{\omega}_\alpha^1 (Y_\alpha + h_\alpha^i X_i) + X_R \dot{\tau}_R + e^{X_R \tau_R} \dot{\omega}_\gamma^2 (Y_\gamma + H_\gamma^j X_j) e^{-X_R \tau_R} = - Q_1^{-1} F_s Q_1. \tag{3.39}$$

System (3.39) allows us to find a system of equations for determining $\dot{\omega}_\alpha^1$, ω_γ^2, and $\dot{\tau}_R$:

$$\begin{aligned} &\dot{\omega}_\alpha^1 f_\alpha^i + \dot{\tau}_R + \dot{\omega}_\beta^2 h_\beta^i + \dot{\omega}_R^2 \tau_R C_{R\beta}^i = \mathrm{Tr}\,(Q_1^{-1} F_s Q_1 Y^i), \\ &\dot{\omega}_\alpha^1 + \dot{\omega}_\alpha^2 = \mathrm{Tr}\,(Q_1^{-1} F_s Q_1 X^\alpha), \end{aligned} \tag{3.40}$$

where the symbol $\mathrm{Tr}(Q_1^{-1} F_s Q_1 X^j)$ denotes the coefficient of the term X_j in the expansion of $Q_1^{-1} F_s Q_1$ with respect to the tangential matrices:

$$Q_1^{-1} F_s Q_1 = \sum_i \mathrm{Tr}\,(Q_1^{-1} F_s Q_1 X^i)\, X_i + \sum_\alpha \mathrm{Tr}\,(Q_1^{-1} F_s Q_1 Y^\alpha)\, Y_\alpha. \tag{3.41}$$

For the generators of the left regular representation we obtain

$$\hat{F}_s = \dot{\omega}_\alpha^1 \hat{\hat{F}}_\alpha^1 + \dot{\omega}_\beta^2 \hat{F}_\beta^2 + \dot{\tau}_R \frac{\partial}{\partial \tau_R}, \tag{3.42}$$

where $\dot{\omega}_\alpha^1$, $\dot{\omega}_\beta^2$, and τ_R are found from solving system (3.40), and $\hat{\hat{F}}_\alpha^1$ $(\hat{F}_\beta^2)$ are generators of the right (left) regular representation of the quotient group. In certain cases it is possible to obtain the representation described above from the regular representation by taking limits with respect to the parameters τ (cf. [16-20]) and thereby relate the regular representation to the representation constructed above.

We consider as an example an interesting nilpotent group, the de Witt group G of even dimension $k = 2n + 2$. Its Lie algebra consists of the vectors $e_1, e_2, \ldots, e_{2n+2}$ satisfying the commutation relations

$$[e_i, e_j] = \begin{cases} (j-i)\, e_{i+j}, & i + j \leqslant 2n + 2, \\ 0 - \text{in the remaining cases}. \end{cases} \tag{3.43}$$

G contains the commutative ideal N consisting of the generators $X_1 = e_{n+1}$, $X_2 = e_{n+2}$, $\ldots$, $X_{2n} = e_{2n+2}$. Introduce the notation $Y_1 = e_1$, $Y_2 = e_2$, $\ldots$, $Y_n = e_n$. The group (3.43) is interesting in that the quotient group F by the normal subgroup corresponding to the ideal N is not constained in the group. The invariance subgroup here is trivial. For the expansion (3.37) we can choose

$$\tau = \exp\,(\tau_{n+1} X_1 + \tau_{2n+2} X_{n+2}). \tag{3.44}$$

Moreover, system (3.40) reduces to the form

$$Q_1^{-1} F_s Q_1 = \dot{\omega}_\alpha^{(1)} (Y_\alpha + h_\alpha^i X_i) + X_{n+1} \dot{\tau}_{n+1} + X_{2n+2} \dot{\tau}_{2n+2} + e^{-\tau_i X_i} \dot{\omega}_\beta^{(2)} (Y_\beta + H_\beta^i X_i) e^{\tau_i X_i}. \tag{3.45}$$

Making some transformations, it is easy to see from system (3.45) that it is possible to take the limit as τ_{n+1} and τ_{2n+2} tend to infinity. We then obtain the limiting expressions for the corresponding operators $\dot{\omega}_\alpha^1$, $\dot{\omega}_\beta^2$, $\dot{\tau}_{n+1}, \ldots, \dot{\tau}_{2n+2}$

$$\begin{aligned} &\dot{\omega}_\alpha^2 = 0, \quad \omega_\alpha^1 = \mathrm{Tr}\,(Q_1^{-1} F_s Q_1 Y^\alpha), \\ &\dot{\tau}_{n+1} = \mathrm{Tr}\,[Q_1^{-1} F_s Q_1 (X^1 - Y^\alpha h_\alpha^{n+1})], \\ &\dot{\tau}_{2n+2} = \mathrm{Tr}\,[Q_1^{-1} F_s Q_1 (X^{n+2} - Y^\alpha h_\alpha^{2n+2})]. \end{aligned} \tag{3.46}$$

Substituting (3.46) into the general equation (3.42), we obtain a representation of the de Witt group in infinitesimal form defined by two arbitrary scalar parameters $\rho_1 \to \partial/\partial \tau_{n+1}$, $\rho_2 \to \partial/\partial \tau_{2n+2}$. Using the general equation (3.36) for the character of the representation $(\rho_1 \rho_2)$, we have

$$\chi^{\rho_1\rho_2}(g) = \sum_{R=1,2} \int dF \exp [J_i \operatorname{Tr}(Q^{-1}X_iQX^R)\rho_R]\cdot\delta(F-1). \tag{3.47}$$

Using the method employed in [16], it is also easy to compute the Plancherel measure for our case,

$$\mu(\rho_1, \rho_2) = (2\pi)^{n+2}\rho_2^n(2n)!! \tag{3.48}$$

In the case of odd dimension k = 2n + 1, the Plancherel measure has the form

$$\mu(\rho) = (2\pi)^{n+1}\rho^n(2n-1)!!, \tag{3.49}$$

where ρ is a single number (Casimir operator) defining the irreducible representation.

4. THE NILPOTENT HEISENBERG GROUP

In order to make the exposition of the canonical transformation method in the theory of Lie group representations as simple as possible, we will demonstrate its main features in the case of the Heisenberg group, which is a very simple but important nilpotent group. The Heisenberg group G is the group of three-dimensional real matrices having the form

$$g_0 = \begin{Vmatrix} 1 & x & z \\ 0 & 1 & y \\ 0 & 0 & 1 \end{Vmatrix}; \qquad -\infty \leqslant x, y, z \leqslant \infty.$$

Now consider the right regular representation $T_r^r(g)$ acting in the space of square-integrable functions on G in accordance with the formula

$$T_r^r(a, b, c)\psi(x, y, z) = \psi(x+a, y+b, z+c+bx), \tag{4.1}$$

or

$$T_r^r(g)\psi(g_0) = \psi(g_0 g), \quad g = \begin{Vmatrix} 1 & a & c \\ 0 & 1 & b \\ 0 & 0 & 1 \end{Vmatrix}, \quad g_0 g = \begin{Vmatrix} 1 & x+a & c+z+xb \\ 0 & 1 & y+b \\ 0 & 0 & 1 \end{Vmatrix}. \tag{4.2}$$

Assuming the parameters a, b, and c to be infinitesimal, we can expand the right-hand side of the equality in Eq. (4.1) into a Taylor series and obtain the infinitesimal operators for infinitely small transformations. We have

$$\begin{gathered}
(aL^a+1)\psi(x, y, z) \simeq T_r^r(a, 0, 0)\psi(x, y, z) \simeq \psi(x, y, z) + a\frac{\partial\psi}{\partial x} = \left(1 + a\frac{\partial}{\partial x}\right)\psi, \\
(bL^b+1)\psi(x, y, z) \simeq T_r^r(0, b, 0)\psi(x, y, z) \simeq \psi(x, y, z) + b\frac{\partial\psi}{\partial y} + bx\frac{\partial\psi}{\partial z} = \left[1 + b\left(\frac{\partial}{\partial y} + x\frac{\partial}{\partial z}\right)\right]\psi, \\
(cL^c+1)\psi(x, y, z) \simeq T_r^r(0, 0, c)\psi(x, y, z) \simeq \psi(x, y, z) + c\frac{\partial\psi}{\partial z} = \left(1 + c\frac{\partial}{\partial z}\right)\psi.
\end{gathered} \tag{4.3}$$

Thus the coefficients of the infinitesimally small parameters a, b, and c give the corresponding generators L^a, L^b, L^c of the right regular representation of the Heisenberg group:

$$L^a = \frac{\partial}{\partial x} = p_x,$$

$$L^b = \frac{\partial}{\partial y} + x\frac{\partial}{\partial z} = p_y + xp_z, \qquad L^c = \frac{\partial}{\partial z} = p_z. \tag{4.4}$$

(We have introduced the notation $\partial/\partial x \equiv p_x$, etc.)

The commutation relations for these operators are as follows:

$$[L^a, L^b] = L^c,$$

$$[L^a, L^c] = [L^b, L^c] = 0. \tag{4.5}$$

We remark that the commutation relations (4.5) are also satisfied if we understand the square brackets to mean the Poisson brackets with respect to the variables $\mathbf{p} = (p_x, p_y, p_z)$ and $\mathbf{q} = (x, y, z)$, that is, for example,

$$[L^a, L^b] = \frac{\partial L^a}{\partial p_x}\frac{\partial L^b}{\partial x} + \frac{\partial L^a}{\partial p_y}\frac{\partial L^b}{\partial y} + \frac{\partial L^a}{\partial p_z}\frac{\partial L^b}{\partial z} - \frac{\partial L^a}{\partial x}\frac{\partial L^b}{\partial p_x} - \frac{\partial L^a}{\partial y}\frac{\partial L^b}{\partial p_y} - \frac{\partial L^a}{\partial z}\frac{\partial L^b}{\partial p_z} = L^c. \tag{4.6}$$

It is easy to verify that the matrix inverse to the matrix g has the form

$$g^{-1} = \begin{Vmatrix} 1 & -a & -c+ab \\ 0 & 1 & -b \\ 0 & 0 & 1 \end{Vmatrix}.$$

The left regular representation $T_l^r(g)$ is defined on the space of square-integrable functions $\psi(g_0)$ on G by the following formula:

$$T_l^r(g)\psi(g_0) = \psi(g^{-1}g_0) = \psi(x-a, y-b, z-c-ay+ab) \tag{4.7}$$

Expanding (4.7) into a Taylor series in the small parameters c, b, and c in the same way as in obtaining Eqs. (4.4), we obtain the infinitesimal operators for the left regular representation

$$\underset{\sim}{L}^a = -\frac{\partial}{\partial x} - y\frac{\partial}{\partial z}, \quad \underset{\sim}{L}^b = -\frac{\partial}{\partial y}, \quad \underset{\sim}{L}^c = -\frac{\partial}{\partial z}, \tag{4.8}$$

which satisfy commutation relations (4.5) with L^i replaced by $\underset{\sim}{L}^i$ $(i = a, b, c)$.

Since the right and left regular representations commute, it is clear from definitions (4.2) and (4.7) that their infinitesimal operators also commute, which can also be checked directly. The operators (4.8) can be written in the "classical" form

$$\underset{\sim}{L}^a = -p_x - yp_z, \quad \underset{\sim}{L}^b = -p_y, \quad \underset{\sim}{L}^c = -p_z. \tag{4.9}$$

The Poisson brackets for these left infinitesimal operators give the same equations (4.5) as for the right infinitesimal operators. The Poisson brackets of any left generator (4.9) with any right generator (4.4) are zero. This suggests that the right and left generators in fact depend on different variables. In order to find these variables, it is sufficient to make a suitable canonical (tangential) transformation which preserves the Poisson brackets of the coordinates and momenta and hence the commutation relations (4.5).

Indeed, consider the change of variables

$$\begin{aligned} &X = x + \frac{p_y}{p_z}, \quad Y = y + \frac{p_x}{p_z}, \quad Z = z - \frac{p_x p_y}{p_z^2}; \\ &\mathcal{P}_X = p_x, \quad \mathcal{P}_Y = p_y, \quad \mathcal{P}_Z = p_z; \\ &x = X - \frac{\mathcal{P}_Y}{\mathcal{P}_Z}, \quad y = Y - \frac{\mathcal{P}_X}{\mathcal{P}_Z}, \quad z = Z + \frac{\mathcal{P}_X\mathcal{P}_Y}{\mathcal{P}_Z^2}. \end{aligned} \tag{4.10}$$

It is easy to see that this transformation is canonical, i.e., the corresponding Poisson brackets have the values

$$\{\mathcal{P}_X, X\} = \{\mathcal{P}_Y, Y\} = \{\mathcal{P}_Z, Z\} = 1, \tag{4.11}$$

and the remaining Poisson brackets are equal to zero.

If we express the generators of the right and left regular representations in terms of the new variables we obtain the equations

$$\begin{aligned} &L^a = \mathcal{P}_X, \quad &&L^b = X\mathcal{P}_Z, \quad &&L^c = \mathcal{P}_Z; \\ &\underset{\sim}{L}^a = -Y\mathcal{P}_Z, \quad &&\underset{\sim}{L}^b = -\mathcal{P}_Y, \quad &&\underset{\sim}{L}^c = -\mathcal{P}_Z. \end{aligned} \tag{4.12}$$

Thus the right and left generators indeed depend on distinct variables: The right generators depend on X and $\mathcal{P}_X$ (as well as $\mathcal{P}_Z$), the left generators on Y and $\mathcal{P}_Y$ (as well as $\mathcal{P}_Z$). It is important that the variable Z in general does not appear in the right and left generators (i.e., is cyclic). Now consider the action of the right and left translations on a point in the group phase space. We have a six-dimensional phase space with coordinates (x, y, z, p_x, p_y, p_z). When the matrix g acts on the matrix g_0 from the right, we obtain the new coordinates x_g, y_g, z_g according to the formula

$$x_g = x + a, \quad y_g = y + b, \quad z_g = z + bx + c. \tag{4.13}$$

In order to obtain the new coordinates $p_{x_g}, p_{y_g}, p_{z_g}$, we proceed as follows. Define $p_{x_g} = \partial/\partial x_g$, $p_{y_g} = \partial/\partial y_g$, $p_{z_g} = \partial/\partial z_g$. We then have

$$\frac{\partial}{\partial x_g} = \frac{\partial}{\partial x} - b\frac{\partial}{\partial z}, \quad \frac{\partial}{\partial y_g} = \frac{\partial}{\partial y}, \quad \frac{\partial}{\partial z_g} = \frac{\partial}{\partial z}. \tag{4.14}$$

Hence

$$\begin{aligned} p_{x_g} &= p_x - bp_z, \\ p_{y_g} &= p_x, \\ p_{z_g} &= p_z, \end{aligned} \qquad \left\| \begin{matrix} p_{x_g} \\ p_{y_g} \\ p_{z_g} \end{matrix} \right\| = \left\| \begin{matrix} 1 & 0 & -b \\ 0 & 1 & 0 \\ 0 & 0 & 1 \end{matrix} \right\| \cdot \left\| \begin{matrix} p_x \\ p_y \\ p_z \end{matrix} \right\|. \tag{4.15}$$

It is easy to check that the transformations (4.13), (4.14) are canonical, i.e., the Poisson brackets are preserved. We now compute the action of right translation on the new variables X, Y, Z, $\mathcal{P}_X$, $\mathcal{P}_Y$, $\mathcal{P}_Z$. This action is determined simply by replacing all the quantities X, Y, Z, $\mathcal{P}_X$, $\mathcal{P}_Y$, $\mathcal{P}_Z$ in Eqs. (4.10) by X_g, Y_g, Z_g, $\mathcal{P}_{X_g}$, $\mathcal{P}_{Y_g}$, $\mathcal{P}_{Z_g}$ and the quantities x, y, z, p_x, p_y, p_z by x_g, y_g, z_g, p_{x_g}, p_{y_g}, p_{z_g}. We then have

$$\begin{aligned} &X_g = x_g + \frac{p_{y_g}}{p_{z_g}}, \quad Y_g = y_g + \frac{p_{x_g}}{p_{z_g}}, \quad Z_g = z_g - \frac{p_{x_g}p_{y_g}}{p_{z_g}}; \\ &\mathcal{P}_{X_g} = p_{x_g}, \quad \mathcal{P}_{Y_g} = p_{y_g}, \quad \mathcal{P}_{Z_g} = p_{z_g}. \end{aligned} \tag{4.16}$$

We now substitute for the right-hand sides of Eqs. (4.16) their expression in terms of the initial points (4.13) and (4.14) and compare similar terms. We then have

$$\begin{aligned} &X_g = X + a, \quad Y_g = Y, \quad Z_g = Z + bX + c; \\ &\mathcal{P}_{X_g} = \mathcal{P}_X - b\mathcal{P}_Z, \quad \mathcal{P}_{Y_g} = \mathcal{P}_Y, \quad \mathcal{P}_{Z_g} = \mathcal{P}_Z. \end{aligned} \tag{4.17}$$

We now carry out the same operations by letting the matrix g^{-1} act on the matrix g_0 from the left, i.e., for the case of a left translation. We have

$$\begin{aligned} &\underset{\sim}{x}_g = x - a, \quad \underset{\sim}{y}_g = y - b, \quad \underset{\sim}{z}_g = z - ay + ab - c; \\ &\underset{\sim}{p}_{x_g} = p_x, \quad \underset{\sim}{p}_{y_g} = \underset{\sim}{p}_y + ap_z, \quad \underset{\sim}{p}_{z_g} = p_z. \end{aligned} \tag{4.18}$$

If we now introduce the points $\underset{\sim}{X}_g$, $\underset{\sim}{Y}_g$, $\underset{\sim}{Z}_g$, $\underset{\sim}{\mathcal{P}}_{X_g}$, $\underset{\sim}{\mathcal{P}}_{Y_g}$, $\underset{\sim}{\mathcal{P}}_{Z_g}$ in accordance with (4.16) and (4.18) and again express them in terms of the initial points X, Y, Z, $\mathcal{P}_X$, $\mathcal{P}_Y$, $\mathcal{P}_Z$, then we have for the left translations

$$\underset{\sim}{X}_g = X_g, \quad \underset{\sim}{Y}_g = Y - b, \quad \underset{\sim}{Z}_g = Z - aY + ab - c;$$
$$\mathcal{P}_{\underset{\sim}{X}_g} = \mathcal{P}_X, \quad \mathcal{P}_{\underset{\sim}{Y}_g} = \mathcal{P}_Y + a\mathcal{P}_Z, \quad \mathcal{P}_{\underset{\sim}{Z}_g} = \mathcal{P}_Z. \tag{4.19}$$

As follows from Eqs. (4.17) and (4.19), the new variables X, Y, Z, $\mathcal{P}_X$, $\mathcal{P}_Y$, $\mathcal{P}_Z$ (the coordinates of a point in group phase space) have the following important properties: 1) the coordinate $\mathcal{P}_Z$; is unchanged under the action of both left and right translations; 2) in addition, the left coordinates Y and $\mathcal{P}_Y$, are invariant under right translations, while the right coordinates X and $\mathcal{P}_X$ are invariant under left translations: 3) all the changes of variables are canonical and hence there exist invariants of the canonical transformations, the Poincaré integral invariants. In particular, the element of phase space volume is invariant under all substitutions of variables,

$$dXdYdZd\mathcal{P}_Xd\mathcal{P}_Yd\mathcal{P}_Z = dxdydzdp_xdp_ydp_z = dX_gdY_gdZ_gd\mathcal{P}_{X_g}d\mathcal{P}_{Y_g}d\mathcal{P}_{Z_g} =$$
$$= d\underset{\sim}{X}_gd\underset{\sim}{Y}_gd\underset{\sim}{Z}_gd\mathcal{P}_{\underset{\sim}{X}_g}d\mathcal{P}_{\underset{\sim}{Y}_g}d\mathcal{P}_{\underset{\sim}{Z}_g}. \tag{4.20}$$

Now consider the equations for the infinitesimal right and left translations in (4.12), and let us realize the representation using the quantization rule

$$X \to \tilde{x}, \quad Y \to \tilde{y}; \quad \mathcal{P}_X \to \frac{\partial}{\partial \tilde{x}}, \quad \mathcal{P}_Y \to \frac{\partial}{\partial \tilde{y}}, \quad \mathcal{P}_Z \to \frac{\partial}{\partial \tilde{z}}. \tag{4.21}$$

Then the infinitesimal operators of the right and left translations have the form

$$L^a = \frac{\partial}{\partial \tilde{x}}, \quad L^b = \tilde{x}\frac{\partial}{\partial \tilde{z}}, \quad L^c = \frac{\partial}{\partial \tilde{z}};$$
$$\underset{\sim}{L}^a = -\tilde{y}\frac{\partial}{\partial \tilde{z}}, \quad \underset{\sim}{L}^b = -\frac{\partial}{\partial \tilde{y}}, \quad \underset{\sim}{L}^c = -\frac{\partial}{\partial \tilde{z}}. \tag{4.22}$$

It is well known that these are generators of irreducible representations of the Heisenberg group; the right representation acts in the space of functions $f(\tilde{x})$, the left in the space of functions $f(\tilde{y})$, where the operator $\partial/\partial\tilde{z}$ must be set equal to the number ρ defining the irreducible representation. We now discuss how to make the formal transition from Eqs. (4.17) and (4.7) to Eqs. (4.22), i.e., from the formulas for the right and left regular representations of the Heisenberg group in the x, y, z representation to the formulas for these same representations in infinitesimal form in the X, Y, Z representation. To do this we construct the kernel G corresponding to the canonical transformation (4.10), i.e., we replace the functions $\psi(x, y, z)$ on the group according to the formula

$$\psi'(X, Y, Z) = \int G(X, Y, Z, x, y, z)\,\psi(x, y, z)\,dxdydz, \tag{4.23}$$

or in operator form,

$$\psi' = \hat{G}\psi.$$

In order to obtain the kernel G it suffices to calculate the generating function of the canonical transformation (4.10) (the action) S(x, y, z, X, Y, Z) and then calculate the Green function according to the Feynman quantization rule by taking the integral over the trajectories of exp (iS). In order to do this, we consider continuous transformations (4.10), viz., a motion of the system in time

$$x(t) = x + t\frac{p_y}{p_z}, \quad y(t) = y + t\frac{p_x}{p_z}, \quad z(t) = z - t\frac{p_xp_y}{p_z^2};$$
$$p_x(t) = p_x, \quad p_y(t) = p_y, \quad p_z(t) = p_z.$$

This motion is defined by the classical Hamiltonian

$$H = p_x p_y / p_z.$$

For t = 1 we recover Eqs. (4.10). The generating function of this transformation (the action) has the form

$$\Phi(x, x(t), y, y(t), z, p_z(t)) = p_z(t)\,[z - (x - x(t))(y - y(t))/t],$$

where

$$\frac{\partial\Phi}{\partial x} = p_x, \qquad -\frac{\partial\Phi}{\partial x(t)} = p_x(t),$$

$$\frac{\partial\Phi}{\partial y} = p_y, \qquad -\frac{\partial\Phi}{\partial y(t)} = p_y(t),$$

$$\frac{\partial\Phi}{\partial z} = p_z, \qquad \frac{\partial\Phi}{\partial p_z(t)} = z(t).$$

We quantize the classical system in the usual way. To do this we may construct a Green function by taking the Feynman integral corresponding to the given Hamiltonian or else, what is equivalent, find the Green function for the Schrödinger equation with a quantum Hermitian Hamiltonian

$$\hat{H} = \hat{p}_x \hat{p}_y / \hat{p}_z.$$

It is easy to find the Green function using functions in the momentum representation normalized in the same way as for the free motion Green function. We obtain

$$G(\mathbf{x}, \mathbf{x}', t) = \frac{1}{(\sqrt{2\pi})^3} \int_{-\infty}^{\infty}\!\!\iint d\mathbf{p} \exp[i\mathbf{p}(\mathbf{x} - \mathbf{x}')] \exp\left(-\frac{ip_x p_y}{p_z} t\right) = $$
$$= \frac{1}{it}\,\delta'\left[z - z' + \frac{(x - x')(y - y')}{t}\right], \qquad \mathbf{x} = (x, y, z). \tag{4.24}$$

The generators of the left and right regular representations expressed in terms of the trajectory momenta and coordinates **x(t)**, **p(t)** satisfy the old commutation relations, and in addition for t = 0 they go over into the generators of the regular representation (4.4) and (4.9), while for t = 1 they go into the generators (4.12). In principle it is necessary to take into account the nonuniqueness of the way of introducing the time" into the change of canonical variables. It is obvious that knowing the two points (initial t = 0 and final t = 1) of the trajectory does not suffice to reconstruct the trajectory completely, but it is these two points which are essential for solving the problem.

The function S as is easily seen has the form

$$S(x, y, z, X, Y, Z) = \rho\,[(Y - y)(X - x) + z - Z] \tag{4.25}$$

(here ρ is a number). It can be seen that the equations

$$\frac{\partial S}{\partial x} = p_x, \quad \frac{\partial S}{\partial y} = p_y, \quad \frac{\partial S}{\partial z} = p_z;$$
$$-\frac{\partial S}{\partial X} = \mathcal{P}_X, \quad -\frac{\partial S}{\partial Y} = \mathcal{P}_Y, \quad \frac{\partial S}{\partial Z} = -\mathcal{P}_Z \tag{4.26}$$

give us the canonical transformation (4.10). The coordinates Z are obtained by the condition that the action S is equal to zero. The kernels of the operators T^r_Γ, T^l_Γ of the right and left regular representations in the x, y, z representation have the form

$$\begin{aligned}
&T_r^r(x,y,z,x',y',z') = \delta(x'-x-a)\,\delta(y'-y-b)\,\delta(z'-z-c-xb),\\
&T_r^l(x,y,z,x',y',z') = \delta(x'-x+a)\,\delta(y'-y+b)\,\delta(z'-z+ay+c-ab),\\
&\widehat{T}_r^r = \exp\left[(c+xb)\frac{\partial}{\partial z}\right]\exp\left[b\frac{\partial}{\partial y}+a\frac{\partial}{\partial x}\right],\\
&\widehat{T}_r^l = \exp\left[-(ab-c+ay)\frac{\partial}{\partial z}\right]\exp\left[-a\frac{\partial}{\partial x}-b\frac{\partial}{\partial y}\right].
\end{aligned}\tag{4.27}$$

If we now construct the operators associated with T_r^r and T_r^l by means of the canonical transformation

$$\begin{aligned}
&\widehat{T}_r^r = \widehat{G}\widehat{T}_r^r\widehat{G}^{-1},\\
&\widehat{T}_r^l = \widehat{G}\widehat{T}_r^l\widehat{G}^{-1},\\
&\widehat{G} = \exp\left[\frac{(\partial/\partial x)(\partial/\partial y)}{\partial/\partial z}\right],
\end{aligned}\tag{4.28}$$

then calculating the matrices of the operator $\hat{T}_r^r$ and $\hat{T}_r^l$ (their kernels) in the X, Y, Z representation, we obtain (using (4.27), (4.28), and the equation $e^{-u}ae^{u} = a + [ua] + (1/2!)[u[ua]] + \ldots$)

$$\begin{aligned}
&T_r^r(X,Y,Z,X',Y',Z') = \delta(X'-X-a)\,\delta(Y'-Y)\,\delta(Z'-Z-bX-c),\\
&T_r^l(X,Y,Z,X',Y',Z') = \delta(X'-X)\,\delta(Y'-Y+b)\,\delta(Z'-Z+aY+c-ab).
\end{aligned}\tag{4.29}$$

This gives a realization of irreducible representations if we consider the Fourier components of the function $\psi(X, Y, Z)$ with respect to the variable Z:

$$\psi(X,Y,\rho) = \frac{1}{2\pi}\int_{-\infty}^{\infty}\psi(X,Y,Z)\,e^{i\rho Z}dZ. \tag{4.30}$$

Then the kernels of the irreducible representations have the form

$$\begin{aligned}
&T_r^r(X,Y,X',Y',\rho) = \delta(Y-Y')\,\delta(X'-X-a)\,e^{i\rho(bX+c)},\\
&T_l^r(X,Y,X',Y',\rho) = \delta(X-X')\,\delta(Y'-Y+b)\,e^{-i\rho(aY+c-ab)}.
\end{aligned}\tag{4.31}$$

Due to the presence of terms with δ-functions which correspond to the invariance of the points Y and X under right and left translation, respectively, irreducible representations are realized, as remarked above, in the spaces of functions $f(X)$ and $f(Y)$ for the right and left representations, respectively. We have for the kernels the expressions (eliminating δ-function terms)

$$\begin{aligned}
&T^r(\rho, X, X') = \delta(X'-X-a)e^{i\rho(bX+c)},\\
&T^l(\rho, Y, Y') = \delta(Y'-Y+b)e^{-i\rho(aY+c-ab)}.
\end{aligned}\tag{4.32}$$

We remark that expressions (4.29) are the kernels of the regular representations in the "good" representation (right and left variables X and Y and cyclic variable Z). Expanding these kernels into a Fourier integral with respect to the cyclic variable we obtain the kernels of the irreducible representations. The character of an irreducible representation is equal to the regularized character of either the right or the left regular representation, the regularization being performed using the following rule: throw away the $\delta(0)$ which is present because of the extraneous variable (Y for a right and X for a left representation), take the Fourier component of the kernel of the representation with respect to the variable Z, and compute the trace with respect to the remaining variables.

Let us consider the classical generating function for the transformation (4.10) without fixing the momentum $\mathcal{P}_z$, i.e., without imposing the condition $\mathcal{P}_z = \rho$. Then there exists no generating function S from the old to the new coordinates for the same reason that there is no such generating function for the identity transformation $x = X$, $\mathcal{P}_x = p_x$. But there does exist a generating function for the identity canonical transformation which depends on the old coordinates and the new momenta $\Phi(x, \mathcal{P}_x) = x\mathcal{P}_x$. A generating function $\Phi(q_i, \mathcal{P}_i)$ of this type satisfies the conditions

$$\frac{\partial\Phi}{\partial q_i} = p_i, \quad \frac{\partial\Phi}{\partial\mathcal{P}_i} = Q_i,$$

where the p_i are the old momenta and the Q_i are the new coordinates. We now construct a "mixed" generating function for the transformation (4.10) depending on these variables: $\Phi = \Phi(x, X, y, Y, z, \mathcal{P}_z)$. Then it is easy to verify that

$$\Phi = \mathcal{P}_z\,[z - (x - X)(y - Y)].$$

If the function of type (4.25) is known, the rule for constructing Φ consists in the following: the number ρ is replaced by the new momentum $\mathcal{P}_z$, and the new coordinate Z is deleted. In order to construct the kernel G of the Green function (4.23) it is necessary to proceed as follows. First, it is easy to solve the system of equations (equations of the Heisenberg type)

$$\hat{G}\hat{x}\hat{G}^{-1} = \hat{x} + \frac{\hat{p}_y}{\hat{p}_z}, \quad \hat{G}\hat{y}\hat{G}^{-1} = \hat{y} + \frac{\hat{p}_x}{\hat{p}_z}, \quad \hat{G}\hat{z}\hat{G}^{-1} = \hat{z} - \frac{\hat{p}_x\hat{p}_y}{\hat{p}_z^2};$$
$$\hat{G}\hat{p}_x\hat{G}^{-1} = \hat{p}_x, \quad \hat{G}\hat{p}_y\hat{G}^{-1} = \hat{p}_y, \quad \hat{G}\hat{p}_z\hat{G}^{-1} = \hat{p}_z.$$

This system has the solution

$$\hat{G} = \exp(\hat{p}_x\hat{p}_y/\hat{p}_z).$$

As usual, division by an operator is understood by passing to the Fourier transform with respect to the variable z. If we calculate the kernel of the operator written out above, it has the form

$$G(x, X, y, Y, z, Z) = \frac{1}{i}\,\delta'\,[(z - Z) - (x - X)(y - Y)].$$

as is easily verified. If we are given a classical generating function Φ, the above kernel can be obtained from the easily proved rule

$$G(x, X, y, Y, z, Z) = \frac{1}{(2\pi)^r}\int d\mathcal{P}_z \exp[i\Phi(x, X, y, Y, z, \mathcal{P}_z)]\exp(-iZ\mathcal{P}_z).$$

This rule holds generally for those groups for which the equations for the functions S and D [cf. (1.19), (1.23)] coincide, i.e., $\partial\Psi_{ik}/\partial q_k = 0$ (e.g., nilpotent groups). In case (1.24) (semisimple groups), the formula for the kernel in terms of the classical generating function Φ changes somewhat, viz.,

$$G(x, X, y, Y, z, Z) = \frac{1}{(2\pi)^r}\int d\mathcal{P}_z \exp[i\Phi(x, X, y, Y, z, \mathcal{P}_z + \mu)]\exp(-iZ\mathcal{P}_z).$$

Here μ has the coordinates μ_α [translates of the momenta in (1.28)], $\alpha = 1, 2, \ldots, r$. By the Fourier integral we mean the integral over the interval $-\infty, \infty$ for the noncompact momenta and a sum in the case of compact variables, when $\mathcal{P}_z$ is discrete. We remark that a classical canonical transformation bringing the generators of the right and left regular representations to maximally convenient form is not uniquely determined. Thus, arbitrary functions, $f(\mathcal{P}_Z)$ of the momenta $\mathcal{P}_Z$ can be added to variables Z. The form of the generators remains the same. This canonical transformation corresponds to the fact that the generating function Φ of a classical transformation has as summand an arbitrary function of the momenta conjugate to the cyclic variables. The quantum function can therefore be viewed as a Fourier integral with arbitrary Fourier coefficients corresponding to this summand. This nonuniqueness is very similar to gauge invariance in quantum mechanics. If we require that the operator $\hat{G}$ be unitary, the nonuniqueness reduces to an arbitrary phase factor in the Fourier component. However, if the identity

$$\hat{G}\hat{G}^{-1} = E$$

is used, which imposes a condition on the function G (rewritten in terms of the kernels this condition gives the representation for the δ-function on the group), we can obtain information on the measure and characters of the irreducible representations starting from the form of the Green function expanded in a Fourier integral with respect to the cyclic variables (or what is the same, from the form of the classical generating function).

5. REPRESENTATIONS OF THE SEMISIMPLE GROUP SL(2, R)

As another example we consider the group of real second-order matrices SL(2, R).

The formulas for the right Z_+, H, Z_- and left $\underset{\sim}{Z}_+$, $\underset{\sim}{H}$, $\underset{\sim}{Z}_-$ generators of the regular representation in the coordinates x, y, z such that

$$\begin{Vmatrix} \alpha & \beta \\ \gamma & \delta \end{Vmatrix} = \begin{Vmatrix} 1 & 0 \\ y & 1 \end{Vmatrix} \begin{Vmatrix} e^{z} & 0 \\ 0 & e^{-z} \end{Vmatrix} \begin{Vmatrix} 1 & x \\ 0 & 1 \end{Vmatrix} = Z_- H Z_+ \tag{5.1}$$

have the form

$$Z_+ = p_x, \quad H = p_z - 2xp_x, \quad Z_- = e^{-2z}p_y + xp_z - x^2p_x;$$

$$\underset{\sim}{Z}_+ = e^{-2z}p_x + yp_z - y^2p_y, \quad \underset{\sim}{H} = p_z - 2yp_y, \quad \underset{\sim}{Z}_- = p_y. \tag{5.2}$$

In the classical case we understand p_x, p_y, p_z to be ordinary c-numbers, while the quantum case is obtained by formally replacing $p_x \to \partial/\partial x$, $p_y \to \partial/\partial y$, $p_z \to \partial/\partial z$. As an illustration of the general constructions (cf. Section 1) of infinitesimal operators in the required number of parameters starting directly from the commutation relations for the right (left) generators, we find the canonical right (left) variables for the group SL(2, R):

$$\{Z_+, Z_-\} = 2H, \quad \{H, Z_+\} = Z_+, \quad \{H, Z_-\} = -Z_-. \tag{5.3}$$

We choose the quantity Z_+ as the canonical momentum:

$$\mathcal{P}_X = Z_+. \tag{5.4}$$

For the conjugate coordinate X we then have the equation

$$2H\frac{\partial X}{\partial Z_-} - Z_+\frac{\partial X}{\partial H} = 1 \tag{5.5}$$

which follows from (5.3). One of the solutions of this equation has the form

$$X = Z_-/(\mathcal{P}_Z + H). \tag{5.6}$$

In addition, the algebra (5.3) has the cyclic momentum

$$\mathcal{P}_Z^2 = H^2 + Z_+Z_-. \tag{5.7}$$

Solving (5.4)-(5.6) in the reverse direction, we find expressions for the generators in terms of the right coordinates

$$Z_+ = \mathcal{P}_X, \quad Z_- = 2X\mathcal{P}_Z - X^2\mathcal{P}_X, \quad H = \mathcal{P}_Z - X\mathcal{P}_X. \tag{5.8}$$

After quantizing, $\mathcal{P}_X \to \partial/\partial X$, Eqs. (5.8) represent one of the possible realizations of a representation of the Lie algebra of SL(2, R), in which the Casimir operator $C = \frac{1}{2}(Z_+Z_- + Z_-Z_+) + H^2$ has the form

$$C = \mathcal{P}_Z(\mathcal{P}_Z + 1). \tag{5.9}$$

From the left generators with the obvious replacement $\underset{\sim}{Z}_- \to \underset{\sim}{Z}_+$ (although this is not necessary), we find analogously the canonical left coordinates and momenta

$$Y = \underset{\sim}{Z}_+/(\mathcal{P}_Z + H). \tag{5.10}$$

The relations analogous to (5.8) for the left generators have the form

$$\underset{\sim}{Z}_+ = 2Y\mathcal{P}_Z - Y^2\mathcal{P}_Y, \quad \underset{\sim}{H} = \mathcal{P}_Z - Y\mathcal{P}_Y, \quad \underset{\sim}{Z}_- = \mathcal{P}_Y. \tag{5.11}$$

Relations (5.8) and (5.11) are connected with (5.2) by a canonical transformation, the generating function of which has the form

$$\Phi(x, X, y, Y, z, \mathcal{P}_Z) = \mathcal{P}_Z \ln[z + e^{-2z}(x - X)(y - Y)]. \tag{5.12}$$

By differentiating this same equation, it is easy to obtain the cyclic variable Z conjugate to the momentum $\mathcal{P}_Z$,

$$Z = z + \ln(1 - v), \tag{5.13}$$

where

$$\frac{v}{(1-v)^2} = \frac{p_x p_y}{p_z^2} e^{-2z}.$$

It is well known (cf. [15]) how to obtain further formulas for the representation theory of the group SL(2, R) from Eqs. (5.8) and (5.11) for the generators.

6. THE CASIMIR OPERATORS OF SEMISIMPLE LIE GROUPS

The general expressions for the infinitesimal operators of the irreducible representations (cf. below) of semisimple Lie groups G obtained in [16-20] can also be found by the technique of canonical transformations described in Section 1. To this end we first consider a canonical transformation from the group variables in group phase space (**q** and **p** for an element g) to new variables (**q'**, **p'** for the element g') such that the generating function $\Phi(\mathbf{q}, \mathbf{q}')$ for this transormation satisfies the equations

$$\begin{aligned} \hat{F}_i\left(\mathbf{q}, \frac{\partial}{\partial \mathbf{q}}\right)\Phi &= \hat{F}_i\left(\mathbf{q}', \frac{\partial}{\partial \mathbf{q}'}\right)\Phi, \\ \underset{\sim}{\hat{F}}_i\left(\mathbf{q}, \frac{\partial}{\partial \mathbf{q}}\right)\Phi &= \underset{\sim}{\hat{F}}_i\left(\mathbf{q}', \frac{\partial}{\partial \mathbf{q}'}\right)\Phi, \end{aligned} \tag{6.1}$$

where $\hat{F}_i$ and $\hat{\tilde{F}}_i$ are generators of the right and left regular representations of the group G. Moreover, system (6.1) is naturally satisfied by any function on G constant on the conjugacy classes of the elements t

$$\Phi(t) \equiv \Phi(gg'^{-1}) = \Phi(g_0 t g_0^{-1}). \tag{6.2}$$

Taking the limit in Eqs. (6.2) as the values of the noncompact parameters in g' tend to infinity and expressing explicitly the dependence on the cyclic momenta conjugate to the noncompact variables in the element g', we get

$$\Phi(\mathbf{q}, \mathbf{q}') = Y_\rho(k_1 g k_2^{-1}), \tag{6.3}$$

where an irreducible right representation of arbitrary type is realized on the elements k_1 of the compact group, an irreducible left representation of arbitrary type is given on the elements k_2, and the function Y_ρ is the analytic continuation of the highest vector of an irreducible representation of the semisimple Lie group to complex parameters ρ, an expression for which is given in [20]. However, we stipulate that the function Φ in Section 1 (defining the transition from the regular representation in the natural form to the regular representation in the most convenient form) is related to the function Y_ρ by means of some complicated integral transformation (equivalent in fact to knowing the expansion of the δ-function on the group in terms of the irreducible unitary components of the regular representation). It is possible that it would be simpler to find this function if we realized the limiting representation in the root variables Z_+ and Z_- (cf. below). The examples of Lie groups which have been considered, in particular, of semisimple Lie groups, show that in the cases when the construction of the irreducible representations of general type is known, either by using the method of induced representations described in Section 2, or else by means of taking asymptotic values of the noncompact parameters (cf. below), the expressions obtained coincide with the expressions which can be obtained by means of a suitable canonical (classical and quantum) transformation in the regular representation in accordance with the rules of Section 1. This points up the geometric nature of both the method of induced representations and the asymptotic method [16-20] and relates the construction of irreducible representations to the structure of the surfaces in group phase space obtained by means of a group motion from some initial point (q, p). The choice of a correct (simplest) coordinate system in phase space in this context plays a most essential role in calculations.

A significant portion of the applications of the theory of group representations to physical problems reduces in the final analysis to finding the spectrum of the eigenvalues and eigenfunctions of the Casimir operators of the corresponding group [8]. An equation for determining the Casimir operators of a group as functions of the generators of the corresponding representation is obtained in the paper [31] of I. M. Gel'fand, although the explicit form of all its solutions has not been determined up to the present time. In the case of the classical semisimple Lie groups and their real forms, the problem simplifies substantially and the Casimir operators can be expressed functionally in terms of the traces of successive powers of the matrix of generators of the representation. In the case of compact groups the problem of finding the spectrum of eigenvalues of the Casimir operators for finite dimensional unitary representations is solved in [32].

In this section we construct in explicit form, following [16-20], the Casimir operators for the irreducible representations (finite and infinite dimensional) of all the classical semisimple Lie groups (complex groups and their real forms) as well as explicit expressions for the eigenvalue spectrum of the Casimir operators for arbitrary irreducible representations (including nonunitary and infinite dimensional ones) of all the classical Lie groups and all their real forms. We do this using the method of passage to limit and root space techniques. We

TABLE 1

G	Infinitesimal operators of the classical simple Lie groups
$L(n, C)$	$F_{ij} = -\sum_{k=1}^{n} \overset{*}{n}{}_i^k n_j^k \frac{\partial}{\partial \tau_k} + \sum_{k,\, l} \mathrm{sgn}\,(l-n)\, \overset{*}{n}{}_i^k n_j^l \tilde{F}^{kl}, \quad 1 \leqslant i, j \leqslant n$
$L(n, R)$	$F_{ij} = -\sum_{k=1}^{n} n_i^k n_j^k \frac{\partial}{\partial \tau_k} + \sum_{k,\, l} \mathrm{sgn}\,(l-k)\, n_i^k n_j^l \tilde{F}^{kl}, \quad 1 \leqslant i \neq j \leqslant n$
$\overset{*}{U}(2n)$	$F_{kl} = -\sum_{s=-n}^{n} (n_{sk} n_l^s - n_{sl} n_k^s) \frac{\partial}{\partial \tau_s} + \frac{1}{2} \sum_{\lvert a \rvert < \lvert b \rvert} (n_{ak} n_l^b - n_{al} n_k^b) \tilde{F}_b -$ $-\frac{1}{2} \sum_{\lvert a \rvert > \lvert b \rvert} (n_{ak} n_l^b - n_{al} n_k^b) \tilde{F}_b^a$
$U(p, q)$ $p < q$	$F_{\alpha i}^{\mu} = i^{\mu-1} A_{\alpha i} - (-i)^{\mu-1} B_{\alpha i}, \quad \mu = 1, 2, \quad 1 \leqslant \alpha \leqslant q, \quad 1 \leqslant i \leqslant p$ $A_{\alpha i} = -\frac{1}{2} \sum_{k=1}^{p} \overset{*}{q}{}_\alpha^k p_i^k \left(\frac{\partial}{\partial \tau_k} + \tilde{Q}^{kk} - \tilde{\mathcal{P}}^{kk} \right) - \sum_{q \geqslant \beta \geqslant k \geqslant 1} \overset{*}{q}{}_\alpha^\beta p_i^k \tilde{Q}^{\beta k} +$ $+ \sum_{1 \leqslant k \leqslant l \leqslant p} \overset{*}{q}{}_\alpha^k p_i^l \tilde{\mathcal{P}}^{kl}, \quad B_{\alpha i} = \overset{*}{A}_{\alpha i}$
$O(n, C)$	$B_{ij}^{2k} = \sum_{1}^{k} \begin{bmatrix} 2\alpha-1 & 2\alpha \\ i & j \end{bmatrix} \frac{\partial}{\partial \tau_{2\alpha}} + \sum_{\alpha > \beta} \left\{ \begin{bmatrix} 2\beta & 2\alpha \\ i & j \end{bmatrix} \tilde{F}^{2\beta-1,\, 2\alpha} + \right.$ $\left. + \begin{bmatrix} 2\beta & 2\alpha-1 \\ i & j \end{bmatrix} \tilde{F}^{2\beta-1,\, 2\alpha-1} - \begin{bmatrix} 2\beta-1 & 2\alpha-1 \\ i & j \end{bmatrix} \tilde{F}^{2\beta,\, 2\alpha-1} - \begin{bmatrix} 2\beta-1 & 2\alpha \\ i & j \end{bmatrix} \tilde{F}^{2\beta,\, 2\alpha}, \right.$ $n = 2k \quad \begin{bmatrix} \theta & \varphi \\ i & j \end{bmatrix} \equiv n_i^\theta n_j^\varphi - n_i^\varphi n_j^\theta$ $B_{ij}^{2k+1} = B_{ij}^{2k} + \sum \left\{ \begin{bmatrix} 2\alpha & 2k \\ i & j \end{bmatrix} \tilde{F}^{2\alpha-1,\, 2k} - \begin{bmatrix} 2\alpha-1 & 2k+1 \\ i & j \end{bmatrix} \tilde{F}^{2\alpha,\, 2k+1} \right\},$ $n = 2k+1$
$\overset{*}{O}(2n)$	$B_{ij}^{2k} = \sum_{1}^{k} \begin{bmatrix} 2\alpha-1 & 2\alpha \\ i & j \end{bmatrix} \left(\frac{\partial}{\partial \tau_{2\alpha}} - i\tilde{F}^{2\alpha-1,\, 2\alpha-1} - i\tilde{F}^{2\alpha,\, 2\alpha} \right) - i \sum \left\{ \begin{bmatrix} 2\alpha & 2\beta \\ i & j \end{bmatrix} \times \right.$ $\times \tilde{F}^{2\alpha-1,\, 2\beta-1} + \begin{bmatrix} 2\alpha & 2\beta \\ i & j \end{bmatrix} \tilde{F}^{2\alpha-1,\, 2\beta} - \begin{bmatrix} 2\alpha+1 & 2\beta-1 \\ i & j \end{bmatrix} \tilde{F}^{2\alpha,\, 2\beta-1} -$ $\left. - \begin{bmatrix} 2\alpha-1 & 2\beta \\ i & j \end{bmatrix} \tilde{F}^{2\alpha,\, 2\beta} \right\}, \quad n = 2k$ $B_{ij}^{2k+1} = B_{ij}^{2k} + \sum_{1}^{k} \left\{ \begin{bmatrix} 2\alpha & 2k \\ i & j \end{bmatrix} \tilde{F}^{2\alpha-1} - \begin{bmatrix} 2\alpha-1 & 2k+1 \\ i & j \end{bmatrix} \tilde{F}^{2\alpha,\, 2k+1} \right\},$ $n = 2k+1$
$O(p, q)$ $p \leqslant q$	$F_{\alpha i} = -\sum_{1}^{p} q_\alpha^k p_i^k \frac{\partial}{\partial \tau_k} - \sum_{q \geqslant \beta > k \geqslant 1} q_\alpha^\beta p_l^k \tilde{Q}^{\beta k} + \sum_{1 \leqslant k < l \leqslant p} q_\alpha^k p_i^l \tilde{\mathcal{P}}^{kl},$ $1 \leqslant \alpha \leqslant q, \quad 1 \leqslant i < p$
$Sp(2n, C)$	$2F_{2,\, ab} = -i \sum_{1}^{2n} (n_{ca} n_b^c + n_{cb} n_a^c) \frac{\partial}{\partial \tau_c} + i \sum_{c,\, d=1}^{2n} \varepsilon(d-c)(n_{ca} n_b^d + n_{cb} n_a^d) \tilde{F}_d^c$

TABLE 1. Continued

G	Infinitesimal operators of the classical simple Lie groups.
$Sp(2n, R)$	$2F_{1,kl} = \widetilde{A}^{lk} - \widetilde{A}^{kl} + \sum_{1}^{n} (n_l^i n_k^i + \overset{*}{n}{}_l^i n_k^i) \frac{\partial}{\partial \tau_i} + \sum (n_l^i n_k^i - \overset{*}{n}{}_l^i \overset{*}{n}{}_k^i) \widetilde{A}^{ii} + \sum_{i>j} (n_l^i n_k^j + n_k^i n_l^j) \widetilde{A}^{ij} - \sum_{i<j} (\overset{*}{n}{}_l^i \overset{*}{n}{}_k^j + \overset{*}{n}{}_k^i \overset{*}{n}{}_l^j) \widetilde{A}^{ij}$
	$F_{2,kl} + F_{3,kl} = -i \sum_{i=1}^{n} (n_k^i n_l^i - \overset{*}{n}{}_k^i \overset{*}{n}{}_l^i) \frac{\partial}{\partial \tau_i} - i \sum_{i=1}^{n} (n_k^i n_l^i + \overset{*}{n}{}_k^i \overset{*}{n}{}_l^i) \widetilde{A}^{ii} - i \sum_{i>j} (n_k^i n_l^j + n_l^i n_k^j) \widetilde{A}^{ij} - i \sum_{i<j} (\overset{*}{n}{}_k^i \overset{*}{n}{}_l^j + \overset{*}{n}{}_l^i \overset{*}{n}{}_k^j) \widetilde{A}^{lj}$
$Sp(2p, 2q)$	$F^{\lambda k} = -\frac{1}{4} \sum_{-q}^{q} (m_{i\lambda} n_k^i - n_{ik} m_\lambda^i) \frac{\partial}{\partial \tau_i} - \frac{1}{4} \sum_{-q}^{q} (m_{\mu\lambda} n_k^\nu + n_{\mu k} m_\lambda^\nu) \times (\widetilde{M}_\nu^\mu - \widetilde{N}_\nu^\mu) - \frac{1}{4} \sum_{-q}^{q} \varepsilon(\lvert\mu\rvert - \lvert\nu\rvert)(m_{\mu\lambda} n_k^\lambda - n_{\mu k} m_\lambda^\nu)(\widetilde{M}_\nu^\mu + \widetilde{N}_\nu^\mu)$

remark that the realization of the representations of semisimple groups in the mathematical literature is given from the start in integral form, and the representation is determined by a set of complex parameters ρ_s and real integral parameters $\varkappa_s$ (cf., for example, the monograph of Gel'fand and Naimark [12]). The question concerning the expression of the Casimir operators of a group in terms of ρ and $\varkappa$ is studied in [8]. In solving this problem we make use of the explicit form of the generators of any representation of a semisimple group obtained in [16-21] by means of passing to the limit in the generators of the regular representation in the asymptotic domain of infinitely large values of the noncompact parameters, which in fact means that a suitable canonical transormation is used (cf. Section 1). For the convenience of the reader we present the corresponding derivation of the formulas for the generators of the semisimple Lie groups.

Moreover, in this section we will consider the compact Lie groups and also find explicit expressions for the Casimir operators of the group of complex n-th order matrices L(n, C) and its real forms L(n, R), $\overset{*}{U}(2n)$, U(p, q); the analogous problem for the case of the orthogonal groups O(n, C), $\overset{*}{O}(2n)$, O(p, q) and the symplectic groups is also solved. Table 1 shows the explicit expressions for the generators of a representation of semisimple Lie groups, while Table 2 gives the expressions for the Casimir operators of the classical groups. These expressions and their derivation were given in [16-19] and in the survey [20].

Every element g of a semisimple group G (up to a set of smaller dimension) can be written in the form

$$g = k_1 \tau k_2, \tag{6.4}$$

where k_1, k_2 are elements of a maximal compact subgroup of G, and τ is an element of a noncompact Cartan subgroup. In the case of the classical Lie groups, Eq. (6.4) can be viewed as a matrix equation; thus, k_1, k_2 are matrices in a maximal compact subgroup and τ is a matrix in a noncompact Cartan subgroup. In order for decomposition (6.4) to be unique, the parameters k_2 associated with the invariance subgroup S of the maximal compact subgroup commuting with τ (i.e., the centralizer) must be assumed to be contained in k_1. In order to find the ex-

TABLE 2

$L(N, C)$	$(A_\pm)_{ab} = \delta_{ab} A_{aa} \Theta(a, b)$ $(A_\pm)_{aa} = m_a + N - a$	$m_a = {}^1/_2(\rho_a \pm \varkappa_a)$ $\Theta(a, b) = \begin{cases} 1, & a < b \\ 0, & a \geqslant b \end{cases}$
$L(N, R)$	$A_{ab} = \delta_{ab} A_{aa} - \Theta(a, b)$ $A_{aa} = m_a + N - a$ $1 \leqslant a \leqslant N$	$m_a = \rho_a$
$\dot{U}(2N)$	$A_{ab} = \delta_{ab} A_{aa} - \Theta(a, b)$ $A_{aa} = m_a + 2N - a$ $1 \leqslant a \leqslant 2N$	$m_{2k} = {}^1/_2(\rho_k - \varkappa_k)$ $m_{2k-1} = {}^1/_2(\rho_k + \varkappa_k)$ $k = 1, 2, \ldots, N$
$U(p, q)$ $p \leqslant q$	$A_{ab} = \delta_{ab} A_{aa} - \Theta(a, b)$ $A_{aa} = m_a + p + q - a$ $1 \leqslant a \leqslant p + q = N$	$m_a = {}^1/_2(\rho_a + \varkappa_a), \quad 1 \leqslant a \leqslant p$ $m_a = l_{q\ldots a+1}, \ \rho + 1 \leqslant a \leqslant q$ $m_a = {}^1/_2(-\rho_{N-a+1} + \varkappa_{N-a+1})$
$Sp(2N, C)$	$(A_\pm)_{ab} = \delta_{ab} A_{aa} - \Theta(a, b)(1 + \delta_{a, 2n-a+1})$ $A_{ii} = m_i + (N+1)\varepsilon_i - i + N + 1$ $i = a, \quad a \leqslant N; \quad i = a - 2N - 1, \quad a \geqslant N$	$m_i^\pm = \rho_i \pm \varkappa_i$ $m_{-i} \equiv m_i, \quad i = 1, 2, \ldots, N$ $\varepsilon_i = 1, \ i > 0, \ \varepsilon_i = -i, \ i < 0$
$Sp(2N, R)$	$A_{ab} = \delta_{ab} A_{aa} - \Theta_{ab}(1 + \delta_{a, 2N-a+1})$ $A_{ii} = m_i + (N+1)\varepsilon_i - i + N + 1$ $i = a, \quad a \leqslant N; \quad i = a - 2N - 1, \quad a \geqslant N$	$m_i = \rho_i$ $m_{-i} = m_i$ $i = 1, \ldots, N$
$Sp(2n, 2m)$ $n \leqslant m$	$A_{ab} = \delta_{ab} A_{aa} - \Theta_{ab}(1 + \delta_{a, 2N-a+1})$ $A_{ii} = m_i + (N+1)\varepsilon_i - i + N + 1$ $i = a, \quad a \leqslant N; \quad i = a - 2N - 1, \quad a \geqslant N;$ $N = n + m$	$m_i = \begin{cases} m_{2k} = {}^1/_2(\rho_k - m_k), \\ k = 1, \ldots, n \\ m_{2k-1} = {}^1/_2(\rho_k + m_k) \\ m_{N-1+j} = l_j, \\ j = 1, \ldots, m - n \end{cases}$
$O(2N, C)$	$A_{ab}^\pm = \delta_{ab}(A_\pm)_{aa} - \Theta_{ab}(1 - \delta_{a, 2N-a+1})$ $(A_\pm)_{ii} = m_i^\pm + N\varepsilon_i - i + N - 1$ $i = a, \quad a \leqslant N; \quad i = a - 2N - 1, \quad a \geqslant N$	$m_i^\pm = \rho_i \pm m_i$ $m_{-i} \equiv - m_i, \quad i = 1, 2, \ldots, N$
$\dot{O}(2N)$ $N = 2k$	$A_{ab} = \delta_{ab} A_{aa} - \Theta_{ab}(1 - \delta_{a, 2N-a+1})$ $A_{ii} = m_i + N\varepsilon_i - i + N - 1$ $i = a, \quad a \leqslant N; \quad i = a - 2N - 1, \quad a \geqslant N$	$m_{2i} = {}^1/_2(\rho_i - m_i), i = 1, 2, \ldots, k$ $m_{2i-1} = {}^1/_2(\rho_i + m_i), i = 1, 2, \ldots, k$ $m_{-i} = - m_i$
$\dot{O}(2N)$ $N = 2k + 1$	$A_{ab} = \delta_{ab} A_{aa} - \Theta_{ab}(1 - \delta_{a, 2N-a+1})$ $A_{ii} = m_i + N\varepsilon_i - i + N - 1$ $i = a, \quad a \leqslant N; \quad i = a - 2N - 1, \quad a \geqslant N$	$m_{2i} = {}^1/_2(\rho_i - m_i), i = 1, 2, \ldots, k$ $m_{2i-1} = {}^1/_2(\rho_i + m_i), i = 1, 2, \ldots, k$ $m_{k+1} = \varkappa_{k+1}, \quad m_{-i} = - m_i,$ $N = 2k + 1$
$O(p, q)$ $q - p = 2k$ $p + q = 2N$	$A_{ab} = \delta_{ab} A_{aa} - \Theta_{ab}(1 - \delta_{a, 2N-a+1})$ $A_{ii} = m_i + N\varepsilon_i - i + N - 1$ $i = a, \ a \leqslant N; \quad i = a - 2N - 1, \ a \geqslant N$	$m_i = p_i, \ i = 1, 2, \ldots, p$ $m_i = l_{N-i+1}$ $m_{-i} \equiv m_i, \quad i = p + 1, \ldots, N$
$O(2N+1, C)$	$(A_\pm)_{ab} = \delta_{ab}(A_\pm)_{aa} - \Theta_{ab}(1 - \delta_{a, 2N-a+2})$ $(A_\pm)_{ii} = m_i^\pm + (N + {}^1/_2)\varepsilon_i - i + N - {}^1/_2$ $i = a, \ a \leqslant N; \quad i = a - 2N - 2, \quad a > N + 1$	$m_i^\pm = {}^1/_2(\rho_i \pm \varkappa_i)$ $m_0 = a, \quad m_{-i} \equiv m_i$ $i = 1, 2, \ldots, N$
$O(p, q)$ $q - p = 2k + 1$	$A_{ab} = \delta_{ab} A_{aa} - \Theta_{ab}(1 - \delta_{a, 2N+2-a})$ $A_{ii} = m_i + (n + {}^1/_2)\varepsilon_i - i + N - {}^1/_2$	$m_i = \rho_i, \quad i = 1, 2, \ldots, p$ $m_i = l_{i-p}, \quad i = p + 1, \ldots$ $\ldots, {}^1/_2(p + q - 1)$

plicit form of the infinitesimal operators of the left translations in the parametrization (6.4), we make use of the fundamental group relation in differential form

$$\delta g = -F_\varepsilon g \delta_\varepsilon;$$
$$k_1^{-1}\delta k_1 + \tau^{-1}\delta\tau + \tau^{-1}\delta k_2 k_2^{-1}\tau = -k_1^{-1}F_\varepsilon k_1 \delta\varepsilon, \tag{6.5}$$

where F_ε is one of the tangent matrices to g^{-1} and $\delta\varepsilon$ is the corresponding parameter. Equation (6.5) allows us to calculate the derivatives $k_1^{-1}\dot{k}_1$, $\dot{k}_2 k_2^{-1}$, $\dot{\tau}$ of interest to us, where $\dot{k}_i \delta\varepsilon = \delta k_i$, $\delta\tau = \dot{\tau}\delta\varepsilon$, $i = 1, 2$, whence by the usual rules we obtain an expression for the infinitesimal operator for the left transition on the group

$$\hat{F}_\varepsilon = \mathrm{Tr}\,(k_1^{-1}\dot{k}_1)\,\hat{\hat{k}}_1 + \sum_s \dot{\tau}_s \frac{\partial}{\partial \tau_s} + \mathrm{Sp}\,(\dot{k}_2 k_2^{-1})\,\hat{k}_2, \tag{6.6}$$

where $\hat{\hat{k}}_1$ ($\hat{k}_2$) denotes the matrices of the generators of the right (left) translations for the maximal compact subgroup k. Expression (6.6) for the infinitesimal operators of the left regular representation in the general case depends in a rather complicated functional manner both on the choice of the noncompact parameters τ_s and the matrices k_1, k_2. The situation simplifies considerably in the asymptotic domain of infinitely large values of the noncompact parameters. In this domain, the dependence on the operators k_2 drops out completely, while at the same time the dependence on the noncompact parameters reduces to the trivial derivative $\partial/\partial\tau_s$. Explicit expressions for the limit generators are obtained in [16-21]. We give the corresponding calculation for the case of complex semisimple Lie groups in terms of root space techniques [15]. The tangent matrices of a semisimple Lie group corresponding to the compact F and noncompact Φ generators are expressed in terms of the positive and negative roots X_α, $X_{-\alpha}$ and the Cartan subalgebra h_s as follows:

$$F_\alpha^1 = X_\alpha - X_{-\alpha}, \quad F_\alpha^2 = i\,(X_\alpha + X_{-\alpha}), \quad F_s = ih_s,$$
$$\Phi_\alpha^1 = i\,(X_\alpha - X_{-\alpha}), \quad \Phi_\alpha^2 = -(X_\alpha + X_{-\alpha}), \quad \Phi_s = h_s. \tag{6.7}$$

The matrices X_α, $X_{-\alpha}$, h_s generally speaking can be the matrices of any representation of the corresponding Lie algebra. We now use lower-dimensional representations for these matrices which are more convenient at the moment. Define the set Y_μ to be the set of all the matrices X_α, $X_{-\alpha}$, h_s and let Y^ν be defined by the relation $\mathrm{Tr}(Y_\mu Y^\nu) = \delta_\mu^\nu$. The set Y^ν contains the matrices X^α, $X^{-\alpha}$, h^s. Corresponding to Eq. (6.4), and using the expansion $\tau = \exp(\tau_i h_i)$ for the matrix τ, where τ_i are real parameters, we have the following equation:

$$e^{-\tau_i h_i}\dot{\omega}e^{\tau_i h_i} + \dot{\tau}_i h_i + \dot{\Omega} = e^{-\tau_i h_i} S e^{\tau_i h_i}, \qquad S = k_1^{-1}F_\varepsilon k_1, \tag{6.8}$$

or

$$e^{\tau_i h_i^+}\dot{\omega}^+ e^{-\tau_i h_i^+} + \dot{\tau}_i h_i^+ + \dot{\Omega}^+ = e^{\tau_i h_i^+} S^+ e^{-\tau_i h_i^+}, \tag{6.9}$$

where the matrices $\dot{\omega} = k_1^{-1}\dot{k}_1$, $\dot{\Omega} = \dot{k}_2 k_2^{-1}$ are anti-Hermitian ($\dot{\omega}^+ = -\dot{\omega}$, $\dot{\Omega}^+ = -\Omega$) since a compact matrix group can always be realized by unitary matrices for which the infinitesimal matrices $\dot{\omega}$, $\dot{\Omega}$ obviously satisfy these conditions. The noncompact real Cartan matrices h_i are Hermitian. Using the usual expansion

$$\dot{\omega} = \sum_{\substack{\beta>0\\ \beta<0}} \omega_\beta X_\beta + \sum_i \omega_i h_i, \qquad \dot{\Omega} = \sum_{\substack{\beta>0\\ \beta<0}} \Omega_\beta X_\beta + \sum_i \Omega_i h_i, \tag{6.10}$$

where the scalar coefficients $\omega_\beta = \mathrm{Tr}\,(\dot{\omega}X^\beta)$; $\Omega^\beta = \mathrm{Tr}\,(\dot{\Omega}X^\beta)$; $\omega_i = \mathrm{Tr}\,(\dot{\omega}h^i)$, $\Omega_i = \mathrm{Tr}\,(\dot{\Omega}h^i)$ satisfy the relations $\dot{\omega}_\beta = -\omega_{-\beta}$, $\dot{\Omega}_\beta = -\Omega_{-\beta}$, we can add the left and right-hand sides of Eqs. (6.9) and get the relation

$$\sum_{\substack{\alpha>0\\ \alpha<0}} (\omega_\alpha e^{-\tau_i h_i} X_\alpha e^{\tau_i h_i} - \omega_\alpha e^{\tau_i h_i} X_\alpha e^{-\tau_i h_i}) + \sum_i 2\dot{\tau}_i h_i = \sum_{\substack{\alpha>0\\ \alpha<0}} S_\alpha^+ e^{\tau_i h_i} X_\alpha e^{-\tau_i h_i} + \\ + \sum_i S_i^+ h_i + \sum_i S_i h_i + \sum_{\alpha>0} (S_\alpha e^{-\tau_i h_i} X_\alpha e^{\tau_i h_i} + S_\alpha^+ e^{\tau_i h_i} X_\alpha e^{-\tau_i h_i}). \tag{6.11}$$

The root vectors X_α satisfy the commutation relations

$$[h_i, X_\alpha] = \alpha_i X_\alpha,$$

and hence

$$e^{-\tau_i h_i} X_\alpha e^{\tau_i h_i} = X_\alpha \exp[\alpha_i \tau_i].$$

We thus have from (6.11), (6.8), and (6.9) that

$$\begin{aligned} \omega_\alpha &= -\frac{S_\alpha \exp(\alpha_i\tau_i) + S_\alpha^+ \exp(-\alpha_i\tau_i)}{2\,\mathrm{sh}\,\alpha_i\tau_i}, \\ S_\alpha &= \mathrm{Tr}\,(X_\alpha k^{-1} F_\varepsilon k), \\ \Omega_\alpha &= \frac{(S+S^+)\,\alpha}{2\,\mathrm{sh}\,(\alpha_i\tau_i)}, \qquad S_\alpha^+ = \mathrm{Tr}\,(X_{-\alpha} k^{-1} F_\varepsilon k), \\ \dot{\tau}_i &= \frac{1}{2}\,\mathrm{Tr}\,(k^{-1}(F_\varepsilon + F_\varepsilon^+)\,kh^i), \\ \dot{\omega}_i + \dot{\Omega}_i &= \mathrm{Tr}\,(k^{-1}(F_\varepsilon - F_\varepsilon^+)\,kh^i) = 0. \end{aligned} \tag{6.12}$$

Equations (6.12) are exact. We can now pass to the limit in them. Consider the derivatives in (6.12) in the domain of infinite values $\tau_i \to \infty$ of the noncompact parameters of the Cartan subgroup. We then obtain from (6.12)

$$\begin{aligned} \dot{\omega}_\alpha &= \Theta(-\alpha)\, S_\alpha - \Theta(\alpha)\, S_\alpha^+, \quad \dot{\Omega}_\alpha = 0, \quad \dot{\Omega}_i = 0, \\ \dot{\omega}_i &= (S_i - S_i^+)/2i, \quad S_\alpha = \mathrm{Tr}\,(k^{-1}F_\varepsilon k X^\alpha), \\ \dot{\tau}_i &= (S_i + S_i^+)/2, \quad S_i = \mathrm{Tr}\,(k^{-1}F_\varepsilon k h^i), \end{aligned} \tag{6.13}$$

where $\Theta(\alpha) = 1$ for $\alpha > 0$, $\Theta(\alpha) = 0$ for $\alpha < 0$.

This limit procedure is a convenient method for finding the generators of the regular representation in most convenient form with the aid of a canonical transformation which is not a point transformation, and it is related to the separation of variables in group phase space. Using the operation described above, we obtain very simple expressions for the left regular representation of the group G. The infinitesimal operator $\hat{F}$ of this representation corresponding to the transformation F has the form

$$\hat{F} = \sum_{\alpha \gtrless} \dot{\omega}_\alpha \hat{\tilde{X}}^\alpha + \sum_i \left(\dot{\tau}_i \frac{\partial}{\partial \tau_i} + \dot{\omega}_i \hat{\tilde{h}}^i\right), \tag{6.14}$$

where the operators $\hat{\tilde{X}}^\alpha$, $\hat{\tilde{h}}^i$ are operators of the right regular representation of the compact subgroup k depending only on the parameters k_1. The differential operator $\tilde{X}_1^\alpha$ corresponds to the compact root vector X_α while the operator $\tilde{h}_i$ corresponds to the matrix h_i in the Cartan subgroup. It would have been possible to carry out the same limiting procedure for the right regular representation of G. The corresponding infinitesimal operators in this case would depend only on the parameters of the subgroup k_2. (This situation is connected with the existence of a coordinate system in which the variables are completely separated.) The trivial dependence on the parameters τ_i in this case is $\sim\partial/\partial\tau_i$, in the same way as in the case of the left regular representation considered above. Expression (6.14) can be considered in which we put $\partial/\partial\tau_i = \rho_i$, where the ρ_i are any complex numbers, since the operators $\partial/\partial\tau_i$ commute with both the left and the right limit regular representations. (The Casimir operators must necessarily commute simultaneously with the left and right representations.) Thus we have the following formulas for an infinitesimal representation of a complex group G:

$$\hat{F}_\varepsilon = \sum_\alpha \operatorname{Tr}(X_\alpha k^{-1} F_\varepsilon k)\,\hat{\hat{X}}^{-\alpha} - \sum_{\alpha>0} \operatorname{Tr}(X_{-\alpha} k^{-1} F_\varepsilon^+ k)\,\hat{\hat{X}}^{\alpha} + \\ + \sum_i \left[\rho_i \operatorname{Tr}\left(k^{-1}\frac{F_\varepsilon^+ + F_\varepsilon}{2} k h^i\right) + \operatorname{Tr}\left(k^{-1}\frac{F_\varepsilon - F_\varepsilon^+}{2} k h^i\right)\hat{\hat{H}}^i\right]. \tag{6.15}$$

Introducing as usual the tangent matrices $F_+ = (\Phi + iF)/2$ and $F_- = (\Phi - iF)/2$ (the matrices of the systems F_+ and F_- commute, $[F_+, F_-] = 0$), we have from (6.15) for the operators $F_\pm$

$$F_+ = \frac{1}{2}\sum_s \operatorname{Tr}(h^s k^{-1}\Phi k)(\rho_s + \varkappa^s) + \sum_{\alpha>0} \operatorname{Tr}(X_\alpha k^{-1}\Phi k)\,\hat{X}^{-\alpha}, \tag{6.16}$$

$$F_- = \frac{1}{2}\sum_s \operatorname{Tr}(h^s k^{-1}\Phi k)(\rho_s - \varkappa^s) - \sum_{\alpha>0} \operatorname{Tr}(X_{-\alpha} k^{-1}\Phi k)\,\hat{X}^{\alpha}. \tag{6.17}$$

The generators of the classical complex groups L(n, C), O(n, C), Sp(n, C) obtained by means of Eq. (6.15) are given in Table 1.

It should be remarked that Eqs. (6.15)-(6.17) are also valid for the exceptional Cartan groups G_2, F_4, E_6, E_7, E_8. The real forms of a given complex semisimple group are determined by the properties of the corresponding involutive automorphisms σ, which are enumerated in [14] for all the series A_n, B_n, C_n, D_n, G_2, F_4, E_6, E_7, E_8. By means of the automorphism σ, a Cartan subalgebra of a compact group splits into two sets: ih_α, the generators of a compact Cartan subalgebra, and h_s, the generators of a noncompact Cartan subalgebra. The complete set of noncompact generators is obtained as a result of commuting the noncompact Cartan subalgebra h_s with the infinitesimal operators of a maximal compact subgroup of the given real form. An infinitesimal transformation of the maximal compact group can be expanded with respect to the root vectors of the compact group, from which we obtain the given real form. Hence the basic equation (6.8) does not undergo any changes in the case of a real form. We remark only that since the h_i in general make up only a portion of the generators of a Cartan subalgebra of some compact group k, the algebra $\Sigma h_s \tau_s$ commutes with some root space X_i, where X_i forms a basis for the Lie algebra of the invariance subgroup. Thus in the real case, (6.8) takes the form

$$\sum_\alpha i\dot{\omega}_\alpha h_\alpha + \sum_\gamma \dot{\omega}_\gamma X_\gamma + \sum_s h_s \dot{\tau}_s + \sum_\beta i\Omega_\beta h_\beta + \sum_\xi X_\xi \dot{\Omega}_\xi e^{\xi(\tau)} + \sum_i (\dot{\omega}_i + \dot{\Omega}_i) X_i = k^+ F k, \tag{6.18}$$

from which we find

$$\begin{aligned} \dot{\omega}_\gamma &= \frac{1}{2\operatorname{sh}\gamma(\tau)}\left[\operatorname{Tr}(X_{-\gamma}k^{-1}Fk)\,e^{-\gamma(\tau)} + \operatorname{Tr}(X_{-\gamma}k^{-1}Fk)\,e^{\gamma(\tau)}\right], \\ \dot{\Omega}_\gamma &= \frac{1}{2\operatorname{sh}\gamma(\tau)}\operatorname{Tr}\left[X_{-\gamma}k^{-1}(F + F^+)k\right], \\ \dot{\tau}_s &= \operatorname{Tr}(k^{-1}Fkh^s), \\ i(\dot{\omega}_\alpha + \dot{\Omega}_\alpha) &= \operatorname{Tr}(k^{-1}Fkh^\alpha), \\ \dot{\omega}_i + \dot{\Omega}_i &= \operatorname{Tr}(X_{-i}k^{-1}Fk), \\ [h, X_\gamma] &= \gamma(h)\,X_\gamma, \end{aligned} \tag{6.19}$$

where $\gamma(h)$ is the linear form corresponding to the root γ on the Cartan subalgebra.

For the generators of a representation in the asymptotic domain, we obtain

$$\hat{F} = \sum_s \mathrm{Tr}\,(k^{-1}Fkh^s)\,\rho^s + \sum_s \mathrm{Tr}(k^{-1}Fkh^s)\,\tilde{H}^s + \sum_{\gamma>0} \mathrm{Tr}\,(k^{-1}FkX_\gamma)\,\underline{\tilde{X}}^{-\gamma} + \sum_{\alpha\gtrless 0} \mathrm{Tr}\,(k^{-1}FkX_{-\alpha})\,\tilde{X}^\alpha, \tag{6.20}$$

where by $\underline{\tilde{X}}^{-\gamma}$ we mean the generators obtained by restricting the original compact group to a maximal compact subgroup of the real form considered: $\underline{\tilde{X}}^{-\gamma}k = k(X^{-\gamma} + \sigma X^\gamma\sigma)$. The definition of $\hat{F}$ (6.20) and the properties of h^σ, h^s, X^i, X^α given above imply that the generators (6.20) commute with the operators $\hat{X}^i$, $\hat{H}^\sigma$ which form a basis for the Lie algebra of the invariance subgroup. Since the operators F are diagonal in the quantum numbers of the invariance subgroup, we may assume that F acts on the highest vector of the invariance subgroup, i.e., $\hat{X}_i\Phi = 0$, $\hat{H}^\sigma\Phi = l_\sigma\Phi$. Thus for $\hat{F}$ we get

$$\hat{F} = \sum_s \mathrm{Tr}\,(k^{-1}Fkh^s)\,\rho_s + \sum_s \mathrm{Tr}\,(k^{-1}Fkh^s)\,l_s + \sum_\alpha \mathrm{Tr}\,(k^{-1}FkX_\alpha)\,\underline{\tilde{X}}^{-\alpha}. \tag{6.21}$$

The last sum in (6.21) extends over all the positive roots of the original compact group G. Comparing (6.15) with (6.21), we see that the generators of a real form of a complex group G are obtained from the generators of the complex group simply by restricting a maximal compact subgroup of G to a maximal compact subgroup of its real form. Consider the matrices G_+ and G_-:

$$G_+ = \sum_\alpha (\hat{F}_\alpha)_+ Y^\alpha, \qquad G_- = \sum_\alpha (\hat{F}_\alpha)_- Y^\alpha.$$

Using expressions (6.16) and (6.17) and taking into account the completeness of the system of root vectors $X^{\pm\alpha}$ and h^s, we have

$$G_\pm = \sum_s kh^sk^{-1}\left(\frac{\rho_s \pm \tilde{H}_s}{2}\right) \pm \sum_{\alpha>0} (kX_{\pm\alpha}k^{-1})\,\tilde{X}^{\mp\alpha}. \tag{6.22}$$

The form for the generators in (6.22) turns out to be convenient for calculating the traces of successive powers of the matrices $G_\pm$.

In the case of complex Lie groups, the algebras $\hat{F}_+$ and $\hat{F}_-$ of the generators in (6.16), (6.17) are by their commutation properties a direct product of the algebras of the corresponding compact group $k \oplus k$. Hence the system of Casimir operators of a semisimple complex group splits into two sets of Casimir operators C_s^+, C_s^- corresponding to the compact group constructed from the generators F_+ and F_-, respectively.

In order to compute the successive powers of the matrices F_+ and F_-, we rewrite the generators (6.22) in the form

$$G_+ = k\left[\sum_s h^s\left(\frac{\rho_s + \varkappa_s}{2}\right) + \sum_{\alpha>0} X_\alpha X_{-\alpha} + \sum_{\alpha>0} X_\alpha \tilde{X}^{-\alpha}\right]k^{-1}. \tag{6.23}$$

In going from (6.22) to (6.23) we have taken into account the definition of right translation $\tilde{X}^\alpha k = kX_\alpha$, which has led to the appearance of an extra summand $\sum_\alpha X_\alpha X_{-\alpha}$ in Eq. (6.23).

Calculation of the trace of the s-th power G_+^s of the matrix G_+ gives

$$C_s^+ \equiv \mathrm{Tr}\,(G_+)^s = \mathrm{Tr}\left\{k\left[\sum_s h^s\left(\frac{\rho_s + \varkappa_s}{2}\right) + \sum_{\alpha>0}\left(X^\alpha X^{-\alpha} + X^\alpha \tilde{X}^{-\alpha}\right)\right]^s k^{-1}\right\}. \tag{6.24}$$

This expression can be simplified. To do this it is necessary to move the matrix k^{-1} over to the left by commuting it with the operators X^α. The result of the commutation can be written as follows:

$$\left[h^s_{ab}\left(\frac{\rho_s+\varkappa_s}{2}\right)+\sum_{\alpha>0}(X_\alpha X_{-\alpha})_{ab}+\sum_{\alpha>0}(X_\alpha)_{ab}\,X^{-\alpha}\right](k^{-1})^i_c=$$
$$=\sum_\alpha (k^{-1})^i_d\left\{\delta_{cd}\left[\sum_s h^s_{ab}\left(\frac{\rho_s+\varkappa_s}{2}\right)+\sum_{\alpha>0}(X_\alpha X_{-\alpha})_{ab}+\sum_{\alpha>0}(X_\alpha)_{ab}X^{-\alpha}\right]-\sum_{\alpha>0}(X_\alpha)_{ab}(X_{-\alpha})_{cd}\right\}. \tag{6.25}$$

Since the trace of a product of any number $n \neq 0$ of positive roots with matrices in the Cartan subgroup is equal to zero, as follows from the commutation relations $[h, X_\alpha] = \alpha(h)X_\alpha$, where $\alpha(h) > 0$, we obtain from (6.24), (6.25)

$$C_k=\sum_{da,\,bc}\left[\sum_s h^s_{da}\left(\frac{\rho_s+\varkappa_s}{2}\right)+\sum_\alpha (X_\alpha X_{-\alpha})_{da}\right](D^k)^{bc}_{da}\left[\sum_l h^l_{bc}\left(\frac{\rho_l+\varkappa_l}{2}\right)\right]. \tag{6.26}$$

The matrix D is defined by the right-hand side of Eq. (6.6) and has two double indices: an initial index *ad* and a final one *bc*

$$D^{bc}_{ad}=\delta_{cd}\left[\sum_s h^s_{ab}\left(\frac{\rho_s+\varkappa_s}{2}\right)+\sum_{\alpha>0}(X_\alpha X_{-\alpha})_{ab}\right]-\sum_{\alpha>0}(X_\alpha)_{ab}(X_{-\alpha})_{cd}. \tag{6.27}$$

We do not restrict ourselves to any concrete representation of the matrices h^s, $X^{\pm\alpha}$ or use their concrete properties in any of the equations displayed above. The representation can be arbitrary. The matrix D acts on a column vector having two indices, i.e., on a matrix, and as a result we obtain a new doubly indexed column vector. The action of the matrix D on a matrix A can be written in the form

$$DA=\left[\sum_s h^s\left(\frac{\rho_s+\varkappa_s}{2}\right)+\sum_\alpha X_\alpha X_{-\alpha}\right]A-\sum_\alpha X_\alpha A X_{-\alpha}. \tag{6.28}$$

Equation (6.28) implies that the matrix DA commutes with the Cartan matrix $H=\sum_s h^s\tau_s$ if we assume that $[A, H] = 0$. This means that the action of the matrix D is equivalent to the action of a square matrix of dimension $\sum_i k_i^2$, where k_i is the multiplicity of the i-th root of the Cartan subalgebra.

If we let S_i denote a complete and orthogonal system of matrices $[\mathrm{Tr}(S_iS^j) = \delta_{ij}]$ commuting with the Cartan subalgebra, then the trace of the k-th power of the matrix of generators of a representation is determined by the sum of the matrix elements of the k-th power of the scalar matrix

$$A_{ij}=\sum_s\left(\frac{\rho_s+\varkappa_s}{2}\right)\mathrm{Tr}\,(S_iS^jh^s)+\sum_{\alpha>0}\mathrm{Tr}\,(S_iS^jX_\alpha X_{-\alpha})+\sum_{\alpha>}\mathrm{Tr}\,(X_\alpha S_iX_{-\alpha}S^j),$$
$$C_n=\sum_{ij}\left[\sum_s h^s\left(\frac{\rho_s+\varkappa_s}{2}\right)+\sum_{\alpha>0}(X_\alpha X_{-\alpha})\right]_{ii}A^{n-2}_{ij}\left[\sum_s h^s\left(\frac{\rho_s+\varkappa_s}{2}\right)\right]_{jj}. \tag{6.29}$$

In particular, if the representation h^s, $X^{\pm\alpha}$ is nondegenerate, i.e., only diagonal matrices commute with the Cartan subalgebra, then the Casimir operators can be described by means of powers of a scalar matrix, the dimension of which equals the dimension of the representation of h^s, $X^{\pm\alpha}$. It is clear from (6.28) that the trace of the n-th power of the matrix G_+ is determined by the sum of the matrix elements of some scalar matrix A^n

$$C_n=\mathrm{Tr}\,G_+^n=\sum_{ab}(A^n)_{ab},$$
$$A_{ab}=\sum_s h^s_{ab}\left(\frac{\rho_s+\varkappa_s}{2}\right)+\sum_\alpha (X_\alpha X_{-\alpha})_{ab}-\sum_\alpha (X_\alpha)_{ab}(X_{-\alpha})_{ba}. \tag{6.30}$$

It should be emphasized that, as follows from (6.28), Eq. (6.30) also holds for finite dimensional unitary representations of a compact group. Thus, Eq. (6.30) contains the spectrum of the Casimir operators for compact groups. This result was previously obtained by another method in [33]. We remark that the method used in [33] to find the spectrum of the Casimir operators of compact groups cannot be applied in the case of a degenerate representation for h_s, $X^{\pm\alpha}$. At the same time, Eq. (6.28) of the present paper also holds in this case. It also follows from Eq. (6.30) that the Casimir operators of the infinite dimensional representations $(\rho, \varkappa)$ of complex semisimple Lie groups can be obtained by formally replacing the weights l_s in the equations for the Casimir operators of compact groups by $(\rho_s \pm \varkappa_s)/2$.

The main steps of the calculation leading to Eq. (6.27) also go through unchanged in the case of real semisimple groups. Indeed, (6.27) follows from the explicit form of the generators (6.20), the main difference from the complex case consisting in the need to restrict the matrices $\tilde{X}^{\alpha}$ of the generators to a maximal compact subgroup of the corresponding real form, which leads instead of (6.20) to the following expression:

$$D_{ad,\,bc} = \{\delta_{cd}\,[h^{s}_{ab}\underline{\rho}_s + \sum_{\alpha} (X_\alpha \underline{X}_{-\alpha})_{ab}] - (X_\alpha)_{ab}(\underline{X}_{-\alpha})_{cd}\}, \tag{6.31}$$

where $\underline{\rho}^s$ denotes the corresponding complex weight and $\underline{X}_{-\alpha}$, the matrix in the relation $\tilde{X}^{\alpha}k = kX_\alpha$. In terms of the involutive automorphism, $\underline{X}_\alpha$ can be expressed in terms of the roots of the complex group as follows: $\underline{X}^{s}_{\alpha} = X_\alpha + \sigma X_\alpha \sigma$, where σ is the matrix of the involutive automorphism of the given real form. Indeed, for the increments $\dot{\omega}_\alpha$ and $\dot{\omega}^{*}_{\alpha}$ we have [cf. (6.19)]

$$\dot{\omega}_\alpha = \mathrm{Tr}\,(FkX_\alpha k^{-1}), \quad \dot{\omega}^{*}_{\alpha} = -\,\mathrm{Tr}\,(Fk\sigma X_\alpha \sigma k^{-1}).$$

The last relation for $\dot{\omega}_\alpha$ is obtained by conjugating the first, taking into account the properties of the involutive automorphisms of the given real form, i.e., $\sigma k = k\sigma$, $\sigma F^{+}\sigma = -F$. The expressions for $\dot{\omega}_\alpha$ and $\dot{\omega}^{*}_{\alpha}$ imply that $\underline{X}_\alpha = X_\alpha + \sigma X_\alpha \sigma$ is valid. If we denote by h_τ a maximal noncompact Cartan subalgebra of the given real form, the properties of the involutive automorphism imply the commutation relation $[h_\tau, \sigma X_\alpha \sigma] = -\alpha(\tau)\sigma X_\alpha \sigma$, while at the same time $[h_\tau, X_\alpha] = \alpha(\tau) X_\alpha$. This means that $\sigma X_\alpha \sigma$ belongs to the system of negative roots with respect to h_τ. The last fact permits us to carry out further transformations in (6.31). We write the matrix A in (6.30) in the form

$$A = \sum_i C_i S_i + \sum_{\alpha,\,\beta,\,\ldots} C_{\alpha\beta\ldots} X_\alpha X_\beta \ldots = \sum_i C_i S_i + A',$$

where the C_i were introduced earlier [cf. (6.29)], S_i is a complete orthonormal system of matrices commuting with the Cartan subalgebra, and $X_\alpha X_\beta \ldots$ is the product of the nonnegative roots with respect to h_τ. The structure of A' is preserved under the action of D on A [cf. (6.28)], the transformed values of C_i being related to C_i by the scalar matrix (6.29) as before [cf. (6.29)]. The structure of the matrix A is immaterial for the final expression for the traces of the successive powers of the matrix of generators of a representation, since the trace of a product of arbitrarily many positive roots is identically equal to zero. Thus the trace of the successive powers of the matrix of generators is determined, as were previously (6.29) in the general case and (6.30) for the classical groups, by the sum of the matrix elements of the S-th power of the scalar matrix A. Table 2 gives the values of the matrix A for all the classical groups and their real forms.

7. COMPACT AND SOLVABLE GROUPS AND GROUPS NOT OF TYPE I

In the case of compact Lie groups we consider as an example the compact rotation group. The generators of its right and left regular representations have the form

$$\begin{aligned} &L_1 + iL_2 = e^{i\varphi}\left[\frac{\partial}{\partial\theta} + i\cot\theta\frac{\partial}{\partial\varphi}\right], \\ &L_1 - iL_2 = e^{-i\varphi}\left[-\frac{\partial}{\partial\theta} + i\cot\theta\frac{\partial}{\partial\varphi}\right], \qquad L_3 = -i\frac{\partial}{\partial\varphi}; \\ &\underset{\sim}{L}_1 + i\underset{\sim}{L}_2 = e^{i\psi}\left[\frac{\partial}{\partial\theta} + i\cot\theta\frac{\partial}{\partial\psi}\right], \\ &\underset{\sim}{L}_1 - i\underset{\sim}{L}_2 = e^{-i\psi}\left[-\frac{\partial}{\partial\theta} + i\cot\theta\frac{\partial}{\partial\psi}\right], \qquad \underset{\sim}{L}_3 = -i\frac{\partial}{\partial\psi}. \end{aligned} \tag{7.1}$$

As generalized right $\mathcal{P}_X$ (left $\mathcal{P}_Y$) momentum we take the generator $L_3(\underset{\sim}{L}_3)$. Then the generalized coordinates X, Y are found from the solution of the equations

$$\{L_3, X(L_1, L_2, L_3)\} = 1, \qquad \{\underset{\sim}{L}_3, Y(\underset{\sim}{L}_1, \underset{\sim}{L}_2, \underset{\sim}{L}_3)\} = 1. \tag{7.2}$$

Taking into account the commutation relations for the rotation group we find

$$X = \arctan\frac{L_2}{L_3} + f(\mathbf{L}^2), \qquad Y = \arctan\frac{\underset{\sim}{L}_2}{\underset{\sim}{L}_3} + f(\underset{\sim}{\mathbf{L}}^2), \tag{7.3}$$

where the Casimir operator of the group is $\mathbf{L}^2$,

$$\mathbf{L}^2 = L_1^2 + L_2^2 + L_3^2 = \underset{\sim}{\mathbf{L}}^2, \qquad \mathbf{L}^2 = \mathcal{P}_Z(\mathcal{P}_Z + 1). \tag{7.4}$$

Solving system (7.2), taking into account the requirement of linearity, leads to explicit expressions for the generators in terms of the generalized coordinates and momenta

$$\begin{aligned} &L_3 = \mathcal{P}_X, \quad L_1 = i(\cos X\mathcal{P}_X + \sin X\mathcal{P}_Z), \quad L_2 = i(-\sin X\,\mathcal{P}_X + \\ &+ \cos X\mathcal{P}_Z); \quad \underset{\sim}{L}_3 = \mathcal{P}_Y, \quad \underset{\sim}{L}_1 = i(\cos Y\mathcal{P}_Y + \sin Y\mathcal{P}_Z), \\ &\underset{\sim}{L}_2 = i(-\sin Y\mathcal{P}_Y + \cos Y\mathcal{P}_Z), \end{aligned} \tag{7.5}$$

which coincides with the well-known realization of the Lie algebra of the rotation group (cf. [15]).

We investigate the results obtained by the formalism developed above for solvable Lie groups, taking as example the simplest solvable Lie group of dimension 2, the group of real matrices $g = \begin{Vmatrix} x & y \\ 0 & 1 \end{Vmatrix}$ considered in [12]. The generators of the right and left regular representations have the form

$$\begin{aligned} &L_1 = xp_x, \quad L_2 = xp_y, \quad \Psi = \begin{Vmatrix} x & 0 \\ 0 & x \end{Vmatrix}, \quad \det\Psi = x^2; \\ &\underset{\sim}{L}_1 = -xp_x - yp_y, \quad \underset{\sim}{L}_2 = -p_y, \quad \underset{\sim}{\Psi} = -\begin{Vmatrix} x & y \\ 0 & 1 \end{Vmatrix}, \quad |\det\underset{\sim}{\Psi}| = x. \end{aligned} \tag{7.6}$$

We make the following canonical transformation which is found using the procedure described above:

$$\mathcal{P}_X = \frac{p_x}{p_y}, \qquad \mathcal{P}_Y = p_y, \qquad X = xp_y, \qquad Y = y + x\frac{p_x}{p_y}. \tag{7.7}$$

In the new variables, we have

$$L_1 = X\mathcal{P}_X, \quad L_2 = X, \quad \underset{\sim}{L}_1 = -Y\mathcal{P}_Y, \quad \underset{\sim}{L}_2 = -\mathcal{P}_Y. \tag{7.8}$$

for the generators of the regular representation. The classical generating function for the canonical transformation (7.7) has the form

$$\Phi(x, y, X, Y) = \frac{y - Y}{x}X. \tag{7.9}$$

The corresponding quantum generating function bringing the generators of the regular representation (7.6) to the form (7.8) is given by

$$G(x, y, X, Y) = \frac{x^2}{X} \exp \frac{(Y-y)X}{x}. \tag{7.10}$$

We remark that in this example the left and right measures on the group do not coincide. The regular representation of this group is irreducible. We now consider the Mautner group as an example of a group which is not of type I. Its representations are discussed in [14]. The Mautner group can be realized as the group of matrices

$$g = \begin{Vmatrix} e^{it} & 0 & a \\ 0 & e^{i\alpha t} & b \\ 0 & 0 & 1 \end{Vmatrix}, \tag{7.11}$$

where t is a real parameter, a and b are complex numbers, and α is an irrational number $(a = a_1 + ia_2,\ b = b_1 + ib_2)$.

The generators of the right and left regular representations have the form

$$\begin{gathered}
\hat{x}_1 = \frac{\partial}{\partial t} = p_x, \quad \hat{x}_2 = \cos t p_{a_1} + \sin t p_{a_2}, \quad \hat{x}_3 = -\sin t p_{a_1} + \cos t p_{a_2}, \\
\hat{x}_4 = \cos \alpha t p_{b_1} + \sin \alpha t p_{b_2}, \quad \hat{x}_5 = -\sin \alpha t p_{b_1} + \cos \alpha t p_{b_2}; \\
\hat{\mathfrak{x}}_1 = -\left[p_x + a_1 \frac{\partial}{\partial a_2} - a_2 \frac{\partial}{\partial a_1} + \alpha\left(b_1 \frac{\partial}{\partial b_2} - b_2 \frac{\partial}{\partial b_1}\right)\right] = \\
= -[p_x + a_1 p_{a_2} - a_2 p_{a_1} + \alpha(b_1 p_{b_2} - b_2 p_{b_1})], \quad \hat{\mathfrak{x}}_2 = -\frac{\partial}{\partial a_1} = -p_{a_1}, \\
\hat{\mathfrak{x}}_3 = -\frac{\partial}{\partial a_2} = -p_{a_2}, \quad \hat{\mathfrak{x}}_4 = -\frac{\partial}{\partial b_1} = -p_{b_1}, \quad \hat{\mathfrak{x}}_5 = -\frac{\partial}{\partial b_2} = -p_{b_2}.
\end{gathered} \tag{7.12}$$

The matrices Ψ and $-\underset{\sim}{\Psi}$ respectively have the form

$$\begin{gathered}
\Psi = \begin{Vmatrix} 1 & 0 & 0 & 0 & 0 \\ 0 & \cos t & \sin t & 0 & 0 \\ 0 & -\sin t & \cos t & 0 & 0 \\ 0 & 0 & 0 & \cos \alpha t & \sin \alpha t \\ 0 & 0 & 0 & -\sin \alpha t & \cos \alpha t \end{Vmatrix}, \\
-\underset{\sim}{\Psi} = \begin{Vmatrix} 1 & -a_2 & a_1 & -\alpha b_2 & \alpha b_1 \\ 0 & 1 & 0 & 0 & 0 \\ 0 & 0 & 1 & 0 & 0 \\ 0 & 0 & 0 & 1 & 0 \\ 0 & 0 & 0 & 0 & 1 \end{Vmatrix}.
\end{gathered} \tag{7.13}$$

The matrix of the adjoint representation $-\Psi\underset{\sim}{\Psi}^{-1} = D$ has the form

$$D = \begin{Vmatrix} 1 & -a_2 & a_1 & -\alpha b_2 & \alpha b_1 \\ 0 & \cos t & \sin t & 0 & 0 \\ 0 & -\sin t & \cos t & 0 & 0 \\ 0 & 0 & 0 & \cos \alpha t & \sin \alpha t \\ 0 & 0 & 0 & -\sin \alpha t & \cos \alpha t \end{Vmatrix}. \tag{7.14}$$

It is easy to verify that $\Psi = -\det \underset{\sim}{\Psi}$, $\det D = 1$, and the left and right Haar measures on the group coincide. Using the recipes applied above, it is possible to construct a canonical tranformation, the classical generating function Φ for which has the following form:

$$\begin{gathered}
\Phi(a_1, a_2, b_1, b_2, p_x, X, Y, \mathcal{P}_{Z_1}, \mathcal{P}_{Z_2}, \mathcal{P}_{Z_3}) = -a_1 \mathcal{P}_{Z_1} \cos Y + \\
+ a_2 \mathcal{P}_{Z_1} \sin Y - p_x (X - Y) + b_2 (\mathcal{P}_{Z_2} \sin \alpha Y - \mathcal{P}_{Z_3} \cos \alpha Y) - \\
- b_1 (\mathcal{P}_{Z_2} \cos \alpha Y + \mathcal{P}_{Z_3} \sin \alpha Y).
\end{gathered} \tag{7.15}$$

In the new variables the generators of the right and left regular representations can be written

$$\begin{aligned}
&\hat{\mathfrak{x}}_1 = \mathscr{P}_Y, \quad \hat{\mathfrak{x}}_2 = \mathscr{P}_{Z_1} \cos Y, \qquad \hat{\mathfrak{x}}_3 = -\mathscr{P}_{Z_1} \sin Y,\\
&\hat{\mathfrak{x}}_4 = \mathscr{P}_{Z_2} \cos \alpha Y + \mathscr{P}_{Z_3} \sin \alpha Y, \quad \hat{\mathfrak{x}}_5 = -\mathscr{P}_{Z_2} \sin \alpha Y + \mathscr{P}_{Z_3} \cos \alpha Y;\\
&\hat{x}_1 = \mathscr{P}_X, \quad \hat{x}_2 = \mathscr{P}_{Z_1} \cos X, \qquad \hat{x}_3 = -\mathscr{P}_{Z_1} \sin X,\\
&\hat{x}_4 = \mathscr{P}_{Z_2} \cos \alpha X + \mathscr{P}_{Z_3} \sin \alpha X, \quad \hat{x}_5 = -\mathscr{P}_{Z_2} \sin \alpha X + \mathscr{P}_{Z_3} \cos \alpha X.
\end{aligned} \tag{7.16}$$

The three Casimir operators of the group are expressed in terms of the generators by the formulas

$$\begin{gathered}
\mathscr{P}^2_{Z_1} = x_2^2 + x_3^2, \quad \mathscr{P}^2_{Z_3} + \mathscr{P}^2_{Z_2} = x_4^2 + x_5^2,\\
\arctan \frac{x_3}{x_2} - \frac{1}{\alpha} \arctan \frac{x_5}{x_4} = f(\pi_{Z_1}, \pi_{Z_2}, \pi_{Z_3}).
\end{gathered} \tag{7.17}$$

If we construct the induced representation, writing g = kr, where

$$k = \begin{Vmatrix} e^{it} & 0 & 0 \\ 0 & e^{i\alpha t} & 0 \\ 0 & 0 & 1 \end{Vmatrix}, \tag{7.18}$$

and induce using the group character $\chi(\mathbf{r}) = \exp[i \operatorname{Re}(\mathbf{k}_1\mathbf{a} + \mathbf{k}_2\mathbf{b})]$, the irreducible representation obtained coincides with the representation which can be derived from Eqs. (7.16) by taking the Fourier integral with respect to the variables Z_i. It is possible to find another canonical transformation to another form which is maximally convenient to study, and such that in the new form the regular representation decomposes into irreducible representations which are not equivalent to the ones just obtained.

8. CONCLUSION

In conclusion we one again briefly discuss the structure of the method considered above for constructing representations of Lie groups and their Lie algebras based on canonical transformations. We find "good" right and left coordinates $\mathbf{X}$, $\mathbf{Y}$, $\mathscr{P}_\mathbf{X}$, $\mathscr{P}_\mathbf{Y}$ in the phase space of the group as functions of the right $\mathbf{L}$ and left $\mathfrak{L}$ generators of the regular representation. We recall that we first dealt with the classical canonical formalism. The fundamental theorem of functional groups guarantees the existence of the correct number of canonically conjugate coordinates and momenta $\mathbf{X}$, $\mathscr{P}_\mathbf{X}$, $\mathbf{Y}$, $\mathscr{P}_\mathbf{Y}$ and cyclic variables $\mathbf{Z}$ and $\mathscr{P}_\mathbf{Z}$, in terms of which the generators of the right and left translations can be expressed. The only point which remains to be proved is that it is possible to find canonical variables with respect to which the generators become linear forms in the momenta. We remark that the linearity of the system in the sense indicated above evidently allows us to quantize the classical system in the simplest traditional way by formally replacing the canonical momenta by first derivatives, which causes no difficulties with the ordering of operators (since the differential operator appears only once and must be located on the right). This is equivalent to proving that the conditions of the main theorem of functional groups (cf. [30]) are consistent with the condition that the second derivatives of all the generators in the new momenta $\mathscr{P}_\mathbf{X}$, $\mathscr{P}_\mathbf{Y}$, $\mathscr{P}_\mathbf{Z}$ are equal to zero.

In essence, we are required to study and prove the consistency of the following system of equations.

Assume given a Lie algebra of an arbitrary Lie group with generators L_i satisfying the commutation relations

$$\{L_i, L_k\} = c^j_{ik} L_j, \qquad i, k, j = 1, 2, \ldots, n. \tag{8.1}$$

Here the brackets denote the classical Poisson brackets and the quantities c^j_{ik} are the structure constants of the group. Consider a second Lie algebra with generators $\underset{\sim}{L}_i$ satisfying the same commutation relations, with moreover $\{L_i, \underset{\sim}{L}_k\} = 0$. It is then required to find quantities $X_\rho(\mathbf{L})$, $\mathcal{P}_{X_\sigma}(\mathbf{L})$, $\mathcal{P}_{Z_\alpha}(\mathbf{L})$ ($\alpha = 1, \ldots, r$), such that

$$\begin{gathered}\{\mathcal{P}_{X_\rho}, X_\sigma\} = \frac{\partial \mathcal{P}_{X_\rho}}{\partial L_s}\frac{\partial X_\sigma}{\partial L_m} c^n_{sm} L_n = \delta_{\rho\sigma},\\ \{\mathcal{P}_{X_\rho}, \mathcal{P}_{X_\sigma}\} = \{X_\rho, X_\sigma\} = 0, \quad \{\mathcal{P}_{Z_\alpha}, X_\rho\} = \{\mathcal{P}_{Z_\alpha}, \mathcal{P}_{X_\rho}\} = 0,\\ \rho, \sigma = 1, 2, \ldots, (n-r)/2.\end{gathered} \tag{8.2}$$

It is shown in [30] that system (8.2) is consistent and has a solution.

An analogous system of equations can be written for the quantities $Y_\rho(\underset{\sim}{L})$, $\mathcal{P}_{Y_\sigma}(\underset{\sim}{L})$, $\mathcal{P}_{Z_\alpha}(\underset{\sim}{L})$, $\alpha = 1, \ldots, r$:

$$\begin{gathered}\{\mathcal{P}_{Y_\rho}, Y_\sigma\} = \frac{\partial \mathcal{P}_{Y_\rho}}{\partial \underset{\sim}{L}_s}\frac{\partial Y_\sigma}{\partial \underset{\sim}{L}_m} c^n_{sm} L_n = \delta_{\rho\sigma},\\ \{\mathcal{P}_{Y_\rho}, \mathcal{P}_{Y_\sigma}\} = \{Y_\rho, Y_\sigma\} = 0, \quad \{\mathcal{P}_{Z_\alpha}, Y_\rho\} = \{\mathcal{P}_{Z_\alpha}, \mathcal{P}_{Y_\rho}\} = 0,\\ \rho, \sigma = 1, 2, \ldots, (n-r)/2.\end{gathered} \tag{8.2'}$$

In order to carry out the construction proposed in this paper, it is necessary to prove the linearity of the generators $L_i(\underset{\sim}{L}_k)$ expressed as functions of the variables $\mathbf{X}$, $\mathcal{P}_\mathbf{X}$, $\mathcal{P}_\mathbf{Z}$ ($\mathbf{Y}$, $\mathcal{P}_\mathbf{Y}$, $\mathcal{P}_\mathbf{Z}$). The linearity of the generators in the momenta is expressed by the relations

$$\{X_\rho\{X_\sigma, L_i\}\} = 0, \quad \{Y_\rho\{Y_\sigma, \underset{\sim}{L}_i\}\} = 0. \tag{8.3}$$

In [30] it is also proved that if the generators L_i and $\underset{\sim}{L}_k$ are expressed as functions of the 2n canonical conjugate variables q_i, p_i ($i = 1, 2, \ldots, n$), then the combined system formed by Eqs. (8.2), (8.2') and the system of equations

$$\begin{gathered}\{\mathcal{P}_{Z_\beta}, Z_\alpha\} = \frac{\partial \mathcal{P}_{Z_\beta}}{\partial p_i}\frac{\partial Z_\alpha}{\partial q_i} - \frac{\partial \mathcal{P}_{Z_\beta}}{\partial q_i}\frac{\partial Z_\alpha}{\partial p_i} = \delta_{\alpha\beta}, \quad \alpha, \beta = 1, \ldots, r,\\ \{Z_\alpha, X_\rho\} = \{Z_\alpha, \mathcal{P}_{X_\rho}\} = \{Z_\alpha, Y_\rho\} = \{Z_\alpha, \mathcal{P}_{Y_\rho}\} = 0\end{gathered} \tag{8.4}$$

has a solution. Here Z_α is a function of the 2n variables $\mathbf{q}$ and $\mathbf{p}$, and the cyclic variable $\mathbf{Z}$ is not expressed as a function of the generators $\mathbf{L}$ and $\underset{\sim}{\mathbf{L}}$. The linearity of the generators $\mathbf{L}$, $\underset{\sim}{\mathbf{L}}$ in the momenta $\mathcal{P}_\mathbf{Z}$ means that

$$\{Z_\rho\{Z_\sigma, L_i\}\} = 0, \quad \{Z_\rho\{Z_\sigma, \underset{\sim}{L}_i\}\} = 0. \tag{8.3'}$$

To complete the construction of the irreducible representations proposed above, it remains to prove that if we add the linearity conditions (8.3), (8.3') to the consistent system of equations (8.2), (8.2'), (8.4) (the consistency of which is proved in [30]) Eqs. (8.2)-(8.4) are consistent.

On the basis of all the examples of Lie groups considered, we convince ourselves directly that this system of equations is consistent, and this evidently should hold for an arbitrary Lie group. The proof that this system of differential equations is consistent is in essence equivalent to proving that any irreducible representation of a Lie group defined by r Casimir operators (the momenta $\mathcal{P}_\mathbf{Z}$) is induced. Indeed, the form of the generators (1.15) is the infinitesimal form of an induced representation. Equations (1.11), (1.12) moreover clearly reveal the geometric meaning of the exponential factor $\exp[i\rho\, \mathbf{f}_g(\mathbf{X})]$ in the matrix element of an irreducible induced Lie group representation [14, 15]. This factor is associated with the shift of the canonical cyclic variable $\mathbf{Z}$ by the vector $\mathbf{f}_g(\mathbf{X})$. Thus the nature of the motion of the points of group phase space under the action of the group itself dictates both the form of the irreducible representations of the group and the number of them. We remark also that this construction can also be extended to the cases where we do not consider the group phase space associated with the action of the regular representation, but the phase associated with the action of the group on a quotient space by a normal subgroup. In this case the right and left group translations again commute. After parametrizing the quotient space, it is possible to add momenta as in the scheme proposed above and to choose suitable canonical variables in which the dependence of the generators on the different coordinates is seen explicitly.

We stress, however, that although it is possible to construct a representation of the Lie algebra by the method described, the question of the possibility of constructing a representation of the whole group (integrating the algebra) requires additional study.

An operator of a group representation can be determined (as a Green function) by solving a first-order equation of the form

$$i\frac{\partial S}{\partial t} = (L_i u_i)\, S(\mathbf{x}, \mathbf{y}, t),$$
$$H = L_i u_i = f_\alpha(\mathbf{x})\frac{\partial}{\partial x_\alpha} + \varphi(\mathbf{x}), \tag{8.5}$$

where L_i is the infinitesimal operator of an infinitely small rotation transformation about the "vector" $\mathbf{u}$ by an "angle" t. A solution of this equation can be found directly: The Green function for a first-order equation can be written in the form of a product of δ-functions in the arguments which can be determined by solving the classical equations of motion, where the group parameter (the variable t) plays the role of time,

$$S(\mathbf{x}, \mathbf{y}, t) = f(t)\prod_{i=1}^{n} \delta(x_i - x_i'(t, \mathbf{y})),$$
$$\dot{x}_\alpha' = f_\alpha(\mathbf{x}), \qquad f(t) = \exp\Big[\sum_{\alpha=1}^{r} \mathcal{P}_{z_\alpha}\int_0^t \varphi_{z_\alpha}(\mathbf{x}'(\mathbf{y}, \tau))\, d\tau\Big], \tag{8.6}$$

with the boundary condition

$$S(0) = \prod_{i=1}^{n} \delta(x_i - y_i), \qquad \mathbf{x}'(\mathbf{y}, 0) = \mathbf{y}. \tag{8.7}$$

Expression (8.6) can be reduced to an integral over the 2n-dimensional phase space by a change of variables; taking into account the Casimir operators, this leads to an expression for the character of a representation in the theory of orbits [14] in those cases where the hypotheses necessary for the theory to be applicable are satisfied. On the other hand, the equation in the form of a phase space integral is always valid. In essence, the study of a matrix element of a representation of a Lie group as Green functions of some finite-dimensional quantum-system with a very simple Hamiltonian linear in the momenta permits the use of methods well developed in quantum mechanics, in particular, the method of coherent states. The expression of a matrix element of a representation as a phase-space integral corresponds to expressing the Green function as an integral over coherent states. Moreover, it is possible to introduce an equation for the matrix element of a representation by using 2n operators which are analogs of the 2n integrals of motion of an n-dimensional quantum system which describe the initial points of the trajectory of the system in phase space. Coherent states, integrals of motion, and their relation to the Green function for quantum systems were considered in [34].

Integrals of motion describing the initial points of a trajectory of a quantum system in a phase space with averaged coordinates and momenta are obtained as operators $\hat{\mathbf{q}}_0 = U\mathbf{x}U^{-1}$ and $\hat{\mathbf{p}}_0 = U(-i\partial/\partial\mathbf{x})U^{-1}$, where U is the evolution operator of the system (the Green function of the system is its kernel) and describes its motion. In order to find the operators $\hat{q}_0$ and $\hat{p}_0$ in explicit form, it is necessary to move the evolution operator past the operator $\mathbf{x}$ and $i\partial/\partial\mathbf{x}$ or, what is technically equivalent, to solve the Heisenberg equation of motion. Sometimes, however, the integrals of motion are known from independent considerations. They can then be used to find the Green function. This method together with examples of physical systems is considered in [34]. Any Lie group is equivalent to such a simple quantum system with linear Hamiltonian, for which the integrals of motion are also immediately known, if the variables $\mathbf{X}_g, \mathbf{Y}_g, \mathbf{Z}_g$, and $\mathcal{P}_{\mathbf{X}_g}, \mathcal{P}_{\mathbf{Y}_g}$ have been found. This is easily seen, since the Heisenberg equations give the evolution of the coordinate and momentum operators, and we know what this evolution is, the answer being given by Eqs. (1.11) and (1.12), where the inverse g^{-1} must replace the element g. The dynamical system corresponding to the right regular representation has 2n integrals of motion (with Hamiltonian $L_i u_i$), while the system corresponding to the left regular representation also has 2n invariants (the Hamiltonian being $\underset{\sim}{L}_i u_i$). The kernel of the right regular representation in the variables X, Y, Z satisfies the system of equations (1.18). The kernel of the left regular representation satisfies the analogous system of 2n equations. As for quantum systems, it is possible [34] to introduce normal coordinates in which the Green function factorizes and has the form of a product of n functions of two variables each. In order to do this, it is necessary to choose changes of variables which uncouple system (1.18), which can always be done. The matrix element of the representation (the Green function) in such a factorized form can turn out to be convenient for further calculations. The operators $\hat{q}_0$ and $\hat{p}_0$ are convenient since by the Stone–von Neumann theorem, they have no nontrivial invariant subspaces. But since any function of invariants is an invariant, other integrals of motion can also be used for writing an equation for the matrix element.

Thus the generators L_i of a representation (in any number of variables) are linear forms in the derivatives. If we carry out a translation in the coordinates and momenta, we obtain a new generator $L_g = D(g)\mathbf{L}$, where the matrix D(g) is the matrix of the adjoint representation. As is easily seen, the operators $\mathbf{L}_{g^{-1}}$, are dependent integrals of motion (functions $\hat{q}_0$ and $\hat{p}_0$) and therefore the matrix element $S(x_1, \ldots, x_p, x_1' \ldots, x_p')$ of a representation (where here the functions on which the representation is realized depend on the coordinates $x_1, \ldots, x_p$) satisfy the system of n equations

$$D_{ik}(g^{-1}) L_k\left(\mathbf{x}, \frac{\partial}{\partial \mathbf{x}}\right) S(\mathbf{x}, \mathbf{x}') = L_i^T\left(\mathbf{x}', \frac{\partial}{\partial \mathbf{x}'}\right) S(\mathbf{x}, \mathbf{x}'), \tag{8.8}$$

where L_i^T is the transposed operator formed using the usual rules, i.e., we have $x_\rho^T = x_\rho$, $(\partial/\partial x_\rho)^T = -\partial/\partial x_\rho$ and $(AB)^T = B^T A^T$. Equation (8.8) is valid for any representation.

Another way to solve Eq. (8.5) for the Green function is to calculate the corresponding Feynman path integral of the expression for a Hamiltonian which is linear in the momenta.

In a recent paper [35] the path integral for an equation of Schrödinger type is calculated for a Hamiltonian which is linear in the momenta. When applied to the matrix element of a representation of a Lie group, the expression for the Green function in [35] coincides with (8.6).

We remark that the approach proposed by us, which is based on the method of canonical transformations, can be applied to construct Lie group representations which are not of type I. An example of this is the five-dimensional Mautner group (cf. [14]).

The important problem of decomposing the regular representation into irreducible representations can apparently be solved by a more precise formulation of the boundary conditions for the system of equations (1.23) for the purpose of selecting the unitary solutions, automatically leading to the decomposition of the regular representation, from the wide class of possible solutions of the system. This means that complete information is contained in $\hat{G}$

concerning the Plancherel measure and the characters of irreducible representations of the Lie groups considered. All the problems listed in this conclusion unquestionably require further development and additional study.

The authors are deeply grateful to D. P. Zhelobenko, A. A. Kirillov, and E. S. Fradkin for helpful discussions.

LITERATURE CITED

1. High Energy Physics and the Theory of Elementary Particles [in Russian], Naukova Dumka, Kiev (1967).
2. A. O. Barut, Proceedings of the Conference on High Energy Physics and Elementary Particles [in Russian], Kiev (1970).
3. A. O. Barut, Phys. Rev., 139:1433 (1965).
4. Y. Neeman, Algebraic Theory of Particle Physics, New York (1967).
5. I. A. Malkin and V. I. Man'ko, Pis'ma Zh. Éksp. Teor. Fiz., 2:230 (1965); Yad. Fiz., 3:373 (1966); 8:1264 (1968).
6. R. Streater, Commun. Math. Phys., 4:217 (1967).
7. I. Segal, Mathematical Problems of Relativistic Physics, Am. Math. Soc., Providence, R. I. (1963).
8. E. Wigner, Ann. Math., 40:149 (1939).
9. N. N. Bogolyubov, A. A. Logunov, and I. T. Todorov, Foundations of Axiomatic Quantum Field Theory, Benjamin, Reading, Mass. (1975).
10. A. Z. Petrov, New Methods in the General Theory of Relativity [in Russian], Nauka, Moscow (1966).
11. G. W. Mackey, Ann. Math., 55:101 (1952).
12. I. M. Gel'fand and M. A. Naimark, Proc. Math. Inst. Acad. Sciences [in Russian], Moscow (1950).
13. G. Frobenius, The Theory of Characters and Representations of Groups [Russian translation], Khar'kov (1937).
14. A. A. Kirillov, Elements of the Theory of Representations [in Russian], Nauka, Moscow (1970).
15. D. P. Zhelobenko, Compact Lie Groups and Their Representations [in Russian], Nauka, Moscow (1970).
16. A. N. Leznov, IFVÉ Preprint, 68-20-K (1968); 71-42 (1971).
17. A. N. Leznov and M. V. Savel'ev, Teor. Mat. Fiz., 2:311 (1970); 4:310 (1971); 8:161 (1974); IFVÉ Preprint, 74-46 (1974).
18. A. N. Leznov and I. A. Fedoseev, Teor. Mat. Fiz., 7 298 (1971).
19. A. N. Leznov, I. A. Malkin, and V. I. Man'ko, FIAN Preprint, No. 139 (1972).
20. A. N. Leznov and M. V. Savel'ev, Fiz. Élem. Chast. At. Yad., 7:55 (1976).
21 A. N. Leznov, I. A. Malkin, and V. I. Man'ko, FIAN Preprint, No. 40 (1971).
22. B. Kostant, Lecture Notes in Mathematics, Vol. 170, Springer-Verlag, New York (1970), p. 87.
23. V. A. Arnol'd, Mathematical Methods of Classical Mechanics [in Russian], Nauka, Moscow (1974).
24. J. M. Souriau, Structures des Systémes Dynamiques, Dunod, Paris (1970).
25. F. Onofri and N. Pauri, J. Math. Phys., 13:533 (1972).
26. D. J. Simms, Z. Naturforsch., 28A:538 (1972).
27. M. Andrié, J. Math. Phys., 17:394 (1976).
28. A. Simoni, F. Zaccaria, and B. Vitale, Nuovo Cimento, 51A:448 (1967).
29. E. B. Aronson, I. A. Malkin, and V. I. Man'ko, Fiz. Élem. Chast. At. Yad., 5:122 (1974).
30. A. Eisenhart, Continuous Transformations [in Russian], Fizmatgiz, Moscow (1946).

31. I. M. Gel'fand, Mat. Sb., 26:103 (1950).
32. M. Micu, Nucl. Phys., 60:353 (1964).
33. A. M. Perelomov and V. S. Popov, Izv. Akad. Nauk SSSR, Ser. Mat., 32:1368 (1968).
34. V. V. Dodonov, I. A. Malkin, and V. I. Man'ko, J. Phys., A8, L19 (1975); Int. J. Theor. Phys., 14:37 (1975); Teor. Mat. Fiz., 24:164 (1975); J. Phys., A9:1791 (1976); Kratk. Soobshch. Fiz., No. 1, p. 3 (1976).
35. W. B. Campbell, P. Finkler, C. E. Jones, and M. M. Misheloff, Ann. Phys., 96:286 (1976).

THE NEWMAN–PENROSE METHOD IN THE THEORY OF GENERAL RELATIVITY

V. P. Frolov

This survey gives a systematic exposition of the Newman–Penrose method and its application to physical problems. The paper discusses the geometric properties of congruences of light rays in a gravitational field, gives an account of the spinor formalism, and considers its use in studying the properties of a gravitational field. A survey is given of the known exact algebraically special solutions of the Einstein equations. A separate chapter is devoted to the asymptotic properties of a gravitational field and the problem of constructing a quantum theory in asymptotically flat spaces. The application of the method treated to the calculation of the energy flux of particles created under conditions of quantum evaporation of black holes is discussed.

INTRODUCTION

The title of this survey is to a certain extent arbitrary and only partly reflects its content. Strictly speaking, the formalism developed by Newman and Penrose [1] (which is also called the method of spin coefficients) consists of the use of complex light tetrads for studying the Einstein equations and the equations of other fields. As was shown by Newman and Penrose, this formalism is completely equivalent to the spinor approach. The expanded system of equations connecting the spinor components of the curvature tensor with the spin coefficients (the components of the spinor connection) has become known as the system of Newman– Penrose equations (or NP equations for short).

At first sight* this system of equations looks rather cumbersome and intractable to study. This usually creates a certain psychological barrier to mastering the NP method and using it. However, in the time elapsed from the appearance of the NP formalism, its viability and convenience have become established. Moreover, at the present time the language of spin coefficients has become generally accepted and widely used in the modern literature. What is the reason for such popularity of the NP formalism? Apparently, the main reason consists in its adequacy and internal adaptability for studying the solutions of the Einstein equations (and the equations of massless fields) possessing certain algebraic properties.

The algebraic classification of the solutions of the Einstein equations was first studied in the papers [2-3] of Petrov and was used by Pirani [4] to study the wave properties of a gravitational field. Subsequently, Penrose [12] used the spinor formalism to simplify Petrov's method in an essential way. Moreover, it turned out that the property for a gravitational field to belong to one or another algebraic type is connected with the fulfillment of certain algebraic conditions on the spinor components of the Weyl tensor. The fact that such components appear linearly in the NP equations permits the successful use of the NP system for studying the solutions of a certain algebraic type.

---

* The complete system of NP equations is given in the Appendix to Chapter 3 of this survey.

The Weyl tensor (coinciding with the curvature tensor in vacuum) determines a set of "eigen" lightlike vectors (called principal lightlike vectors), the mutual position of which is directly related to the algebraic type of the gravitational field. The integral curves of principal lightlike vector fields form a family (congruence) of light curves on the spacetime manifold. The geometric characteristics of such congruences can serve to give a more detailed classification of gravitational fields (Jordan, Ehlers, and Sachs [6]; Sachs [7, 8]). If a field of complex light tetrads is chosen so that one of the two real lightlike vectors of the tetrad coincides with a principal light vector, then the individual spin coefficients coincide with differential invariants (the so-called optical scalars) characterizing the properties of the principal light congruence. In this case, the NP equations which relate different optical scalars among themselves take on a completely definite geometric meaning and are often amenable to solution.

A successful choice of coordinate system is of great importance both for solving the NP equations and for studying the properties of the solutions. The most convenient coordinates turn out to be the ones associated with one of the principal light congruences of the gravitational field being studied. In these coordinates, which are intrinsically related to the properties of the class of solutions being studied, the NP system also simplifies substantially when complex light tetrads adapted to a principal light congruence are used. If we compare such coordinate systems with the coordinates usually used and associated with imposing certain conditions on the metric (e.g., comoving coordinates), the transition from one set of coordinates to another usually involves a complicated nonlinear transformation which often cannot be described explicitly. This nonlinear change of variables leads to the fact that although the solution can be obtained explicitly in the coordinates associated with one of the principal light congruences, and may have a simple form, in other coordinates (e.g., comoving coordinates) it is not possible to find an explicit solution. Both the choice of light coordinates associated with a principal light congruence and the use of complex light tetrads make the NP formalism a convenient one for describing and studying massless fields (including gravitational fields).

As is also true of other possible classifications of gravitational fields (e.g., in terms of symmetry groups or holonomy groups), the algebraic classification* is a method for obtaining particular solutions of the corresponding field equations. The property for a solution to belong to a given algebraic class can be connected with the vanishing of certain spinor components of the Weyl tensor and of certain spin coefficients. Under such assumptions, the NP system simplifies and frequently becomes solvable. It turns out that the solutions thus obtained usually possess small symmetry groups, and it is possible to obtain solutions having no symmetry at all.

At the present time a significant number of exact solutions have been obtained. In this paper, we have therefore deemed it necessary to give a survey of the known algebraically special solutions.

The NP equations have also turned out to be convenient in studying the asymptotic properties of gravitational (and other massless) fields. Using these equations together with the definition of asymptotically flat space introduced by Penrose [9-11], it is possible to prove a number of theorems concerning the asymptotic behavior of the fields at infinity. Penrose's idea of conformal infinity and the existence of a group of asymptotic symmetries (Bondi, van der Burg, and Metzner [13]; Sachs [14]) in an asymptotically flat space, together with the property of conformal invariance of massless fields, allow the construction of classical and quantum scattering theories in asymptotically flat spaces. The application of these methods enabled

---

* Here and in the sequel, by an algebraic classification we mean a division of gravitational (and other massless) fields into classes, where two fields belong to the same class if they have the same Petrov type and their principal light congruences have the same properties.

Hawking to obtain the well-known result on the instability of the vacuum in the gravitational field of a black hole [15].

Another interesting direction is the attempt developed by Newman and co-workers [16] to relate asymptotic properties of a gravitational field to the nature of the motion of a body generating the field. The integration of the NP equations in the asymptotic domain plays an important role in this approach. The NP formalism is also convenient for studying initial condition problems and for giving initial data on lightlike surfaces (Sachs [17, 18]). Even if we mention the use of the NP equations in investigating the development of trapping surfaces and horizons [19, 20], or in solving the wave equations by separation of variables in the Kerr metric [21-23], we obtain an enumeration of the possible applications of the NP formalism which is far from complete.

The original idea of using the spinor method, or the method of complex light tetrads, in the Einstein theory of relativity has turned out to be organically related to a wealth of other directions (the algebraic approach, theory of light congruences, study of the asymptotic behavior of fields, etc.). This has led to the circumstance that at the present time the Newman–Penrose method is understood to subsume a wider class of problems than just the writing down of the Einstein equations in complex light tetrads.

Although at the present time the language of the NP formalism is generally adopted, the Newman–Penrose method is still used comparatively little in the papers of Soviet authors. Apparently the main reason for this is the lack of accessible surveys devoted to this problem. Translations of the papers of Sachs [8] and Penrose [11], and in particular Penrose's book, have substantially improved the situation. This survey is written with the purpose of giving an idea of the methods used in the NP formalism and of the principal results achieved through its use; in particular, we want to give an idea of the application of this formalism to finding exact solutions of the Einstein equations. We have also taken the opportunity to include a chapter in this survey devoted to quantum theory in asymptotically flat spaces. This decision is motivated by the fact that quantum processes in black holes are widely discussed at present in the literature. The application in quantum theory of methods related to the Newmann–Penrose method makes the study of such processes substantially simpler.

In referring to the literature we have in many cases restricted ourselves to mentioning only one or a few of a number of existing papers on a given subject, assuming that the references to the literature or the historical remarks given in these papers will make it possible to obtain complete information. Therefore, the bibliography listed at the end of this paper can in no way presume to be a complete survey of the literature devoted to the NP method. We hope, however, that the main papers on the problem we consider have been properly reflected in the bibliography.

In this survey we employ a double enumeration of the equations appearing in a single chapter. Equation (3.5), for example, denotes Eq. (5) of Section 3. In referring to equations in another chapter, as well as in the notation for propositions, a triple enumeration is used. For example, Proposition 2.5.4 denotes the fourth proposition of the fifth section of the second chapter. Each chapter of the survey begins with a short introduction in which general remarks and some historical references and references to the literature relating to the chapter are given. The two Appendices contain a number of useful formulas and relations. The reference (A1.3) refers to the third formula of the Appendix to Chapter 1.

We adopt the following notation in this paper. The Greek indices $\alpha, \beta, \gamma, \ldots$ take the values from 0 to 3; the Latin indices in the middle of the alphabet i, j, k, ... denote tensor components and vary from 1 to 3; Latin indices $a, b, c, \ldots$ serve to denote spin indices and take values 0 and 1; the Latin indices m, n, l, ... are usually used to denote the tetrad components of tensors and vary from 1 to 4; the capital Latin letters A, B, C, ... denote abstract spinor indices and do not take numerical values; the metric $g_{\mu\nu}$ has the signature +, −, −, −; the curvature tensor $R^{\alpha}_{\beta\gamma\delta}$ and Ricci tensor $R_{\beta\gamma}$ are defined as follows:

$$\xi_{\beta;\gamma;\delta} - \xi_{\beta;\delta;\gamma} = -R^{\alpha}_{\beta\gamma\delta}\xi_{\alpha}, \qquad R_{\beta\gamma} = -R^{\alpha}_{\beta\gamma\alpha},$$

and the Einstein equations have the form

$$R_{\alpha\beta} - {}^{1}/_{2}g_{\alpha\beta}R = -\frac{8\pi G}{c^4}T_{\alpha\beta}.$$

Symmetrization and antisymmetrization are indicated as usual by round and square brackets respectively, e.g.,

$$\xi_{(\alpha\beta)} = {}^{1}/_{2}(\xi_{\alpha\beta} + \xi_{\beta\alpha}), \quad \xi_{[\alpha\beta]} = {}^{1}/_{2}(\xi_{\alpha\beta} - \xi_{\beta\alpha});$$

$\varepsilon_{\alpha\beta\gamma\delta}$ denotes the completely antisymmetric symbol $\varepsilon_{\alpha\beta\gamma\delta} = \varepsilon_{[\alpha\beta\gamma\delta]}$, where $\varepsilon_{0123} = (-g)^{1/2}$.

CHAPTER 1

LIGHT RAYS IN A GRAVITATIONAL FIELD

1.1. Introduction

One of the ways to obtain information on the structure of a gravitational field is by studying the behavior of different test particles in the field. If massless particles are considered (photons, neutrinos), then the trajectories corresponding to them are light rays (isotropic geodesics). It turns out that the space-time metric can be recovered uniquely up to a conformal transformation from the known trajectories of massless test particles (photons) ([3, p. 319]). This chapter is devoted to the study of the geometric properties of congruences (families) of light rays in a gravitational field.

The importance of the concept of light ray is connected with the fact that the propagation of various physical fields (including the gravitational field itself) is described by a system of equations of hyperbolic type, and the light rays play the role of bicharacteristics for these equations. Essential information on the characteristic surfaces and bicharacteristics is given in Section 1.2. General information on the theory of hyperbolic differential equations and the role of characteristic surfaces and bicharacteristics in solving the Cauchy problem can be found in [27, 28].

Although in the general case it is not completely accurate to think of a light ray as being the trajectory of a real physical massless particle, this idealization is nevertheless completely justified when the wavelength of the corresponding radiation is vanishingly small. The approximation based on the smallness of the wavelength of the radiation is called the approximation of geometrical optics. In geometrical optics, propagation of energy takes place along the light rays while the null surfaces are eikonal, i.e., surfaces of constant phase. Necessary information on the approximation of geometrical optics in a flat space and additional references to the literature can be found in [29-31]. The approximation of geometrical optics in a curved space [32-34] is considered in Section 1.3.

Finally, we remark that light rays and congruences of light rays arise naturally in the study of the algebraic classification of a gravitational, electromagnetic, or other massless field with spin.* All of this makes it important to study the geometric properties of congruences of null curves.

There exists an important property which differentiates null curves from spacelike and timelike curves. The point is that although the null curves form a submanifold of a space possessing a metric, the notion of distance along them is not defined. The parallel transport of vectors along null curves is induced by imbedding them into space-time and thus, since they have no metric properties, the null curves are one-dimensional affine spaces. This leads to certain rather unaccustomed and peculiar properties of null curves and their congruences.

The study of the properties of null congruences was initiated in [6, 7], where it was in particular proved that there exists a set of invariants (now called optical scalars) determining the geometric properties of congruences and having a simple physical meaning. Section 1.4 and all the subsequent sections of this chapter are devoted to the study of the geometrical properties of congruences of null curves in a curved space and to the description of these properties in the formalism of complex null tetrads and spin coefficients [1, 5, 6–8, 10, 35, 36].

1.2. Characteristic Surfaces and Bicharacteristic Rays

The propagation of an electromagnetic or some other field in a given gravitational field is described by hyperbolic equations. A basic role in the study of the Cauchy problem and the radiation problem is played by the characteristic surfaces and bicharacteristic rays (along which perturbations are propagated). We recall the main concepts of the associated theory using the example of a scalar massless field.

We write the equation for a scalar massless field φ

$$L[\varphi] \equiv g^{\mu\nu}\varphi_{;\mu;\nu} \equiv \frac{1}{\sqrt{-g}}\partial_\mu(\sqrt{-g}\,g^{\mu\nu}\partial_\nu\varphi) = 0 \tag{2.1}$$

in the following form which is conveniently studied:

$$g^{00}\partial_0^2\varphi + 2g^{0i}\partial_0\partial_i\varphi + g^{ij}\partial_i\partial_j\varphi + \frac{1}{\sqrt{-g}}\partial_\mu(\sqrt{-g}\,g^{\mu\nu})\,\partial_\nu\varphi = 0. \tag{2.2}$$

If $g^{00} \neq 0$, the initial data φ and $\partial_0\varphi$ (Cauchy data) on the surface Σ: x^0 = const permit one to find $\partial_0^2\varphi$ on the surface and thereby obtain a unique solution φ, at least in some neighborhood of the surface Σ. If in the general case the surface Σ is defined by an equation $u = u(x^\mu)$, then using new coordinates $x^{0'} = u(x^\mu)$, $x^{i'} = f^i(x^\mu)$ it can be seen that the Cauchy problem with data φ and $\partial_u\varphi$ on Σ is well posed and admits a unique solution if and only if $g^{uu} \equiv g^{0'0'} = \partial_\mu u \partial_\nu u g^{\mu\nu} \neq 0$. If on the other hand

$$g^{uu} = g^{\mu\nu}\partial_\mu u \partial_\nu u = 0, \tag{2.3}$$

Eq. (2.2) represents an additional condition imposed on the initial data and does not make it possible to determine $\partial_u^2\varphi$. In this case the initial data φ and $\partial_u\varphi$ on Σ do not determine the value of the field φ outside the surface Σ. Such hypersurfaces, on which the Cauchy data do not determine the second derivatives, are called characteristic surfaces.

The characteristic surfaces play the role of a wave front, i.e., on these surfaces the solutions of Eq. (2.1) may suffer discontinuities, e.g., discontinuities of the second derivatives. Indeed, assume that φ and $\partial_\mu\varphi$ are continuous on some not necessarily characteristic surface Σ, and let us show under what conditions $\varphi_{,\mu\nu}$ can have a jump in crossing this surface. Denote

* See Sections 2.6 and 3.5 of this survey concerning the algebraic classification.

the jump of a quantity A on the surface Σ by the symbol $\Delta[A]$. We then have

$$\Delta[\varphi] = \Delta[\varphi_{,\mu}] = 0, \quad \Delta[\varphi_{,\mu\nu}] = \Psi_{\mu\nu} = \Psi_{\nu\mu}. \tag{2.4}$$

Consider two nearby points x^μ and $x^\mu + dx^\mu$ on a surface Σ, defined by the equation $u(x^\mu) = c$. Since dx^μ is a translation vector along the surface, we have $u_{,\mu}dx^\mu = 0$. For each such translation the function $\partial_\mu \varphi(x + dx) = \varphi_{,\mu}(x) + \varphi_{,\mu\nu}dx^\nu$ is continuous on Σ and hence

$$\Delta[\varphi_{,\mu}(x + dx)] = \Delta[\varphi_{,\mu}(x)] + \Delta[\varphi_{,\mu\nu}(x)\, dx^\nu] = 0,$$

or

$$\Delta[\varphi_{,\mu\nu}(x)\, dx^\nu] = \Psi_{\mu\nu}dx^\nu = 0. \tag{2.5}$$

Fixing the index μ and writing $\Psi_{\mu\nu} = P_{(\mu)\nu}$, we obtain that $u_{,\nu}$ and $P_{(\mu)\nu}$ are orthogonal to the same three-dimensional surface elements formed by the translation vectors along Σ, and therefore they are parallel,

$$\Psi_{\mu\nu} = P_{(\mu)\nu} = P_{(\mu)}\, u_{,\nu}. \tag{2.6}$$

By the symmetry of $\Psi_{\mu\nu}$, we have

$$\Delta[\varphi_{,\mu\nu}] = \Psi_{\mu\nu} = \Psi u_{,\mu}u_{,\nu}. \tag{2.7}$$

On the other hand, Eq. (2.1) is satisfied in the entire space. Considering this equation in a neighborhood of Σ and calculating the jump of both sides of (2.1) taking Eq. (2.4) into account, we obtain

$$g^{\mu\nu}\Delta[\varphi_{,\mu\nu}] = \Psi g^{\mu\nu}\, u_{,\mu}u_{,\nu} = 0. \tag{2.8}$$

It follows that a jump in $\varphi_{,\mu\nu}$ is possible only on characteristic surfaces. Hence instead of defining characteristics to be surfaces on which Cauchy data cannot be given, it is possible to define characteristics equivalently as surfaces on which physically meaningful discontinuities of the solutions are possible. Such surfaces are called wave fronts. Such a wave front arises, for example, on a boundary beyond which there is no disturbance at time t. The solution describing the disturbance vanishes identically on one side of this surface and is not equal to zero on the other side.

It is clear from the example given that the equation of a characteristic surface (2.3) is in no way related to the specific properties of the scalar field, but is completely determined by structure of space-time, or, more accurately, its geometry. It is therefore not surprising that the surfaces u = const, where u is subject to Eq. (2.3), are also characteristic for the Maxwell equations and for the equations describing the propagation of weak perturbations of a gravitational field [5, 37]. Moreover, it turns out in studying the nonlinear Einstein equations that if we assume the continuity of $g_{\mu\nu}$ and $\partial_\gamma g_{\mu\nu}$, then the curvature tensor $R_{\alpha\beta\gamma\delta}$ containing the second derivatives of $g_{\mu\nu}$ can have discontinuities only on the characteristic surface [38, 39]. Moreover, in view of the continuity of $g_{\mu\nu}$, Eq. (2.3) has a completely determined meaning.

Disturbances of physical fields in a vacuum propagate with a velocity equal to the velocity of light, and therefore the characteristics describing the location of the wave front at various times are lightlike surfaces. The physical discontinuities of the solutions move over these characteristics along the bicharacteristic rays, and the propagation of discontinuities is described by a simple ordinary differential equation.

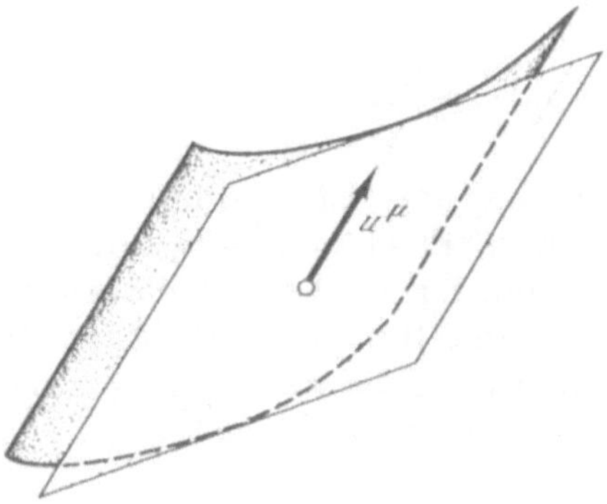

Fig. 1.1

In order to show this, let us consider in more detail the concept of bicharacteristic rays and the main properties of bicharacteristics. Assume that the characteristic surface Σ is defined by the equation $u(x) = c$. Then $\partial_\mu u$ is a normal covector to the surface Σ and $u^\mu = g^{\mu\nu}\partial_\nu u$ is a normal vector to Σ. The vectors ξ^μ tangent to Σ satisfy the equation $\xi^\mu u_{,\mu} = 0$. On the other hand, Eq. (2.3) shows that $u^\mu u_{,\mu} = g^{\mu\nu} u_{,\mu} u_{,\nu} = 0$ and therefore the vector u^μ itself is tangent to the surface Σ. Hence the vector u^μ is lightlike. All the remaining vectors of the space tangent to Σ which are not parallel to u^μ and are determined by the equation $\xi^\mu u_{,\mu} = 0$ are spacelike (Fig. 1.1), as is easily verified. Therefore the characteristic surface is lightlike.*

Any characteristic surface $u(x^\mu) = 0$ can be included in a one-parameter family of characteristic surface $u(x^\mu) = c$. Such a family generates a lightlike vector field u^μ in some region of space-time. The integral curves $x^\mu(r)$ of this field defined by the equation

$$dx^\mu/dr = g^{\mu\nu}u_{,\nu} \tag{2.9}$$

are called bicharacteristic rays (or simply bicharacteristics, or rays). We have the following proposition.

Proposition 1.2.1. Bicharacteristics have the following properties:

1) Every solution of Eq. (2.9) which passes through a point of the surface $u(x) = c$ is completely contained in it. Any characteristic surface is generated by a two-parameter family of bicharacteristic curves.
2) The bicharacteristics are lightlike curves, i.e., the tangent vectors to them are isotropic.
3) The bicharacteristics are null geodesics.

Proof. The proof of the first assertion can be found, for example, in [27, Chapter II]. The second assertion is obvious if we recall that the vector u^μ satisfies Eq. (2.3). In order to prove the third assertion it suffices to verify that the vector u^μ is parallel transported along a bicharacteristic curve, i.e.,

$$Du^\mu/dr \equiv u^\nu u^\mu_{;\nu} = 0. \tag{2.10}$$

* A hypersurface is said to be lightlike if its tangent space at every point has exactly one lightlike direction. If there are two distinct lightlike directions at every point of the tangent space, the hypersurface is said to be timelike, while if there is no lightlike direction, it is spacelike.

This equality is established as follows:

$$u^\nu u^\mu_{;\nu} = u^\nu (g^{\mu\alpha} u_{,\alpha})_{;\nu} = u^\nu g^{\mu\alpha} u_{;\nu;\alpha} = {}^1/_2\, g^{\mu\alpha} (u^\nu u_\nu)_{;\alpha} = 0.$$

We now show that the discontinuities of the solutions propagate along the rays. To do this we choose a coordinate system x^μ in which $x^0 = u$, i.e., the surfaces $x^0 = \mathrm{const}$ are characteristic for Eq. (2.1). It follows from Eq. (2.7) that the quantity $\varphi_{,00}$ experiences a jump on the surface Σ: $x^0 = \mathrm{const}$ while at the same time φ itself, its first derivatives, and all the second derivatives except $\varphi_{,00}$ are continuous. If we write $\varphi_{,00} = w$ and use the fact that $g^{00} = g^{uu} = 0$, we obtain by differentiating (2.2) with respect to x^0 that

$$2g^{0i}\partial_i w + \frac{1}{\sqrt{-g}}\partial_\mu (\sqrt{-g}\, g^{\mu 0})\, w + J = 0, \tag{2.11}$$

where J denotes a sum of terms not containing $\varphi_{,00}$.

If we denote the jump by $\Delta[\varphi_{,00}] = \Delta[w] = \Psi$ and take into account that J in Eq. (2.11) is continuous on Σ, we obtain an equation for the propagation of Ψ,

$$2g^{0i}\partial_i \Psi + \frac{1}{\sqrt{-g}}\partial_\mu (\sqrt{-g}\, g^{\mu 0})\Psi = 0. \tag{2.12}$$

In order to rewrite (2.12) in an arbitrary coordinate system, we observe that $g^{0i}\partial_i\Psi = g^{\mu\nu}u_{,\mu}\Psi_{,\nu} = u^\nu\Psi_{,\nu}$ and hence the first term in (2.12) is a derivative along the direction of a bicharacteristic. The second term is equal to L[u], where the operator $L[\varphi]$ is defined by Eq. (2.1). We therefore have

$$d\Psi/dr + P\Psi = 0, \tag{2.13}$$

where d/dr indicates differentiation along a bicharacteristic ray on the surface Σ, and $P \equiv {}^1/_2 L[u]$ is known on Σ.

Equation (2.13) gives the law of propagation along rays lying on a characteristic discontinuity surface. It has the form of an ordinary differential equation and shows, in particular, that the size of the jump cannot vanish at any point of the ray, provided the jump is different from zero somewhere on the ray.

It has already been pointed out above that the family of characteristics $u(x) = c$ generates a lightlike vector field $u^\mu = g^{\mu\nu}u_{,\nu}$. The integral curves (rays) of this field generate null surfaces (characteristics). We now assume that a lightlike vector field ξ^μ is given in some region of space-time. It turns out that in the general case the integral curves of this field do not lie on null surfaces. Indeed, we have the next proposition.

Proposition 1.2.2. The condition

$$\xi_{[\lambda}\xi_{\mu,\nu]} = 0 \tag{2.14}$$

is necessary and sufficient for there to exist a family of null hypersurfaces $u(x) = c$ such that the integral curves

$$dx^\mu/dr = \xi^\mu \tag{2.15}$$

of the vector field ξ^μ lie completely on these surfaces. If Eq. (2.14) is satisfied, then the integral curves of the vector field ξ^μ are null geodesics.

Proof. In order to prove the necessity of condition (2.14), assume that there exists a family of null hypersurfaces $u(x) = c$ and that the integral curves lie on the hypersurfaces. Since ξ^μ is a tangent vector to the integral curve and lightlike, it must be proportional to the vector u^μ which is the only lightlike vector in the space tangent to the null surface $u(x) = c$. Therefore

$$\xi_\mu = f u_{,\mu}. \tag{2.16}$$

In this case, it follows from the equality $u_{,[\mu\nu]} = 0$ that $\xi_{[\mu,\nu]} = f^{-1}\xi_{[\mu}f_{,\nu]}$. Multiplying both sides of this equality by ξ_λ and antisymmetrizing, we get (2.14).

The sufficiency of condition (2.14) follows from the following considerations. Consider the Pfaffian equation

$$\xi_\mu dx^\mu = 0. \tag{2.17}$$

By the Frobenius theorem [40, 41], condition (2.14) is necessary and sufficient for the complete integrability of Eq. (2.17), i.e., for the existence of a family of hypersurfaces $u(x) = \text{const}$ passing through each point such that $\xi_\mu = f u_{,\mu}$. Since $g^{\mu\nu}u_{,\mu}u_{,\nu} = f^{-2}\xi_\mu\xi^\mu = 0$, the surfaces $u(x) = \text{const}$ are null surfaces.

We now observe that if the vectors ξ^μ and u^μ are proportional, then geometrically, the integral curves of both vector fields coincide. It follows from Proposition 1.2.1 that the integral curves of the vector field u^μ are null geodesics, and therefore the integral curves of ξ^μ are also null geodesics.

A congruence $\Gamma(\xi)$ of null curves generated by a vector field ξ satisfying condition (2.14) will be called a normal congruence.*

It seems appropriate to point out that even in the case where (2.14) does not hold, the integral curves (2.15) can be used to construct families of hypersurfaces. In the general case, however, these hypersurfaces will be timelike, and only turn out to be lightlike when (2.14) is satisfied.

1.3. Geometric Optics in a Gravitational Field

The concepts of characteristic surface (wave front) and bicharacteristic rays are also essential in the study of the propagation of high-frequency electromagnetic waves (or waves of other fields) in twisted space-time. In the case where the wavelength of the radiation is small compared to the characteristic parameters of variation of the gravitational field, one can expect to get a good first approximation for the laws of propagation of radiation if the finiteness of the wavelengths is ignored altogether.

The approximation in which the finiteness of the wavelength of the radiation is completely ignored (this corresponds to the limit $\lambda \to 0$) is called the approximation of geometric optics. This terminology is connected with the fact that the optical laws describing the propagation of radiation can be formulated in the language of geometry in this approximation. Within the framework of geometric optics it is assumed that radiation propagates along light rays (bicharacteristics). With each such ray there is associated a polarization state, and variation in radiation intensity is associated with variation in the cross section of a light beam.

In order to formulate more precisely the conditions under which the approximation of geometric optics is valid, we consider the following characteristic parameters for the di-

* In the literature the term congruence orthogonal to a hypersurface is also used.

mension of length which arise, for example, in the problem of the propagation of an electromagnetic wave in a given gravitational field:

1. λ is the characteristic wavelength of radiation, $\lambdabar = \lambda/2\pi$;
2. L is a characteristic length over which there is significant variation of the amplitude, polarization, or wavelength of radiation, e.g., the radius of curvature of the wave front or the dimensions of a wave packet;
3. R is the characteristic radius of curvature of space-time in the region in which the wave propagates.

The approximation of geometric optics corresponds to the situation where

$$\lambdabar \ll L, \ \lambdabar \ll R. \tag{3.1}$$

If we write l = min (L, R), then in a region of space-time of dimension d: $\lambdabar \ll d \ll l$, electromagnetic waves propagate like plane waves in an almost flat space-time. Let us consider the propagation in this region of a monochromatic wave.* We write the vector potential A_μ of

$$A_\mu = \mathrm{Re}\,(a_\mu e^{i\Theta}), \tag{3.2}$$

where a_μ is a slowly varying amplitude and Θ a rapidly varying phase (the change in the phase Θ over a distance l is a quantity of the order $l/\lambdabar$). If we write $\varepsilon = \lambdabar/l$, then under the above assumption we have

$$\begin{gathered} a_\mu = O\,(1), \quad a_{\mu;\,\nu} = O\,(1), \quad k_\mu \equiv \partial_\mu \Theta = O\,(\varepsilon^{-1}), \\ k_{\mu;\,\nu} = O\,(\varepsilon^{-1}), \quad R_{\mu\nu} = O\,(1), \end{gathered} \tag{3.3}$$

where $R_{\mu\nu}$ is the Ricci tensor of the gravitational field.

The Maxwell equations in a gravitational field

$$F^{\mu\nu}_{;\nu} = 0, \qquad F_{[\mu\nu,\lambda]} = 0 \tag{3.4}$$

are equivalent to the following equations for the vector potential A_μ ($F_{\mu\nu} = \partial_\mu A_\nu - \partial_\nu A_\mu$):

$$A^{;\alpha}_{\mu;\alpha} + R_{\mu\sigma} A^\sigma = 0, \tag{3.5a}$$

$$A^{\mu}_{;\mu} = 0. \tag{3.5b}$$

Substituting expression (3.2) for the potential into Eq. (3.5) and using (3.3), we have

$$[-k_\alpha k^\alpha a_\mu] + i\,[2k^\alpha a_{\mu;\alpha} + k^{\alpha}_{;\alpha} a_\mu] + [a^{;\alpha}_{\mu;\alpha} + R_{\mu\alpha} a^\alpha] = 0, \tag{3.6a}$$

$$ik_\mu a^\mu + a^{\mu}_{;\mu} = 0. \tag{3.6b}$$

The expressions in (3.6a) in the square brackets have the order ε^{-2}, ε^{-1}, and 1, respectively. Equating the term of the order ε^{-2} to zero, we have

$$k_\alpha k^\alpha \equiv g^{\alpha\beta} \Theta_{,\alpha} \Theta_{,\beta} = 0. \tag{3.7}$$

In lowest approximation, Eq. (3.6b) gives

$$k_\mu a^\mu = 0. \tag{3.8}$$

* The propagation of an arbitrary wave can be described using the usual Fourier transform.

If we write $a_\mu = a e_\mu$, where e_μ is the polarization vector, $e_\mu e^\mu = -1$, and $a = (-a_\mu a^\mu)^{1/2}$ is the scalar amplitude, then the term of order ε^{-1} in (3.6a) leads to the relations

$$e^\mu_{;\alpha} k^\alpha = 0, \tag{3.9}$$

$$(a^2 k^\alpha)_{;\alpha} = 0. \tag{3.10}$$

The light rays are defined as the curves normal to the surface of constant phase Θ. Therefore, the equation of the light rays is

$$dx^\mu/dr = g^{\mu\nu}\Theta_{,\nu}. \tag{3.11}$$

The light rays are evidently bicharacterics for the characteristic surfaces Θ = const and therefore by Proposition 1.2.1 are null geodesics.

The results obtained above permit us to state the following proposition.

Proposition 1.3.1. In geometric optics, propagation of light takes place along null geodesics. The polarization vector is perpendicular to the light rays and is parallel transported along rays. The scalar amplitude a varies along a ray in accordance with Eq. (3.10).

Relation (3.10) shows that the quantity $N = \int_\Sigma a^2 k^\mu d\sigma_\mu$, calculated for an arbitrary spacelike surface Σ does not depend on Σ.

From the quantum point of view, $N/8\pi\hbar$ is the number of photons in a light beam [32]. It is not an absolute invariant, but rather represents what is called an adiabatic invariant, i.e., a quantity which does not change its value under a change in the parameters of the system which is slow compared to the characteristic period.

If expression (3.2) for the potential is substituted into the expression for the energy-momentum tensor of the electromagnetic field and we restrict ourselves to the leading approximation with respect to ε (terms of order ε^{-2}), then after averaging over a space-time domain of dimension d ($\lambda \ll d \ll l$) we obtain the following expression for the effective energy-momentum tensor [33, 34]:

$$T_{\mu\nu} = q l_\mu l_\nu, \tag{3.12}$$

where q is the effective energy density and $l_\mu = \varepsilon\Theta_{,\mu}$ the null vector giving the propagation of energy flux. An analogous expression is obtained for the effective energy-momentum tensor for high-frequency radiation and for ether massless fields.

1.4. Propagation of Shadows. Optical Scalars

It was shown in the preceding sections that the null curves play an important role in the propagation of radiation. It turns out also that when the problem of the algebraic classification of the electromagnetic field and the Weyl tensor of a gravitational field* is considered, with every suitable tensor there is associated the set of its "eigen" lightlike vectors. Therefore, the corresponding tensor fields (electromagnetic and gravitational) generate lightlike vector fields in space-time. The integral curves of a lightlike vector field form a congruence† of null curves.

* Sections 2.6 and 3.5 are devoted to questions of algebraic classification.

† The term "congruence" is used in this paper to indicate a family of curves (or surfaces, etc.) such that there exists an element of the family passing through an arbitrarily given nonsingular point of space.

In this section we consider the geometric properties of a congruence of null geodesics. An important property of null curves is the fact that although they lie in a space possessing a metric, the concept of distance along them is not defined. Such curves are one-dimensional affine spaces (only parallel tranport is defined; there is no notion of distance). This fact is indeed reflected in a number of unusual properties of null curves.

The null geodesic congruences can be defined parametrically in the form $x^\alpha = x^\alpha(r, y^i)$ $(i = 1, 2, 3)$, an individual curve of the congruence being defined by the equation $y^i = c^i = \text{const}$, and r is an affine parameter along the curve. We will assume that r is a canonical parameter along the null geodesic, i.e., if $l^\alpha = \partial x^\alpha/\partial r$, then

$$l_\alpha l^\alpha = 0, \qquad Dl^\alpha/dr \equiv l^\alpha{}_{;\beta}\, l^\beta = 0. \tag{4.1}$$

The canonical parameter r along each null geodesic is defined uniquely up to an inhomogeneous linear transformation. Therefore, the canonical parameter r for a congruence of null geodesics is determined up to a transformation

$$r \to r' = (A^0)^{-1} r + R^0, \qquad l^\alpha \to l'^\alpha = A^0 l^\alpha, \tag{4.2}$$

where A^0 and R^0 depend only on y^i.

The key to understanding the geometric properties of a congruence of null geodesics is obtained by considering the following Gedanken experiment. Assume that at a given moment of time an observer places an opaque object in the path of light rays. If another observer places a screen in the path of the rays, the shape, dimension, and position of the shadow on the screen are determined not only by the characteristics of the object, but also depend essentially on the geometric properties of the congruence of light rays.

Let us consider such an experiment in more detail. Assume that the world line of the first observer is $x^\mu = x^\mu(\tau)$ (Fig. 1.2). If the parameter τ is the proper time, then the velocity vector $V^\mu = dx^\mu/d\tau$ of the observer is timelike and normalized, $V_\mu V^\mu = 1$. The set of events

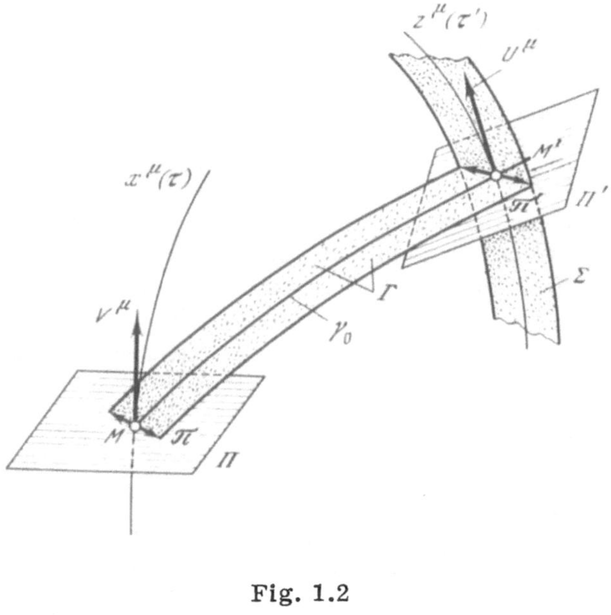

Fig. 1.2

simultaneous with the event M, which has coordinates $x^{\mu}(\tau_0)$, forms a three-dimensional area Π orthogonal to the vector V^{μ}.*

It is easy to verify that the tensor

$$h_{\nu}^{\mu}(V) = \delta_{\nu}^{\mu} - V_{\nu}V^{\mu} \tag{4.3}$$

is a projection operator projecting four-dimensional vectors onto the area Π.

Let γ_0 be a light geodesic in the congruence which passes through the world point M, and let l^{μ} be the vector tangent to γ_0 at this point. Then the vector $\lambda^{\nu} = h_{\mu}^{\nu}(V)l^{\mu}$ lies in the three-dimensional space Π and defines the direction of propagation of light in ordinary three-dimensional space. We place a two-dimensional object π, which is opaque at time τ_0 perpendicular (in the three-dimensional sense) to the direction of propagation of this light ray. If we let π denote an arbitrary vector joining M with the point of the object ζ^{μ}, then we have the equality

$$\lambda^{\mu}\zeta_{\mu} = \zeta_{\mu}h_{\nu}^{\mu}(V)\, l^{\nu} = 0. \tag{4.4}$$

The propagation of the shadow created by the object π is determined by these rays of the congruence which pass through π at the moment τ_0 when it is opaque. The set of these rays forms a two-parameter family of light geodesics Γ.

Assume now that $z^{\mu}(\tau')$ is the world line of an observer studying the shadow from the object π and let its velocity be $U^{\mu} = dz^{\mu}/d\tau'$, $U^{\mu}U_{\mu} = 1$. Let the ray γ_0 intersect the world line of this observer at the point M' at some time τ_0'. The three-dimensional area Π' passing through M' and spanned by the vectors orthogonal to U^{μ} defines the set of events simultaneous with M', and the projection operator onto this set is

$$h_{\nu}^{\mu}(U) = \delta_{\nu}^{\mu} - U_{\nu}U^{\mu}. \tag{4.5}$$

The vector l'^{μ} tangent to the ray γ_0 at M' defines a three-dimensional direction $\lambda'^{\mu} = h_{\nu}^{\mu}(U)l'^{\nu}$ in Π'. We position the screen, i.e., a two-dimensional area π', perpendicular (in the three-dimensional sense) to the light ray λ'^{μ}. If ζ'^{μ} is the vector joining M' with a point of the screen π' at time τ_0', then we have

$$\lambda'^{\mu}\zeta'_{\mu} = \zeta'_{\mu}h_{\nu}^{\mu}(U)\, l'^{\nu} = 0. \tag{4.6}$$

Let Σ denote the three-dimensional hyperplane formed by the world lines of the points of the screen π' as time passes. It is clear that the rays in the family Γ lying near γ_0 intersect Σ, i.e., they fall on the screen, but it is not at all obvious that they intersect Σ at the same moment of time τ_0', i.e., pass through the two-dimensional surface π'. However, it turns out that this holds, and we have the following proposition.

Proposition 1.4.1 (Ehlers – Sachs Theorem [6-7]). In the experiment considered above the following conditions hold:

1. All parts of the shadow reach the screen at the same time.
2. The dimension, shape, and orientation of the shadow depend only on the position of the screen and do not depend on the velocity of motion of the observer.
3. If the screen is positioned at a small distance δr from the object, the shadow is magnified, rotated, and deformed by the quantities $\theta\,\delta r$, $\omega\,\delta r$, and $|\hat{\sigma}|\,\delta r$, where

$$\theta = {}^1/_2 l^{\alpha}_{;\alpha}, \qquad \omega = [{}^1/_2 l_{[\alpha;\beta]}l^{\alpha;\beta}]^{1/2}, \qquad |\hat{\sigma}| = [{}^1/_2 l_{(\alpha;\beta)}\, l^{\alpha;\beta} - \theta^2]^{1/2}. \tag{4.7}$$

* The definition of simultaneity and a description of the associated procedure for synchronization of clocks can be found, for example, in [29, p. 302].

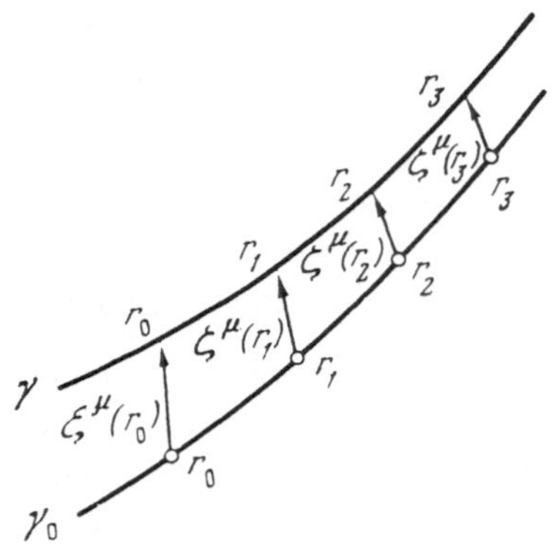

Fig. 1.3

Proof. In order to prove the first assertion we consider along with γ_0 another arbitrary ray γ in the family Γ. We denote by r a canonical parameter along these curves. Let $\zeta^\mu(r)$ be a vector joining the points γ_0 and γ with the same value of the canonical parameter r (Fig. 1.3). By construction, the vector field $\zeta^\mu(r)$ is parallel in the sense of Lie* along γ_0, and therefore its Lie derivative along $l^\mu = dx^\mu/dr$ (a vector tangent to γ_0) vanishes:

$$\mathcal{L}_l \zeta^\mu = l^\nu \zeta^\mu_{;\nu} - \zeta^\nu l^\mu_{;\nu} = 0. \tag{4.8}$$

The variation of the product $l^\mu \zeta_\mu$ along γ_0 is determined by the relation

$$\frac{d}{dr}(l^\mu \zeta_\mu) = l^\nu (l^\mu \zeta_\mu)_{;\nu} = l^\nu l^\mu_{;\nu} \zeta_\mu + l^\nu l^\mu \zeta_{\mu;\nu}. \tag{4.9}$$

The second term on the right-hand side of (4.9) can be rewritten using (4.8) in the form $\zeta^\nu l^\mu l_{\mu;\nu}$, and it vanishes since $l^\mu l_\mu = 0$. The first term in (4.9) is equal to zero because γ_0 is geodesic. Hence the quantity $l^\mu \zeta_\mu$ is constant along γ_0. We remark also that if the canonical parametrization along γ is changed, the vector $\zeta^\mu(r)$ goes into $\zeta^\mu(r) + \alpha(r) l^\mu(r)$ and the value of the product $l^\mu(r)\zeta_\mu(r)$ does not change.

We choose canonical parameters along the rays of the family Γ in such a way that for all rays γ of Γ, the value of r is equal to r_0 at the points of the object π and the parameter r takes the value r_1 for the points of incidence onto the screen. Now let $\zeta^\mu(r_0)$ be any vector joining M with some point of π. Since $\zeta^\mu(r_0)$ belongs to Π, Eq. (4.4) gives

$$l^\mu(r_0)\zeta_\mu(r_0) = 0. \tag{4.10}$$

Therefore, since the product $l^\mu(r)\zeta_\mu(r)$ is constant along the curve we also have

$$l^\mu(r_1)\zeta_\mu(r_1) = 0. \tag{4.11}$$

We remark that since the screen is positioned perpendicular to the ray γ_0, Eq. (4.6) holds and allows us to write

$$\lambda'^\mu(r_1)\zeta_\mu(r_1) \equiv \zeta_\mu(r_1)(\delta^\mu_\nu - U^\mu U_\nu) l^\nu(r_1) = 0. \tag{4.12}$$

* The definition of parallel transport in the sense of Lie can be found, for example, in [42, p. 112].

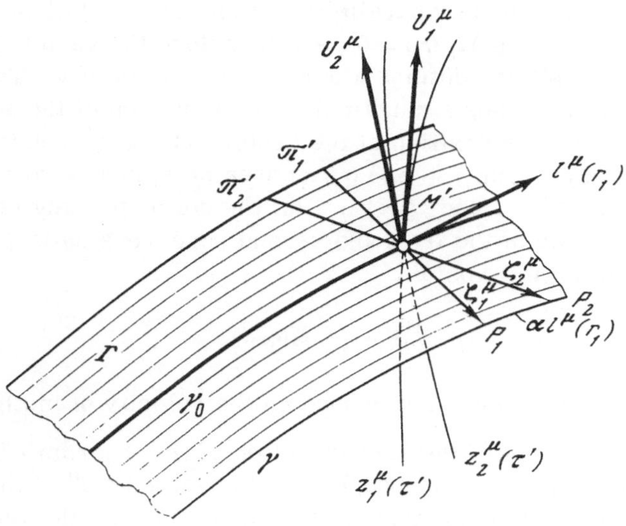

Fig. 1.4

Comparing Eqs. (4.11) and (4.12) and taking into account that the scalar product of the lightlike vector l^ν with the timelike vector U_ν cannot be zero, we get

$$\zeta_\mu (r_1)\, U^\mu = 0. \tag{4.13}$$

The fulfillment of this condition means that all parts of the shadow reach the screen simultaneously.

To prove the second part of the theorem, consider the world lines $z_1^\mu(\tau')$ and $z_2^\mu(\tau')$ passing through the point M' (Fig. 1.4). Let the velocities of these observers at the point M' be U_1^μ and U_2^μ, respectively, and let $\Pi_1'(\Pi_2')$ be the three-dimensional surface of events simultaneous with M' from the point of view of the first (second) observer. We denote the two-dimensional screens lying in Π_1' and Π_2' and perpendicular to the light ray γ_0 by π_1' and π_2', respectively. According to what was proved above, the light rays of the family Γ reach a screen positioned perpendicular to the light in the reference frame of an arbitrary observer at the same time from the point of view of this observer, and therefore π_1' and π_2' lie completely in the hypersurface Γ. Let the ray γ of Γ intersect the screen π_1' at the point P_1. If we then extend it, it will meet the screen π_2' at the point P_2 and hence the vectors ζ_1^μ and ζ_2^μ recording the points of incidence of the same ray γ on the screens π_1' and π_2', differ only by a quantity proportional to $l^\mu(r_1)$:

$$\zeta_2^\mu = \zeta_1^\mu + \alpha l^\mu (r_1). \tag{4.14}$$

The dimension and shape of the shadow are determined by the values of the scalar products of the vectors ζ^μ joining the point M' and the shadow boundary on the screen. However, it follows from Eq. (4.14) and the conditions $\zeta_1^\mu l_\mu (r_1) = \zeta_2^\mu l_\mu (r_1) = l_\mu (r_1) l^\mu (r_1) = 0$ that the equality

$$\zeta^\mu (P_1)\, \zeta_\mu (Q_1) = \zeta^\mu (P_2) \zeta_\mu (Q_2)$$

holds for points P_1 and Q_1 of the shadow on the first screen and corresponding points P_2 and Q_2 on the second screen. Hence the dimensions and the shape of the shadow on a screen do not depend on the velocity and direction of motion of the observer. We now choose a unit vector $e^\mu(r_0)$ at the point M which lies in the surface π and determines a direction in this surface.

Let $e^{\mu}(r_1)$ denote the vector obtained by parallel transport of $e^{\mu}(r_0)$ along γ_0 to the point M'.* Since $e^{\mu}(r_0)\, l_{\mu}(r_0)=0$, we have $e^{\mu}(r_1)\, l_{\mu}(r_1)=0$, and therefore the vector $e(r_1)$ lies in the tangent space to the family Γ at M' and defines a curve γ' in this family. Therefore, this curve γ' and hence also $e^{\mu}(r_1)$ distinguishes a certain direction on each of the screens π_1' and π_2'. The orientation of the shadow, i.e., the position of the vectors ζ_1^{μ} and ζ_2^{μ} relative to the directions $n_{(1)}^{\mu}=e^{\mu}(r_1)+\beta_1 l^{\mu}(r_1)$ and $n_{(2)}^{\mu}=e^{\mu}(r_1)+\beta_2 l^{\mu}(r_1)$, chosen on π_1' and π_2', are determined by the value of the scalar products $\zeta_1^{\mu} n_{(1)\mu}$ and $\zeta_2^{\mu} n_{(2)\mu}$. But the corresponding vectors of the boundary of the shadow on the screen of the first and second observers satisfy Eq. (4.14), and by the orthogonality $l^{\mu}(r_1)e_{\mu}(r_1)=0$ we obtain

$$\zeta_1^{\mu} n_{(1)\mu}=\zeta_2^{\mu} n_{(2)\mu},$$

i.e., the orientation of the shadow does not depend on the velocity of motion of the observer.

Finally, we turn to the proof of the last assertion of the theorem. Since the orientation, shape, and dimension of the shadow do not depend on the velocity U^{μ} of the observer, for the sake of convenience we take U^{μ} to coincide with the vector V^{μ} parallel transported along γ_0 to the point M'. Having chosen an arbitrary vector $\zeta^{\mu}(r_0)$ corresponding to a point of the object, we parallel transport it along γ_0 to M'. The vector $\eta^{\mu}(r_1)$ obtained in this way is as before orthogonal to the light ray and to the velocity vector U^{μ}, and hence it joins M' with one of the points of the shadow on π'. If we let γ denote the ray passing through $\zeta^{\mu}(r_0)$, this ray determines a vector $\zeta^{\mu}(r_1)$ as it falls onto the screen π'. It was already remarked above that this vector is obtained from $\zeta^{\mu}(r_0)$ by means of a parallel transport in the sense of Lie. We therefore have

$$\frac{D\zeta^{\mu}(r)}{dr}\equiv l^{\nu}\zeta^{\mu}_{;\,\nu}=\zeta^{\nu}l^{\mu}_{;\,\nu};\qquad \frac{D\eta^{\mu}(r)}{dr}=0. \tag{4.15}$$

If the screen is situated at a small distance δr from the object, then putting $r_1 = r_0 + \delta r$, we have up to a quantity of second order in δr

$$\zeta^{\mu}(r_1)=\eta^{\mu}(r_1)+\eta^{\nu}(r_1)\, l^{\mu}_{;\nu}\delta r. \tag{4.16}$$

This relation shows that if $l_{\mu;\nu}$ does not vanish, then the shadow undergoes a deformation described by the linear operator $\delta^{\mu}_{\nu}+l^{\mu}_{;\nu}\delta r$. This operator, which takes vectors of the two-dimensional surface π' into themselves, differs from the identity operator by the infinitesimal quantity $l^{\mu}_{;\nu}\delta r$. It turns out to be convenient to split the infinitesimal deformation $l_{\mu;\nu}\delta r$ into three irreducible components: the dilation $\frac{1}{2}l^{\alpha}_{;\alpha}\delta r$, rotation $l_{[\alpha;\beta]}\delta r$, and translation $[l_{(\alpha;\beta)}-\frac{1}{2}g_{\alpha\beta}l^{\gamma}_{;\gamma}]\,\delta r$.

If we introduce the notation

$$\theta=\tfrac{1}{2}l^{\alpha}_{;\,\alpha},\qquad \omega=[\tfrac{1}{2}l_{[\alpha;\,\beta]}l^{\alpha;\,\beta}]^{1/2},$$
$$|\hat{\sigma}|=[\tfrac{1}{2}l_{(\alpha;\,\beta)}l^{\alpha;\,\beta}-\theta^2]^{1/2}, \tag{4.17}$$

then starting from the usual interpretation of the dilation, rotation, and translation operators, it is possible to depict the action of each of these operators as in Fig. 1.5.

Attention should be addressed to the fact that although the quantities $\theta\delta r$, $\omega\delta r$, and $|\hat{\sigma}|\delta r$ are uniquely determined by the geometry of the congruence, the values of the rate of dilation θ, rotation ω, and translation $|\hat{\sigma}|$ depend on the choice of canonical parameter, and

*We can choose the vector $e^{\mu}(r)$ to be the polarization vector which, as was shown in Section 1.3, is parallel transported along a light ray in the approximation of geometric optics.

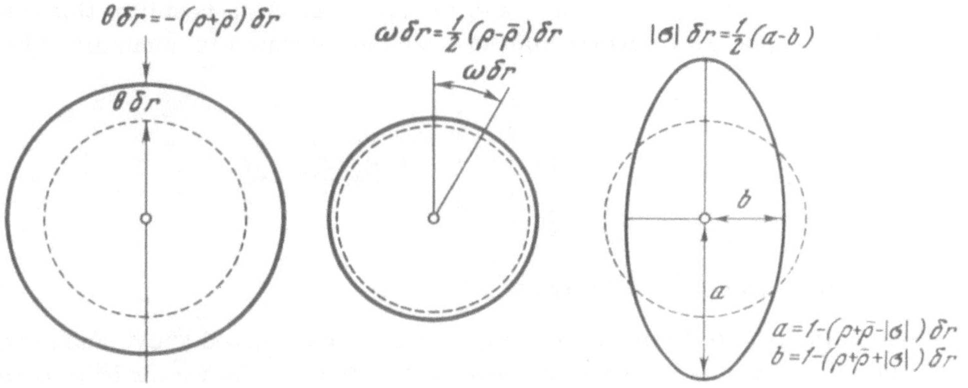

Fig. 1.5

when this parameter is changed (4.2) they transform as follows:

$$\theta' = A^0\theta, \qquad \omega' = A^0\omega, \qquad |\hat{\sigma}'| = A^0\,|\hat{\sigma}|. \tag{4.18}$$

In other words, the ratio $\theta : \omega : |\hat{\sigma}|$ is an invariant. The quantities θ, ω, and $|\hat{\sigma}|$ are called optical scalars.

1.5. Complex Null Tetrads and Spin Coefficients

In considering many questions, and in particular in the study of the geometric properties of congruences of null curves (not necessarily geodesic), it turns out to be convenient to introduce a suitably chosen field of tetrads which is adapted to the given congruence. To this end we construct at an arbitrary point P of space-time a lightlike vector l^μ which is tangent to a null curve of the congruence passing through P. Any other lightlike vector l^μ distinct from n^μ and directed into the future is linearly independent of l^μ and $n^\mu l_\mu > 0$. Multiplying the vector n^μ by a suitable factor, we can arrange that $n^\mu l_\mu = 1$. The pair of vectors l^μ and n^μ determines a two-dimensional subspace orthogonal to them (Fig. 1.6). We introduce an orthonormal basis in this subspace consisting of the pair of vectors a^μ and b^μ:

$$a_\mu a^\mu = -1, \; b_\mu b^\mu = -1, \; a_\mu b^\mu = 0,$$

and thereby obtain a basis $(l^\mu, n^\mu, a^\mu, b^\mu)$ in the tangent space. It turns out to be more convenient to work not with this basis but with a complex modification of it. Put $m^\mu = (a^\mu + ib^\mu)/\sqrt{2}$

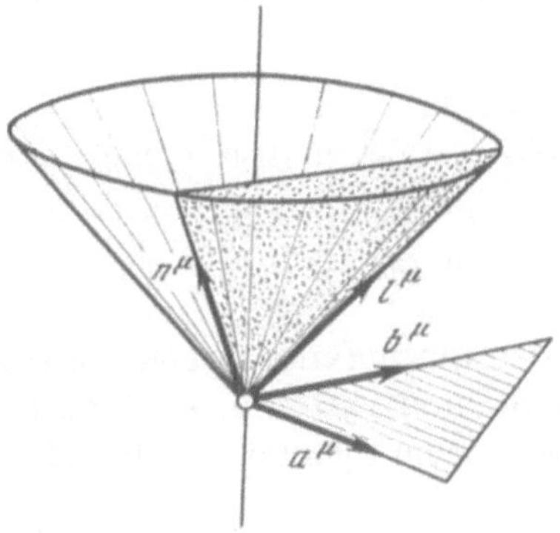

Fig. 1.6

and $\bar{m}^\mu = (a^\mu - ib^\mu)/\sqrt{2}$. Then $m_\mu m^\mu = \bar{m}_\mu \bar{m}^\mu = 0$, $m^\mu \bar{m}_\mu = -1$. As a result of this a tetrad $z^\mu_m = (l^\mu, n^\mu, m^\mu, \bar{m}^\mu)$ appears at every point of space-time which is normalized by the condition

$$z^\mu_m z_{n\mu} = \eta_{mn},$$
$$\eta_{mn} = \begin{Vmatrix} 0 & 1 & 0 & 0 \\ 1 & 0 & 0 & 0 \\ 0 & 0 & 0 & -1 \\ 0 & 0 & -1 & 0 \end{Vmatrix} = \eta^{mn}. \tag{5.1}$$

Such tetrads are called complex null tetrads.

It is obvious that the field of complex null tetrads constructed above and associated with a congruence of null curves is far from uniquely determined. The vector l^μ is defined by the congruence up to a transformation $l'^\mu = Al^\mu$ corresponding to a change of parametrization of the curves of the congruence. If a congruence of null geodesics is being considered and a canonical parametrization is chosen, then $A = A^0(y^i)$ does not depend on r. The normalization condition leads to the transformation $n^\mu \to n'^\mu = A^{-1}n^\mu$. The remaining arbitrariness is related to the nonuniqueness of the choice of the lightlike direction n^μ. For fixed l^μ, each such choice corresponds to choosing a two-dimensional subspace orthogonal to l^μ and n^μ. Such a subspace, being orthogonal to l^μ, lies in a hyperplane tangent to the light cone along the direction l^μ. A basis in the new two-dimensional subspace obtained by changing n^μ can be obtained from the basis a^μ, b^μ in the old subspace by means of the transformation

$$a'^\mu = a^\mu + \beta_1 l^\mu, \qquad b'^\mu = b^\mu + \beta_2 l^\mu$$

or for complex vectors m^μ and $\bar{m}^\mu$

$$m'^\mu = m^\mu + Bl^\mu, \qquad \bar{m}'^\mu = \bar{m}^\mu + \bar{B}l^\mu, \qquad B = (\beta_1 + i\beta_2)/\sqrt{2}.$$

The conditions that n^μ be orthogonal to m^μ and $\bar{m}^\mu$ and the normalization $l_\mu n^\mu = 1$ allow us to determine the corresponding transformation for n^μ,

$$n'^\mu = n^\mu + \bar{B}m^\mu + B\bar{m}^\mu + B\bar{B}l^\mu.$$

The remaining arbitrariness in the choice of tetrad is connected with the nonuniqueness of the choice of basis a^μ, b^μ in the two-dimensional space. This basis is determined up to a rotation, and hence the associated arbitrariness in the choice of m^μ is $m'^\mu = e^{iC}m^\mu$.

Summarizing what has been said above, we obtain that the most general arbitrariness in choosing complex null tetrads based on a congruence of null curves is described by the transformations

$$\begin{aligned} l'^\mu &= Al^\mu, \\ n'^\mu &= A^{-1}(n^\mu + \bar{B}m^\mu + B\bar{m}^\mu + B\bar{B}l^\mu), \\ m'^\mu &= e^{iC}(m^\mu + Bl^\mu), \\ \bar{m}'^\mu &= e^{iC}(\bar{m}^\mu + \bar{B}l^\mu). \end{aligned} \tag{5.2}$$

More precisely, Eq. (5.2) describes the maximal group of transformations preserving the direction of the vector l^μ and the normalization conditions (5.1). These transformations form a four-parameter subgroup of the Lorentz group.* Therefore, the overall arbitrariness in

* Each transformation of a basis is determined by four parameters: two real A, C, and one complex B. It is easy to convince oneself that these transformations form a group.

the choice of the field of complex null tetrads is described by four arbitrary functions of four variables. In the case of a congruence of null geodesics with canonical parametrization, one of the three arbitrary functions (A) depends only on three coordinates.

If we denote by z^m_μ the basis of forms dual to z^μ_m then we have

$$z^m_\mu z^\mu_n = \delta^m_n. \tag{5.3}$$

Using this equality and the normalization conditions (5.1), it is easy to obtain that

$$z^m_\mu = g_{\mu\nu}\eta^{mn}z^\nu_n, \qquad z^m_\mu = (n_\mu, l_\mu, -\bar{m}_\mu, -m_\mu). \tag{5.4}$$

Multiplying the equality $z^\mu_m z_{n\mu} = \eta_{mn}$ by z^n_ν, we obtain the relation $z^\mu_m z_{n\mu} z^n = z_{m\nu}$ valid for all vectors z^μ_m. This equality permits us to conclude that

$$g_{\mu\nu} = z_{m\mu}z^m_\nu = \eta_{mn}z^m_\mu z^n_\nu = 2\,[l_{(\mu}n_{\nu)} - m_{(\mu}\bar{m}_{\nu)}], \tag{5.5}$$

$$g^{\mu\nu} = \eta^{mn}z^\mu_m z^\nu_n = 2\,[l^{(\mu}n^{\nu)} - m^{(\mu}\bar{m}^{\nu)}]. \tag{5.5a}$$

Giving a field of complex null tetrads in space-time permits the use of the tetrad formalism for describing geometric objects. As usual, the operation of covariant differentiation is determined by giving the Ricci rotation coefficients γ_{mnl}

$$\gamma_{mnl} = z^\nu_n z^\lambda_l\, z_{m\nu;\,\lambda} = -\gamma_{nml}. \tag{5.6}$$

The last equality follows from the relation $z^\nu_n z_{m\nu} = \eta_{nm}$ after applying the operator $z^\lambda_l \nabla_\lambda$ to it. By virtue of this antisymmetry property there are 24 independent real quantities γ_{mnl}, which in the case of the complex light tetrads used above are grouped into 12 complex quantities.

It is customary to denote these quantities (more precisely, the following linear combinations of them) using particular letters,

$$\begin{aligned}
&k = \gamma_{131} = l_{\mu;\,\nu}m^\mu l^\nu, \qquad \pi = -\gamma_{241} = -n_{\mu;\,\nu}\bar{m}^\mu l^\nu,\\
&\varepsilon = {}^1\!/_2(\gamma_{121} - \gamma_{341}) = {}^1\!/_2(l_{\mu;\,\nu}n^\mu l^\nu - m_{\mu;\,\nu}\bar{m}^\mu l^\nu),\\
&\rho = \gamma_{134} = l_{\mu;\,\nu}m^\mu\bar{m}^\nu, \qquad \lambda = -\gamma_{244} = -n_{\mu;\,\nu}\bar{m}^\mu\bar{m}^\nu,\\
&\alpha = {}^1\!/_2(\gamma_{124} - \gamma_{344}) = {}^1\!/_2(l_{\mu;\,\nu}n^\mu\bar{m}^\nu - m_{\mu;\,\nu}\bar{m}^\mu\bar{m}^\nu),\\
&\sigma = \gamma_{133} = l_{\mu;\,\nu}m^\mu m^\nu, \qquad \mu = -\gamma_{243} = -n_{\mu;\,\nu}\bar{m}^\mu m^\nu,\\
&\beta = {}^1\!/_2(\gamma_{123} - \gamma_{343}) = {}^1\!/_2(l_{\mu;\,\nu}n^\mu m^\nu - m_{\mu;\,\nu}\bar{m}^\mu m^\nu),\\
&\nu = -\gamma_{242} = -n_{\mu;\,\nu}\bar{m}^\mu n^\nu, \qquad \tau = \gamma_{132} = l_{\mu;\,\nu}m^\mu n^\nu,\\
&\gamma = {}^1\!/_2(\gamma_{122} - \gamma_{342}) = {}^1\!/_2(l_{\mu;\,\nu}n^\mu n^\nu - m_{\mu;\,\nu}\bar{m}^\mu n^\nu).
\end{aligned} \tag{5.7}$$

These quantities have become known as the spin coefficients.* If the field of complex null tetrads is determined by a congruence of null curves, the spin coefficients carry information concerning both the geometric properties of the congruence and the concrete choice of tetrads.

It is easy to verify that the relation

$$\gamma_{m[nl]} = {}^1\!/_2 z^\nu_n z^\lambda_l(z_{m\nu;\,\lambda} - z_{m\lambda;\,\nu}) = {}^1\!/_2 z^\nu_n z^\lambda_l(z_{m\nu,\,\lambda} - z_{m\lambda,\,\nu}) \tag{5.8}$$

---

* The name "spin coefficients" indicates that these quantities arise in a natural way in the spinor formalism. This connection is considered in more detail in Chapter 3.

and Eq. (5.6) allow us to write

$$\gamma_{mnl} = \gamma_{m[nl]} + \gamma_{n[lm]} - \gamma_{l[mn]}. \tag{5.9}$$

Therefore, calculation of the quantities γ_{mnl} in terms of a given z^μ_m only requires application of the operation of partial differentiation along with algebraic operations.

As usual, in the tetrad formalism it is convenient to carry out all the calculations using tetrad components of tensors defined by the relation

$$T_{\alpha_1\ldots\alpha_p}{}^{\beta_1\ldots\beta_q} \leftrightarrow T_{m_1\ldots m_p}{}^{n_1\ldots n_q} = z^{\alpha_1}_{m_1}\ldots z^{\alpha_p}_{m_p} z^{n_1}_{\beta_1}\ldots z^{n_q}_{\beta_q} T_{\alpha_1\ldots\alpha_p}{}^{\beta_1\ldots\beta_q}, \tag{5.10}$$

with moreover

$$\begin{aligned} T_{\alpha_1\ldots\alpha_p}{}^{\beta_1\ldots\beta_q}{}_{;\alpha} \leftrightarrow T_{m_1\ldots m_p}{}^{n_1\ldots n_q}{}_{;m} &= z^{\alpha}_{m} z^{\alpha_1}_{m_1}\ldots z^{\alpha_p}_{m_p} z^{n_1}_{\beta_1}\ldots z^{n_q}_{\beta_q} T_{\alpha_1\ldots\alpha_p}{}^{\beta_1\ldots\beta_q}{}_{;\alpha} = \\ &= T_{m_1\ldots m_p}{}^{n_1\ldots n_q}{}_{,m} + \gamma^{l}{}_{\cdot m_1 m} T_{l m_2\ldots m_p}{}^{n_1\ldots n_q} + \ldots + \gamma^{l}{}_{\cdot m_p m} T_{m_1\ldots m_{p-1} l}{}^{n_1\ldots n_q} + \\ &+ \gamma_{l\cdot m}^{n_1} T_{m_1\ldots m_p}{}^{l n_2\ldots n_q} + \ldots + \gamma_{l\cdot m}^{n_q} T_{m_1\ldots m_p}{}^{n_1\ldots n_{q-1} l}, \end{aligned} \tag{5.11}$$

where $T_{m_1\ldots m_p}{}^{n_1\ldots n_q}{}_{,m} = z^{\nu}_{m} (T_{m_1\ldots m_p}{}^{n_1\ldots n_q})_{,\nu}$.

The Ricci identity

$$z_{m\mu[;\alpha;\beta]} = -{}^1/_2 z_{m\nu} R^{\nu}{}_{\cdot\mu\alpha\beta} \tag{5.12}$$

serves to define the curvature tensor, and after suitable contraction with the quantities $z^n_\mu z^p_\alpha z^q_\beta$, it can be used to find the tetrad components of the curvature tensor

$$R^{mnqp} = \gamma^{mnp,q} - \gamma^{mnq,p} + \gamma_l{}^{mq}\gamma^{lnp} - \gamma_l{}^{mp}\gamma^{lnq} + \gamma^{mnl}(\gamma_l{}^{pq} - \gamma_l{}^{qp}). \tag{5.13}$$

In tetrad notation the Bianchi identities $R_{\alpha\beta[\gamma\delta;\mu]} = 0$ take the form

$$R_{mn[pq,r]} - \gamma_{m\cdot[r}^{\ l} R_{pq]ln} + \gamma_{n\cdot[r}^{\ l} R_{pq]lm} - 2R_{mnl[p}\gamma_{r\cdot q]}^{\ l} = 0. \tag{5.14}$$

1.6. The Geometric Meaning of the Spin Coefficients

In this section we return once more to the discussion of the geometric properties of congruences of null curves with the purpose of clarifying the geometric meaning of the separate spin coefficients and finding their relation to the optical scalars introduced previously.

We again consider the Gedanken experiment on the propagation of a shadow, which was previously discussed in Section 1.4 (cf. Fig. 1.2). Choose a lightlike vector $n^\mu(r_0)$ at the point M and let it lie in the two-dimensional plane spanned by $l^\mu(r_0)$ and V^μ and be normalized by the condition $n^\mu(r_0)l^\mu(r_0) = 1$. In the two-dimensional surface π orthogonal to n^μ and l^μ choose a basis $a^\mu(r_0)$ and $b^\mu(r_0)$ and define the complex vectors $m^\mu = (a^\mu + ib^\mu)/\sqrt{2}$, $\bar{m}^\mu = (a^\mu - ib^\mu)/\sqrt{2}$. An arbitrary real vector ζ^μ in π can be written in the form

$$\zeta^\mu = \zeta\bar{m}^\mu + \bar{\zeta}m^\mu. \tag{6.1}$$

We parallel transport the complex null tetrad z^μ_m constructed at the point M along γ_0 to the point M' and denote the corresponding basis by $z^\mu_m(r_1)$. Since the dimension, shape, and orientation of the shadow do not depend on the velocity U^μ of the observer studying the shadow, we will assume for convenience that U^μ coincides with the vector obtained by parallel transport of V^μ from M to M'. In this case the vectors $m^\mu(r_1)$ and $\overline{m}^\mu(r_1)$ span the two-dimensional space in which the screen π' lies.

The vector $\eta^\mu(r_1)$ obtained from $\xi^\mu(r_0)$ by parallel transport has the same components in the basis $m^\mu(r_1)$, $\overline{m}^\mu(r_1)$ as the vector $\xi^\mu(r_0)$ does in the basis $m^\mu(r_0)$, $\overline{m}^\mu(r_0)$:

$$\eta^\mu(r_1) = \zeta\overline{m}^\mu(r_1) + \overline{\zeta}m^\mu(r_1). \tag{6.2}$$

A ray γ passing through the point $\xi^\mu(r_0)$ of the object determines a vector $\xi^\mu(r_1)$ on the screen

$$\xi^\mu(r_1) = (\zeta + \delta\zeta)\overline{m}^\mu(r_1) + (\overline{\zeta} + \delta\overline{\zeta})m^\mu(r_1). \tag{6.3}$$

If we make use of Eq. (4.16) after multiplying both sides by $m^\mu(r_1)$ and use the definition of the spin coefficients (5.7), it is possible to obtain

$$\delta\zeta = -(\rho\zeta + \sigma\overline{\zeta})\delta r. \tag{6.4}$$

Thus the transformation

$$\zeta \to \zeta' = \zeta + \delta\zeta = \zeta(1 - \rho\delta r) - \overline{\zeta}\sigma\delta r \tag{6.5}$$

establishes a relation between the shape of the object and the shape of the shadow. If we take the object to be a disk with boundary $\zeta = \exp(i\Phi)$, the boundary of the shadow is determined by the equation

$$\zeta' = (1 - \rho\delta r)\exp(i\Phi) - \sigma\delta r\exp(-i\Phi), \tag{6.6}$$

which describes an ellipse with semiaxes $a_\pm = 1 - (\rho + \overline{\rho} \mp |\sigma|)\delta r$. The direction of the principal axis makes an angle Φ_0 with OX, where

$$e^{2i\Phi_0} = (\overline{\sigma}/\sigma)^{1/2}. \tag{6.7}$$

The area of the ellipse is $\pi a_+ a_- = \pi(1 - (\rho + \overline{\rho})\delta r)$, and hence the factor $-\mathrm{Re}\,\rho = -\frac{1}{2}(\rho + \overline{\rho})$ determines the increase in the linear scale. The ellipse (6.6) is rotated with respect to the circle by an angle determined by the average value of (ζ'/ζ) and is equal to $(\rho - \overline{\rho})\delta r/2i$. The modulus of the shear $|\sigma|$ is determined by the ratio of the semiaxes

$$a_+ a_- = 1 + 2|\sigma|\delta r.$$

Comparing all these results with the expressions for the optical scalars θ, ω, and $|\hat{\sigma}|$ given in Section 1.4, it is easy to obtain that

$$\rho = -(\theta + i\omega), \qquad |\sigma| = |\hat{\sigma}|. \tag{6.8}$$

We now show how the optical scalars vary along light rays. To do this we observe that the Ricci dentity (5.12) permits us to write

$$l_{\mu;\alpha;\beta} = l_{\mu;\beta;\alpha} - R^{\nu}_{\mu\alpha\beta}l_\nu.$$

Therefore

$$\frac{D}{dr}(l_{\mu;\alpha}) = l^\beta l_{\mu;\alpha;\beta} = l_{\mu;\beta;\alpha}l^\beta - R^{\nu}_{\mu\alpha\beta}l_\nu l^\beta = -l_{\mu;\beta}l^{\beta}_{;\alpha} - R_{\nu\mu\alpha\beta}l^\nu l^\beta. \tag{6.9}$$

The last equality is obtained after using $l_{\mu;\beta}l^{\beta} = 0$. If we now multiply both sides of (6.9) by $m^{\mu}\bar{m}^{\alpha}$ and choose basis vectors m^{μ}, $\bar{m}^{\mu}$ parallel transported along γ_0, it is easy to verify that Eq. (6.9) leads to the relation

$$D\rho = \partial\rho/\partial r = \rho^2 + \sigma\bar{\sigma} - R_{\nu\mu\alpha\beta}l^{\nu}l^{\beta}m^{\mu}\bar{m}^{\alpha}. \tag{6.10}$$

We remark that

$$R_{\nu\beta}l^{\nu}l^{\beta} = -g^{\mu\alpha}R_{\nu\mu\alpha\beta}l^{\nu}l^{\beta} = 2R_{\nu\mu\alpha\beta}l^{\nu}l^{\beta}m^{\mu}\bar{m}^{\alpha}. \tag{6.11}$$

In deriving this equality we used the symmetry properties of the curvature tensor $R_{(\nu\mu)\alpha\beta} = 0$, $R_{\nu\mu\alpha\beta} = R_{\alpha\beta\nu\mu}$ and expression (5.5a) for the metric tensor $g^{\mu\alpha}$. On the basis of (6.10) and (6.11) we have finally

$$D\rho = \rho^2 + \sigma\bar{\sigma} + \Phi_{00}, \tag{6.12}$$

where $\Phi_{00} = -{}^1/_2 R_{\nu\beta}l^{\nu}l^{\beta}$. In analogous fashion we obtain the equation*

$$D\sigma = \sigma(\rho + \bar{\rho}) + R_{\alpha\beta\gamma\delta}l^{\alpha}l^{\gamma}m^{\beta}m^{\delta}. \tag{6.13}$$

In order to clarify the physical meaning of Eq. (6.12) we choose an arbitrary two-dimensional spacelike surface S. Let a^{μ} and b^{μ} be two linearly independent vector fields tangent to S. By the Frobenius theorem [40, 41] the condition

$$a^{\nu}b^{\mu}_{,\nu} - b^{\nu}a^{\mu}_{,\nu} = \alpha a^{\mu} + \beta b^{\mu} \tag{6.14}$$

is satisfied for these fields. We say that a pair of vector fields satisfying condition (6.14) are vector fields generating the two-dimensional surface S.

If $m^{\mu} = (a^{\mu} + ib^{\mu})/\sqrt{2}$ and $\bar{m}^{\mu} = (a^{\mu} - ib^{\mu})/\sqrt{2}$, Eq. (6.14) allows us to write

$$m^{\nu}\bar{m}^{\mu}_{,\nu} - \bar{m}^{\nu}m^{\mu}_{,\nu} = i(\zeta\bar{m}^{\mu} + \bar{\zeta}m^{\mu}). \tag{6.15}$$

At each point of S, choose one of the two null directions orthogonal to S and draw the null geodesics, the tangent vector l^{μ} to which coincides with the chosen direction on S. We show that the quantity $\rho - \bar{\rho}$ for this family of null curves vanishes on the surface S. In order to do this, we use the condition $m_{\mu}l^{\mu} = \bar{m}_{\mu}l^{\mu} = 0$ to write $\rho - \bar{\rho}$ in the form

$$\rho - \bar{\rho} = (l_{\mu,\nu} - l_{\nu,\mu})m^{\mu}\bar{m}^{\nu} = -l_{\mu}(m^{\mu}_{,\nu}\bar{m}^{\nu} - \bar{m}^{\mu}_{,\nu}m^{\nu}).$$

Therefore by Eq. (6.15) we have $\rho = \bar{\rho}$ on S. Equation (6.12) shows that ρ remains a real quantity along light rays even outside the surface S.

The equality $\rho = \bar{\rho}$ just proved means that the family of null geodesics emitted orthogonally to the two-dimensional surface S forms a null surface.†

We now consider the manner in which the cross section of a beam of light geodesics orthogonal to the surface S varies. It was shown in the analysis of the propagation of a shadow

* Equations (6.12) and (6.13) in the case considered coincide with the first two equations of the complete Newman–Penrose system of equations (A3.4).

† This is a consequence of Proposition 1.6.3 (cf. the end of this section).

that the variation of the area s of the cross section of a light beam is described by the equation

$$\frac{1}{s}\frac{ds}{dr} = -(\rho + \bar{\rho}). \tag{6.16}$$

Since ρ is a real quantity, we can also write

$$\frac{d(\sqrt{s})}{dr} = -\rho\sqrt{s}. \tag{6.17}$$

Differentiating again with respect to r we get

$$\frac{d^2}{dr^2}(\sqrt{s}) = -(\sigma\bar{\sigma} + \Phi_{00})\sqrt{s} \leqslant 0. \tag{6.18}$$

Thus if $\rho > 0$ at some point of a light curve, (6.17) shows that s decreases, while if $\Phi_{00} \geq 0$,[16*] it follows from (6.18) that s decreases to zero. Thus a light beam inevitably reaches a focal point. It is here that the focusing action of a gravitational field manifests itself.

To conclude this section we prove a number of assertions concerning the geometric properties of congruences of null curves.

Let $\Gamma(\mathbf{l})$ denote a congruence of null curves such that the tangents to it form a field of lightlike vectors l^μ.

Proposition 1.6.1. The null congruence $\Gamma(\mathbf{l})$ is geodesic if and only if $k = 0$ and the condition $\varepsilon + \bar{\varepsilon} = 0$ can be achieved by a suitable choice of affine parameter along $\Gamma(\mathbf{l})$.

Proof. Relation (A1.11a) gives

$$Dl^\mu \equiv l^\nu l^\mu_{;\nu} = (\varepsilon + \bar{\varepsilon})\, l^\mu - \bar{k} m^\mu - k\bar{m}^\mu.$$

The curves $\Gamma(\mathbf{l})$ are geodesics in the case where $Dl^\mu \sim l^\mu$, and therefore $k = 0$. Choosing the parameter along $\Gamma(\mathbf{l})$ in a suitable way, we can get $\varepsilon + \bar{\varepsilon} = 0$.

Proposition 1.6.2. The complex null tetrad z^μ_m is parallel along $\Gamma(\mathbf{l})$ if and only if $k = \pi = \varepsilon = 0$.

Proof. The proof follows from Eqs. (A1.11)-(A1.14) if we bear in mind that under parallel transport along $\Gamma(\mathbf{l})$, $Dl^\mu = Dn^\mu = Dm^\mu = D\bar{m}^\mu = 0$.

Proposition 1.6.3. Let $\Gamma(\mathbf{l})$ be a congruence of null geodesics. Then the condition $l_{[\lambda}\nabla_\mu l_{\nu]} = 0$ is equivalent to $\rho = \bar{\rho}$, i.e., the condition $\mathrm{Im}(\rho) = 0$ is a necessary and sufficient condition for normality of a congruence of null geodesics. The field l^μ is a gradient field, $l_\mu = \partial_\mu u$, if and only if the following equalities are satisfied:

$$\rho = \bar{\rho}, \quad \tau = \bar{\alpha} + \beta. \tag{6.19}$$

Proof. It follows from Proposition 1.6.1 that $k = 0$. The condition $\mathrm{Im}\,\rho = 0$ can be obtained if Eq. (A1.6) for $\nabla_{[\mu} l_{\nu]}$ is used in calculating $l_{[\lambda}\nabla_\mu l_{\nu]}$. The field is a gradient field if the equality $\nabla_{[\mu} l_{\nu]} = 0$ is satisfied. Equation (A1.6) shows that Eq. (6.19) holds in this case.

Proposition 1.6.4. Assume that $\Gamma(\mathbf{l})$ is a congruence of null geodesics and that the parameter along it is chosen such that $\varepsilon + \bar{\varepsilon} = 0$. Then the angular velocity of rotation $\Omega^\mu \equiv {}^1\!/_2\, \varepsilon^{\mu\alpha\beta\gamma} l_\alpha \nabla_{[\beta} l_{\gamma]}$ of the congruence $\Gamma(\mathbf{l})$ is equal to $\Omega^\mu = -\mathrm{Im}(\rho) l^\mu$.

Proof. We observe that

$$\varepsilon^{\mu\alpha\beta\gamma} n_\mu l_\alpha m_\beta \bar{m}_\gamma = -i\varepsilon^{\mu\alpha\beta\gamma}\left(\frac{n_\mu + l_\mu}{\sqrt{2}}\right)\left(\frac{l_\alpha - n_\alpha}{\sqrt{2}}\right) a_\beta b_\gamma = i, \tag{6.20}$$

* This inequality is called the weak energy inequality.

where $m_\beta = (a_\beta + ib_\beta)/\sqrt{2}$, $\bar{m}_\beta = (a_\beta - ib_\beta)/\sqrt{2}$. The last equality reflects the fact that the volume of the four-dimensional rectangular parallelepiped constructed from the unit vectors $(n_\mu + l_\mu)/\sqrt{2}$, and $(l_\mu - n_\mu)/\sqrt{2}$, is equal to unity. The equality

$$\varepsilon^{\mu\alpha\beta\gamma} l_\alpha m_\beta \bar{m}_\gamma = i l^\mu \tag{6.21}$$

is a consequence of Eq. (6.20). The desired equality $\Omega^\mu = -\mathrm{Im}\,(\rho) l^\mu$ is a consequence of Eqs. (A1.6) and (6.21).

Proposition 1.6.5. The assertions stated above concerning the congruence $\Gamma(\mathbf{l})$ go over completely to the case of a congruence $\Gamma(\mathbf{n})$ if we make the substitution

$$k \leftrightarrow -\nu, \quad \varepsilon \leftrightarrow -\gamma, \quad \pi \leftrightarrow -\tau, \quad \rho \leftrightarrow -\mu, \quad \alpha \leftrightarrow -\beta, \; \lambda \leftrightarrow -\sigma. \tag{6.22}$$

Proof. The proof is evident if we remark that when the replacement $l^\mu \longleftrightarrow n^\mu$ is made, the spin coefficients (5.7) undergo the transformation described by Eq. (6.22)

Proposition 1.6.6. (Sach's theorem [7]). If the quantities $\Phi_{00} = -{}^1/_2 R_{\nu\beta} l^\nu l^\beta$ and $\Psi_0 = R_{\alpha\beta\gamma\delta} l^\alpha l^\gamma m^\beta m^\delta$ vanish and r is a canonical parameter along the congruence $\Gamma(\mathbf{l})$ of null geodesics, then the spin coefficients ρ and σ have the following form:

$$\rho = -(r - i\Sigma)/(r^2 + \Sigma^2 - \Omega^2), \quad \sigma = \Omega/(r^2 + \Sigma^2 - \Omega^2), \quad \text{if} \quad \sigma\bar{\sigma} \neq \rho\bar{\rho}; \tag{6.23}$$

$$\rho = -(1 + i\Omega)/2r, \quad \sigma = (1 + i\Omega)/2r, \quad \text{if} \quad \sigma\bar{\sigma} = \rho\bar{\rho}, \; \mathrm{Re}\rho \neq 0; \tag{6.24}$$

$$\rho = -i\Omega, \quad \sigma = i\Omega, \quad \text{if} \quad \sigma\bar{\sigma} = \rho\bar{\rho}, \quad \mathrm{Re}\rho = 0, \tag{6.25}$$

where Σ and Ω are real functions independent of r.

Proof. The proof is based on the system of equations (6.12) and (6.13), which under the assumptions made above have the form

$$\partial\rho/\partial r = \rho^2 + \sigma\bar{\sigma}, \quad \partial\sigma/\partial r = \sigma(\rho + \bar{\rho}). \tag{6.26}$$

Consider the matrix Z

$$Z = \begin{Vmatrix} -\rho & \sigma \\ \bar{\sigma} & -\bar{\rho} \end{Vmatrix}.$$

It is easy to see that system (6.26) is equivalent to the following matrix condition:

$$\partial Z/\partial r = -Z^2. \tag{6.27}$$

If $\det Z = \rho\bar{\rho} - \sigma\bar{\sigma} \neq 0$, then multiplying both sides of (6.27) by Z^{-2} we get

$$\partial Z^{-1}/\partial r = I$$

and therefore $Z^{-1} = \begin{Vmatrix} r + i\Sigma & -\Omega \\ -\Omega & r - i\Sigma \end{Vmatrix}$, where Ω and Σ do not depend on r. Using the arbitrariness in choosing the origin of the parameter r, it is always possible to achieve that Ω and Σ are real functions. Inverting the matrix Z^{-1}, we get

$$Z = \begin{Vmatrix} (r - i\Sigma)/R & \Omega/R \\ \Omega/R & (r + i\Sigma)/R \end{Vmatrix}, \quad R = r^2 + \Sigma^2 - \Omega^2,$$

from which Eq. (6.23) follows. If $\det Z = 0$ we can put $\sigma = -\rho$. Simple integration of (6.26) in this case leads to Eqs. (6.24) and (6.25), and the proof of the proposition is thereby completed.

If we restrict ourselves to normal congruences ($\mathrm{Im}\,\rho = 0$), the light rays of these congruences lie on the null surfaces $u = u_0$. The geometry of the two-dimensional sections t = const of the null surfaces $u = u_0$ (wave-front surface) is determined by the values of the optical scalars $\theta = -\rho$ and $|\sigma|$. Indeed, we have the following proposition [72].

Proposition 1.6.7. Let H be the mean curvature of the wave front and K its Gaussian curvature. Then

$$H = -\rho, \quad K = \rho^2 - |\sigma|^2, \tag{6.28}$$

and if r is a canonical parameter and $\Phi_{00} = \Psi_0 = 0$, the quantities H and K have the form

$$H = r/(r^2 - \Omega^2), \quad K = 1/(r^2 - \Omega^2), \quad |\sigma|^2 \neq \rho^2 \text{ (spherical waves)}, \tag{6.29}$$

$$H = 1/2r, \quad K = 0, \quad |\sigma|^2 = \rho^2, \quad \rho \neq 0 \text{ (cylindrical waves)}, \tag{6.30}$$

$$H = 0, \quad K = 0, \quad |\sigma|^2 = \rho^2 = 0 \text{ (plane waves)}, \tag{6.31}$$

where Ω is a real function not depending on r.

Proof. We denote by S_t the position of the wave front at time t (i.e., the two-dimensional surface given by the intersection of the null surface $u = u_0$ with the hyperplane Π_t: t = const), and by $S_{t+\delta t}$ the position of the wave front at time $t + \delta t$ (Fig. 1.7). Project the surface S_t along the time lines onto the hyperplane $\Pi_{t+\delta t}$, and denote the corresponding surface by S_t'. It is easy to verify that the surface S_t' is parallel to $S_{t+\delta t}$, i.e., is the geometric locus of the endpoints of normals of length δt drawn toward the surface S_t'. Choose a canonical parameter along the rays such that $\delta r = \delta t$. If dS, H, and K are the surface element, mean, and Gaussian curvature of the surface S, then we have for a surface parallel to it and at a distance r away [73]

$$dS_r = (1 + 2rH + r^2K)\,dS, \tag{6.32}$$

$$H_r = (1 + 2rH + r^2K)^{-1}(H + rK), \tag{6.33}$$

$$K_r = (1 + 2rH + r^2K)^{-1}K. \tag{6.34}$$

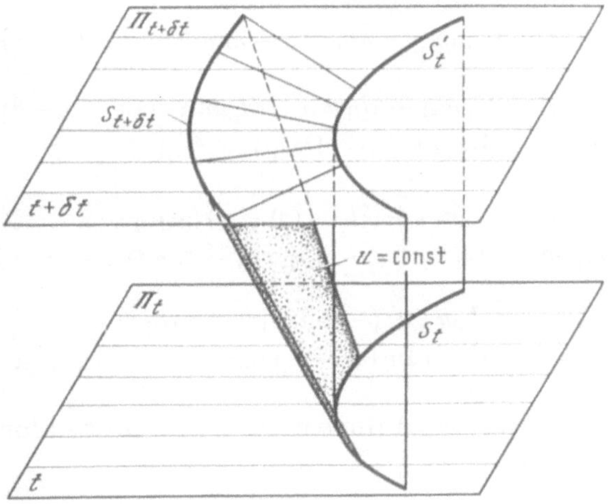

Fig. 1.7

Since $\theta = -\rho = {}^1\!/_2 \lim\limits_{r\to 0} [(dS_r - dS)/rdS]$, we have $H = -\rho = \theta$. Equation (6.33) gives $\left.\frac{\partial H}{\partial r}\right|_{r=0} = -2H^2 + K$. Therefore Eq. (6.12) for $\Phi_{00} = \Psi_0 = 0$ allows us to obtain $|\sigma|^2$ in the form

$$|\sigma|^2 = \frac{\partial\rho}{\partial r} - \rho^2 = \left(-\frac{\partial H}{\partial r} - H^2\right)\Big|_{r=0} = H^2 - K,$$

from which relation (6.28) follows. Equations (6.29)-(6.31) follow directly from (6.28) if we use expressions (6.23)-(6.25) for ρ and σ in the case where $\mathrm{m}\,(\rho) = 0$.

APPENDIX TO CHAPTER 1

In this appendix we list a number of useful formulas containing the spin coefficients [36].

Definition (1.5.7) of the spin coefficients allows us to obtain the following relations:

$$l_{\nu;\,\mu} = (\gamma + \bar{\gamma})\, l_\mu l_\nu - \bar{\tau} l_\mu m_\nu - \tau l_\mu \bar{m}_\nu + (\varepsilon + \bar{\varepsilon})\, n_\mu l_\nu - \bar{k} n_\mu m_\nu - k n_\mu \bar{m}_\nu - (\alpha + \bar{\beta})\, m_\mu l_\nu + \bar{\sigma} m_\mu m_\nu + + \rho m_\mu \bar{m}_\nu - (\bar{\alpha} + \beta)\, \bar{m}_\mu l_\nu + \bar{\rho} \bar{m}_\mu m_\nu + \sigma \bar{m}_\mu \bar{m}_\nu; \tag{A1.1}$$

$$n_{\nu;\,\mu} = -(\gamma + \bar{\gamma})\, l_\mu n_\nu + \nu l_\mu m_\nu + \bar{\nu} l_\mu \bar{m}_\nu - (\varepsilon + \bar{\varepsilon})\, n_\mu n_\nu + \pi n_\mu m_\nu + \bar{\pi} n_\mu \bar{m}_\nu + (\alpha + \bar{\beta})\, m_\mu n_\nu - \lambda m_\mu m_\nu - - \bar{\mu} m_\mu \bar{m}_\nu + (\bar{\alpha} + \beta)\, \bar{m}_\mu n_\nu - \mu \bar{m}_\mu m_\nu - \bar{\lambda} \bar{m}_\mu \bar{m}_\nu; \tag{A1.2}$$

$$m_{\nu;\,\mu} = \bar{\nu} l_\mu l_\nu - \tau l_\mu n_\nu + (\gamma - \bar{\gamma})\, l_\mu m_\nu + \bar{\pi} n_\mu l_\nu - k n_\mu n_\nu + (\varepsilon - \bar{\varepsilon})\, n_\mu m_\nu - \bar{\mu} m_\mu l_\nu + \rho m_\mu n_\nu + + (\bar{\beta} - \alpha)\, m_\mu m_\nu - \bar{\lambda} \bar{m}_\mu l_\nu + \sigma \bar{m}_\mu n_\nu + (\bar{\alpha} - \beta)\, \bar{m}_\mu m_\nu; \tag{A1.3}$$

$$\bar{m}_{\nu;\,\mu} = \nu l_\mu l_\nu - \bar{\tau} l_\mu n_\nu + (\bar{\gamma} - \gamma)\, l_\mu \bar{m}_\nu + \pi n_\mu l_\nu - \bar{k} n_\mu n_\nu + (\bar{\varepsilon} - \varepsilon)\, n_\mu \bar{m}_\nu - \mu \bar{m}_\mu l_\nu + \bar{\rho} \bar{m}_\mu n_\nu + + (\beta - \bar{\alpha})\, \bar{m}_\mu \bar{m}_\nu - \lambda m_\mu l_\nu + \bar{\sigma} m_\mu n_\nu + (\alpha - \bar{\beta})\, m_\mu \bar{m}_\nu. \tag{A1.4}$$

These equations give after contraction

$$\text{a) } l^\alpha_{;\alpha} = -(\rho + \bar{\rho}) + \varepsilon + \bar{\varepsilon}; \quad \text{b) } n^\alpha_{;\alpha} = -(\gamma + \bar{\gamma}) + \mu + \bar{\mu};$$
$$\text{c) } m^\alpha_{;\alpha} = -\bar{\alpha} + \bar{\pi} - \tau + \beta; \quad \text{d) } \bar{m}^\alpha_{;\alpha} = -\alpha + \pi - \bar{\tau} + \bar{\beta} \tag{A1.5}$$

Antisymmetrization of Eqs. (A1.1)-(A1.4) leads to the following inequalities:

$$l_{[\nu;\,\mu]} = -2\,\mathrm{Re}\,(\varepsilon)\, l_{[\mu} n_{\nu]} - (\bar{\tau} - \alpha - \bar{\beta})\, l_{[\mu} m_{\nu]} - - (\tau - \bar{\alpha} - \beta)\, l_{[\mu} \bar{m}_{\nu]} - \bar{k} n_{[\mu} m_{\nu]} - k n_{[\mu} \bar{m}_{\nu]} + 2i\,\mathrm{Im}\,(\rho)\, m_{[\mu} \bar{m}_{\nu]}; \tag{A1.6}$$

$$n_{[\nu;\,\mu]} = -2\,\mathrm{Re}\,(\gamma)\, l_{[\mu} n_{\nu]} + \nu l_{[\mu} m_{\nu]} + \bar{\nu} l_{[\mu} \bar{m}_{\nu]} + (\pi - \alpha - \bar{\beta})\, n_{[\mu} m_{\nu]} + + (\bar{\pi} - \bar{\alpha} - \beta)\, n_{[\mu} \bar{m}_{\nu]} + 2i\,\mathrm{Im}\,(\mu)\, m_{[\mu} \bar{m}_{\nu]}; \tag{A1.7}$$

$$m_{[\nu;\,\mu]} = -(\bar{\pi} + \tau)\, l_{[\mu} n_{\nu]} + (2i\,\mathrm{Im}\,(\gamma) + \bar{\mu})\, l_{[\mu} m_{\nu]} + + \bar{\lambda} l_{[\mu} \bar{m}_{\nu]} + (2i\,\mathrm{Im}\,(\varepsilon) - \rho)\, n_{[\mu} m_{\nu]} - \sigma n_{[\mu} \bar{m}_{\nu]} - (\bar{\alpha} - \beta)\, m_{[\mu} \bar{m}_{\nu]}; \tag{A1.8}$$

$$\bar{m}_{[\nu;\,\mu]} = -(\pi + \bar{\tau})\, l_{[\mu} n_{\nu]} + (-2i\,\mathrm{Im}\,(\gamma) + \mu)\, l_{[\mu} \bar{m}_{\nu]} + + \lambda l_{[\mu} m_{\nu]} + (-2i\,\mathrm{Im}\,(\varepsilon) - \bar{\rho})\, n_{[\mu} \bar{m}_{\nu]} - \bar{\sigma} n_{[\mu} m_{\nu]} - (\alpha - \bar{\beta})\, \bar{m}_{[\mu} m_{\nu]}. \tag{A1.9}$$

If we use the definition of the covariant differentiation operators along the direction of the vectors of a complex null tetrad

$$D = l^\mu \nabla_\mu, \quad \Delta = n^\mu \nabla_\mu, \quad \delta = m^\mu \nabla_\mu, \quad \bar{\delta} = \bar{m}^\mu \nabla_\mu, \tag{A1.10}$$

we can rewrite Eqs. (A1.1)-(A1.4) in the following form:

$$\begin{aligned}
&\text{a) } Dl^\mu = (\varepsilon + \bar{\varepsilon})\, l^\mu - \bar{k} m^\mu - k \bar{m}^\mu,\\
&\text{b) } \Delta l^\mu = (\gamma + \bar{\gamma})\, l^\mu - \bar{\tau} m^\mu - \tau \bar{m}^\mu,\\
&\text{c) } \delta l^\mu = (\bar{\alpha} + \beta)\, l^\mu - \bar{\rho} m^\mu - \sigma \bar{m}^\mu,\\
&\text{d) } \bar{\delta} l^\mu = (\alpha + \bar{\beta})\, l^\mu - \bar{\sigma} m^\mu - \rho \bar{m}^\mu;
\end{aligned} \tag{A1.11}$$

$$\begin{aligned}
&\text{a) } Dn^\mu = -(\varepsilon + \bar{\varepsilon})\, n^\mu + \pi m^\mu + \bar{\pi} \bar{m}^\mu,\\
&\text{b) } \Delta n^\mu = -(\gamma + \bar{\gamma})\, n^\mu + \nu m^\mu + \bar{\nu} \bar{m}^\mu,\\
&\text{c) } \delta n^\mu = -(\bar{\alpha} + \beta)\, n^\mu + \mu m^\mu + \bar{\lambda} \bar{m}^\mu,\\
&\text{d) } \bar{\delta} n^\mu = -(\alpha + \bar{\beta})\, n^\mu + \lambda m^\mu + \bar{\mu} \bar{m}^\mu;
\end{aligned} \tag{A1.12}$$

$$\begin{aligned}
&\text{a) } Dm^\mu = \bar{\pi} l^\mu - k n^\mu + (\varepsilon - \bar{\varepsilon})\, m^\mu,\\
&\text{b) } \Delta m^\mu = \bar{\nu} l^\mu - \tau n^\mu + (\gamma - \bar{\gamma})\, m^\mu,\\
&\text{c) } \delta m^\mu = \bar{\lambda} l^\mu - \sigma n^\mu + (\beta - \bar{\alpha})\, m^\mu,\\
&\text{d) } \bar{\delta} m^\mu = \bar{\mu} l^\mu - \rho n^\mu + (\alpha - \bar{\beta})\, m^\mu;
\end{aligned} \tag{A1.13}$$

$$\begin{aligned}
&\text{a) } D\bar{m}^\mu = \pi l^\mu - \bar{k} n^\mu + (\bar{\varepsilon} - \varepsilon)\, \bar{m}^\mu,\\
&\text{b) } \Delta \bar{m}^\mu = \nu l^\mu - \bar{\tau} n^\mu + (\bar{\gamma} - \gamma)\, \bar{m}^\mu,\\
&\text{c) } \delta \bar{m}^\mu = \mu l^\mu - \bar{\rho} n^\mu + (\bar{\alpha} - \beta)\, \bar{m}^\mu,\\
&\text{d) } \bar{\delta} \bar{m}^\mu = \lambda l^\mu - \bar{\sigma} n^\mu + (\bar{\beta} - \alpha)\, \bar{m}^\mu.
\end{aligned} \tag{A1.14}$$

The commutation relations for the operators D, Δ, δ, and $\bar{\delta}$ are given in the appendix to Chapter 3 [cf. (A3.1)].

Formulas describing the transformation of the spin coefficients (1.5.7) when a new choice of complex null tetrads (1.5.2) is made are given in [49].

CHAPTER 2

CONGRUENCES OF LIGHT RAYS IN MINKOWSKI SPACE AND PROPERTIES OF AN ELECTROMAGNETIC FIELD

2.1. Introduction

In this chapter we consider congruences of light rays in Minkowski space. This simplest case permits a better understanding of many properties and peculiarities of light congruences in curved space. Congruences of null curves in flat space are moreover of independent interest, in particular in connection with the so-called Kerr–Schild [43] solutions of the Einstein equations.*

In order to describe null congruences it is necessary to have a convenient parametric representation for an arbitrary lightlike vector. It is shown in Section 2.2 how in a given reference frame light rays can be placed in correspondence with points on the sphere S^2 so that they are uniquely determined by giving the complex stereographic coordinate ζ on the sphere. The motion of rays along the surface of the light cone which arises by means of a Lorentz transformation in going from one reference frame to another generates a conformal transfor-

* See Section 4.7 of this survey concerning Kerr–Schild metrics.

mation of the sphere S^2 onto itself [44]. The concepts of spin weight and conformal weight [44-46] considered in Section 2.2 and the operators $\eth_0$ and $\bar{\eth}_0$ introduced there [44, 47] turn out to be useful not only in describing null congruences, but also and primarily in describing the asymptotic properties of asymptotically flat spaces [47-49] and in particular, in studying the motion of sources of a gravitational field in an asymptotically flat space [16].

In Section 2.3 a parametric representation is given for an arbitrary congruence of null geodesics with dilation $\theta \neq 0$ in a flat space [50]. Parametrizations of null congruences are used in [43, 51-54, 74] which differ from the one considered in this section. In the simplest case, considered in Section 2.4, when the congruence has no shear or rotation it turns out that every similar congruence is formed from generators of light cones, the vertices of which lie on some world line. This assertion, the proof of which is given in Section 2.4, is well known, although an explicit derivation of it apparently does not exist in the literature.*

The last sections of this chapter are devoted to the algebraic classification of the electromagnetic field [56-62]. It is shown in these sections how the electromagnetic field generates principal null congruences in space which are associated with it. There exists a close relation between the properties of the field itself and the properties of the congruences it generates. The Mariot–Robinson theorem stating that a principal light congruence of an algebraically special solution of the Maxwell equation is shear-free and geodesic and the property of sequential degeneration of the electromagnetic field which are given in Section 2.7 have a direct and immediate analog in the more complicated case of a gravitational field.

In complete analogy with the case of an electromagnetic field, an algebraic classification of the Yang–Mills fields and of a nonlinear electrodynamic field can be given. The geometric properties of the principal light congruences of these fields can evidently be used to obtain new exact classical solutions in these theories.

2.2. Complex Stereographic Coordinates on the Sphere. Spin Weight, Conformal Weight, and the Operators $\eth_0$ And $\bar{\eth}_0$

Before turning to the description of congruences of null geodesics in flat space-time, we consider one of the possible ways of defining null vectors. Choose an orthonormal tetrad e^μ_m: $e^\mu_m e_{n\mu} = \eta_{mn}$ = diag (1, −1, −1, −1) at an arbitrary point O in Minkowski space. Let S be the sphere of unit radius in the three-dimensional space orthogonal to e^μ_0. On the surface of S we introduce the complex stereographic coordinates ζ, $\bar{\zeta}$ by means of the following construction (Fig. 2.1). Through an arbitrary point P of the surface S pass a straight line emanating from the point O' (the point of intersection of S with the vector e^μ_3) and consider the intersection of this line with plane Π spanned by the vectors e^μ_1 and e^μ_2. The point P of S is thereby put into correspondence with the pair of coordinates (x, y) on the plane Π. It turns out to be convenient to introduce the complex coordinate $\zeta = x + iy$ on Π. As a result of this construction, a smooth bijective correspondence is established between the unit sphere S and the extended complex plane, the point O' corresponding to the point at infinity.

The relation between the ordinary polar coordinates (ϑ, φ) on S and the complex stereographic coordinates introduced above is given by the equation

$$\zeta = \cot (\vartheta/2) e^{i\varphi}. \tag{2.1}$$

*On this point see Remark 5 in [55].

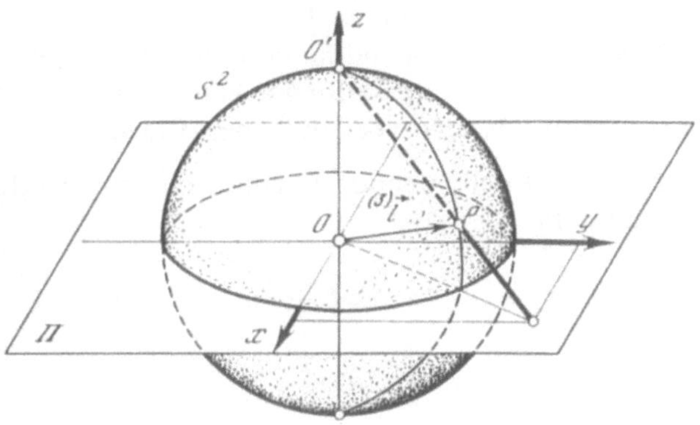

Fig. 2.1

We remark for what follows that the element of length on the surface of the unit sphere has the form

$$ds^2 = d\vartheta^2 + \sin^2\vartheta\, d\varphi^2 = d\zeta d\bar{\zeta}/P_0^2, \qquad P_0 = \frac{1}{2}(1 + \zeta\bar{\zeta}). \tag{2.2}$$

By means of a simple geometric argument it can be shown that the three-dimensional vector ${}^{(3)}\mathbf{l}$, extending from the center O of S to a point P with stereographic coordinates ζ, $\bar{\zeta}$ is given by the expression

$${}^{(3)}\mathbf{l} = \frac{1}{2P_0}(\zeta + \bar{\zeta}, i(\bar{\zeta} - \zeta), -1 + \zeta\bar{\zeta}), \tag{2.3}$$

so that an arbitrary lightlike 4-vector $\hat{l}^\mu$ normalized by the condition $\hat{l}_\mu e_0^\mu = 1/\sqrt{2}$, has the form

$$\hat{l}^\mu = \frac{1}{2\sqrt{2}P_0}(1 + \zeta\bar{\zeta}, \zeta + \bar{\zeta}, i(\bar{\zeta} - \zeta), -1 + \zeta\bar{\zeta}). \tag{2.4}$$

The lightlike vector $\hat{l}^\mu$ just constructed can be completed to a complex null tetrad $\hat{z}_m^\mu$ as follows:

$$\begin{aligned}
\hat{m}^\mu &= 2P_0\partial_\zeta\hat{l}^\mu = \frac{1}{2\sqrt{2}P_0}(0, 1 - \bar{\zeta}^2, -i(1 + \bar{\zeta}^2), 2\bar{\zeta}),\\
\hat{\bar{m}}^\mu &= 2P_0\partial_{\bar{\zeta}}\hat{l}^\mu = \frac{1}{2\sqrt{2}P_0}(0, 1 - \zeta^2, i(1 + \zeta^2), 2\zeta),\\
\hat{n}^\mu &= \hat{l}^\mu + 4P_0^2\partial_\zeta\partial_{\bar{\zeta}}\hat{l}^\mu = \frac{1}{2\sqrt{2}P_0}(1 + \zeta\bar{\zeta}, -(\zeta + \bar{\zeta}), i(\zeta - \bar{\zeta}), 1 - \zeta\bar{\zeta}).
\end{aligned} \tag{2.5}$$

The vectors Re $(\hat{m}^\mu)$ and Im $(\hat{m}^\mu)$ are tangent to the curves Im ζ = const and Re ζ = const on the surface of the sphere S, and therefore the vectors $\hat{m}^\mu$ and $\hat{\bar{m}}^\mu$ span the real two-dimensional space tangent to S at the point $(\zeta, \bar{\zeta})$. The vector $\hat{n}^\mu$ lies in the plane formed by the vectors e_0^μ and $\hat{l}^\mu$, and satisfies the condition $\hat{n}^\mu\hat{l}_\mu = 1$.

We remark that for a given choice of orthonormal tetrad e_m^μ, the complex stereographic coordinates on S and the complex null tetrads $\hat{z}_m^\mu(\zeta, \bar{\zeta})$ are uniquely determined. Assume now that in addition to the initial tetrad e_m^μ at the point O we are given a new orthonormal tetrad $e_m'^\mu$ obtained from e_m^μ by means of a Lorentz rotation, $e_m'^\mu = \Lambda_\nu^\mu e_m^\nu$ and let ζ', $\bar{\zeta}'$ be complex stereographic coordinates and $\hat{z}_m'^\mu(\zeta', \bar{\zeta}')$ the complex null tetrads corresponding to $e_m'^\mu$. If we choose an arbitrary lightlike vector l^μ, then

$$\hat{l}^\mu = l^\mu/(\sqrt{2}\, l_\mu e_0^\mu), \qquad \hat{l}'^\mu = l^\mu/(\sqrt{2}\, l_\mu e_0'^\mu) \tag{2.6}$$

are null vectors having the same direction as l^{μ} and appearing in the tetrads $\hat{z}^{\mu}_{m}$ and $\hat{z}'^{\mu}_{m}$. The vectors $\hat{l}^{\mu}$ and $\hat{l}'^{\mu}$ determine corresponding complex stereographic coordinates $(\zeta, \bar{\zeta})$ and $(\zeta', \bar{\zeta}')$, and hence an arbitrary Lorentz transformation Λ^{μ}_{ν} generates a certain transformation of the complex plane onto itself. It turns out that the following proposition holds.

Proposition 2.2.1. Every proper Lorentz transformation generates a transformation S → S' given by the formula

$$\zeta' = (a\zeta + b)/(c\zeta + d), \quad ad - bc = 1, \tag{2.7}$$

and moreover

$$\hat{l}'^{\mu} = Kl^{\mu}, \quad \hat{m}'^{\mu} = e^{i\lambda}(\hat{m}^{\mu} + H\hat{l}^{\mu}), \tag{2.8}$$

where

$$K = (1 + \zeta\bar{\zeta})[(a\zeta + b)(\bar{a}\bar{\zeta} + \bar{b}) + (c\zeta + d)(\bar{c}\bar{\zeta} + \bar{d})]^{-1}, \tag{2.9}$$

$$e^{i\lambda} = (c\zeta + d)/(\bar{c}\bar{\zeta} + \bar{d}), \quad \dot{H} = (1 + \zeta\bar{\zeta})\,\partial_{\zeta}(\ln K). \tag{2.10}$$

Moreover, the rotation subgroup SO(3) of the Lorentz group corresponds to the transformations

$$\zeta' = (a\zeta + b)/(\bar{a} - \bar{b}\zeta), \quad |a|^2 + |b|^2 = 1. \tag{2.11}$$

Proof. An arbitrary proper Lorentz transformation Λ^{ν}_{μ} can be written in the form [63]

$$\Lambda^{\nu}_{\mu} = \left\| \begin{matrix} \frac{1}{2}(a\bar{a} + b\bar{b} + c\bar{c} + d\bar{d}) & \mathrm{Re}\,(a\bar{b} + c\bar{d}) & -\mathrm{Im}\,(a\bar{b} + c\bar{d}) & \frac{1}{2}(a\bar{a} - b\bar{b} + c\bar{c} - d\bar{d}) \\ \mathrm{Re}\,(a\bar{c} + b\bar{d}) & \mathrm{Re}\,(a\bar{d} + b\bar{c}) & -\mathrm{Im}\,(a\bar{d} - b\bar{c}) & \mathrm{Re}\,(a\bar{c} - b\bar{d}) \\ \mathrm{Im}\,(a\bar{c} + b\bar{d}) & \mathrm{Im}\,(a\bar{d} + b\bar{c}) & \mathrm{Re}\,(a\bar{d} - b\bar{c}) & \mathrm{Im}\,(a\bar{c} - b\bar{d}) \\ \frac{1}{2}(a\bar{a} + b\bar{b} - c\bar{c} - d\bar{d}) & \mathrm{Re}\,(a\bar{b} - c\bar{d}) & -\mathrm{Im}\,(a\bar{b} - c\bar{d}) & \frac{1}{2}(a\bar{a} - b\bar{b} - c\bar{c} + d\bar{d}) \end{matrix} \right\|. \tag{2.12}$$

Taking the vector l^{μ} to be the vector $l^{\mu} = \hat{l}^{\mu}(\zeta, \bar{\zeta})$, we obtain that the components of l^{μ} in a new basis e'^{μ}_{m} have the form $l'^{\mu} = \Lambda^{\mu}_{\nu} l^{\nu}$. Therefore, using Eq. (2.4) to express $\hat{l}'^{\mu}$ in terms of the coordinates ζ', $\bar{\zeta}'$, we can obtain

$$\zeta' = (l'^1 + il'^2)/(l'^0 - l'^3). \tag{2.13}$$

Substituting the explicit form of the matrix Λ^{μ}_{ν} (2.12) into the expression for l'^{μ} and in terms of $\hat{l}^{\nu}$, it is possible to get from Eq. (2.13) that

$$\zeta' = (a\zeta + b)/(c\zeta + d).$$

We now observe that the matrices (2.12) describe a pure rotation if and only if [63]

$$a\bar{a} + b\bar{b} = 1, \quad c\bar{c} + d\bar{d} = 1, \quad a\bar{c} + b\bar{d} = 0,$$

i.e., if $d = \bar{a}$, $c = -\bar{b}$, and therefore the rotation subgroup SO(3) of the Lorentz group corresponds to the transformations (2.11). Equalities (2.8)-(2.10) are verified by direct calculation [44].

We remark that Eq. (2.7) describes a conformal transformation of the extended complex plane onto itself. Moreover, the element of length (2.2) is transformed into

$$ds'^2 = d\zeta' d\bar{\zeta}'/(K^2 P'^2_0), \tag{2.14}$$

where $P'_0 = \frac{1}{2}(1 + \zeta'\bar{\zeta}')$ and K is given by Eq. (2.9). The relation established in Proposition 2.2.1 between the Lorentz group and the group of transformations (2.7) is a concrete manifestation of the well-known fact that the group SL(2, C) is a twofold covering group of the Lorentz group.

Now consider a complex function $\eta(\zeta, \bar{\zeta})$ on the sphere S. We say that the function $\eta(\zeta, \bar{\zeta})$ has spin weight s and conformal weight w if under the coordinate transformation (2.7) the function η transforms as follows:

$$\eta(\zeta, \bar{\zeta}) \to \eta'(\zeta', \bar{\zeta}') = K^w e^{is\lambda} \eta(\zeta, \bar{\zeta}), \tag{2.15}$$

where K and $e^{i\lambda}$ are defined by Eqs. (2.8) and (2.9)

A representation of the Lorentz group is realized in the space of functions with fixed spin and conformal weights. A description of the properties of this representation is given in [44, 46].

If we restrict ourselves to the study of the transformations (2.11) corresponding to a pure rotation, then K = 1 and hence the components of the vectors $\hat{m}^\mu$ and $\hat{\bar{m}}^\mu$ viewed as sets of complex functions on the sphere have spin weight +1 for $\hat{m}^\mu$ and −1 for $\hat{\bar{m}}^\mu$. The components of the vectors $\hat{l}^\mu$ and $\hat{n}^\mu$ viewed as functions on the sphere have spin weight 0.

The differential operators ∂_ζ and $\partial_{\bar{\zeta}}$ can be used to construct the differential operators $\eth_0$ and $\bar{\eth}_0$ [44, 45, 47]; they act on functions with a given spin weight to give a function of spin weight one unit greater (for $\eth_0$) or one unit smaller (for $\bar{\eth}_0$) than the original spin weight. The action of $\eth_0$ and $\bar{\eth}_0$ on a quantity $\eta_s(\zeta, \bar{\zeta})$ of spin weight s is defined by the formula

$$\eth_0\eta_s(\zeta, \bar{\zeta}) = 2P_0^{1-s}\partial_\zeta(P_0^s\eta_s); \qquad \bar{\eth}_0\eta_s = 2P_0^{1+s}\partial_{\bar{\zeta}}(P_0^{-s}\eta_s). \tag{2.16}$$

These operators $\eth_0$ and $\bar{\eth}_0$ have the following easily verified properties:

$$\begin{aligned} &\text{a)}\ (\eth_0\bar{\eth}_0 - \bar{\eth}_0\eth_0)\,\eta_s = -2s\eta_s,\\ &\text{b)}\ \eth_0(A_sB_t) = \eth_0A_sB_t + A_s\eth_0B_t,\\ &\text{c)}\ \overline{(\eth_0A_s)} = \bar{\eth}_0(\bar{A}_s),\\ &\text{d)}\ \eth_0^2A_0 \equiv \eth_0\eth_0A_0 = 0, \qquad \bar{\eth}_0^2A_0 = 0. \end{aligned} \tag{2.17}$$

Using the operators $\eth_0$ and $\bar{\eth}_0$, it is convenient to rewrite Eqs. (2.5) in the following equivalent form:

$$\hat{m}^\mu = \eth_0\hat{l}^\mu, \qquad \hat{\bar{m}}^\mu = \bar{\eth}_0\hat{l}^\mu, \qquad \hat{n}^\mu = \hat{l}^\mu + \eth_0\bar{\eth}_0\hat{l}^\mu. \tag{2.18}$$

It moreover follows from Eq. (2.17d) that

$$\eth_0\hat{m}^\mu = \bar{\eth}_0\hat{\bar{m}}^\mu = 0. \tag{2.19}$$

2.3. Congruences of Light Geodesics in Minkowski Space

In this section we will give a description of an arbitrary congruence of light rays in flat space-time. The light geodesics in flat Minkowski space are the straight lines. Consider an arbitrary congruence C of light rays and choose some timelike curve γ: $x^{\mu} = z^{\mu}(u)$, where u is the proper time along the curve. Choose an orthonormal tetrad $e^{\mu}_{m}(u_0)$ at the point $x^{\mu}_0 = z^{\mu}(u_0)$ and using the parallel transport operator construct the field of tetrads $e^{\mu}_{m}(u)$ along the world line γ. The generators of the light cones with vertices on the line γ make up a congruence $\tilde{C}$ of special null geodesics which in the general case does not coincide with the given congruence C. Each generator of a light cone in the family $\tilde{C}$ is uniquely determined by giving the value u which fixes the position of the vertex of the light cone on the lightlike curve γ, and by giving the parameters ζ, $\bar{\zeta}$ defining the direction of the ray.* Each such generator is uniquely associated with the null surface tangent to the light cone along the generator, and therefore the parameters u, ζ, $\bar{\zeta}$ uniquely define such a hyperplane.

Two arbitrary null hyperplanes either do not meet (and then they are parallel) or else intersect along a two-dimensional spacelike plane. Therefore, an arbitrary null geodesic cannot simultaneously belong to two light hyperplanes. On the other hand, there exists a light hyperplane (u, ζ, $\bar{\zeta}$) which contains a given light line λ. This can be seen as follows. Construct a field of vectors parallel to λ along γ and consider the family of null hyperplanes defined by this vector field. This family is formed by nonintersecting hyperplanes and fills up all of space-time.† Therefore the ray λ intersects one of the null hyperplanes and hence is completely contained in it.

The position of a light ray $\lambda \in C$ can be described by giving a vector η^{μ} lying in a hyperplane and joining a point on the ray λ with the vertex of a light cone lying on γ (Fig. 2.2). This vector is fixed uniquely by the condition $\eta_{\mu} e^{\mu}_{0}(u) = 0$. Thus a null geodesic λ in the family C is

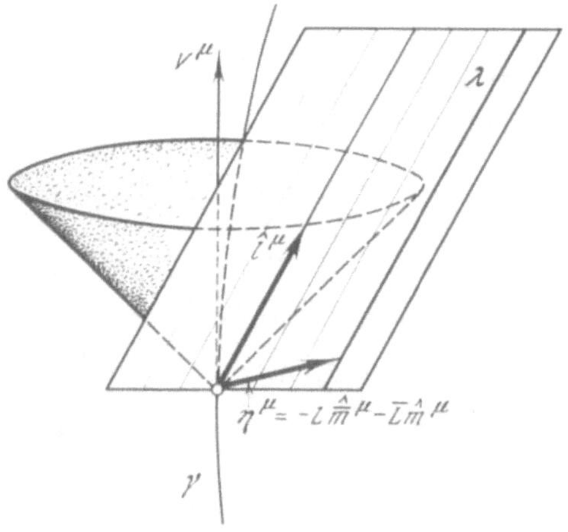

Fig. 2.2

---

* The complex stereographic coordinates ζ, $\bar{\zeta}$ are uniquely determined by the orthonormal tetrad $e^{\mu}_{m}(u)$.

† This is valid for timelike curves γ for which the past $I^{-}[\gamma]$ [10] coincides with all of Minkowski space. It is precisely such curves γ that we are considering.

given parametrically in the form

$$x^{\mu} = z^{\mu}(u) + \eta^{\mu} + (r - r_0)\hat{l}^{\mu}(\zeta, \bar{\zeta})/V, \tag{3.1}$$

where $V = V^{\mu}(u)\hat{l}_{\mu}(\zeta, \bar{\zeta})$, $V^{\mu} = dz^{\mu}/du$, r is a canonical parameter along λ, and $r_0 = r_0(u, \zeta, \bar{\zeta})$ determines the choice of canonical parameters. The vector η^{μ} lies in the two-dimensional space spanned by the vectors $\hat{m}^{\mu}$ and $\hat{\bar{m}}^{\mu}$, and therefore

$$\eta^{\mu} = -L\hat{\bar{m}}^{\mu} - \bar{L}\hat{m}^{\mu}.$$

A congruence of light rays is completely determined if we describe the way in which the vector η^{μ} depends on the hyperplane, i.e., if we give the complex function $L(u, \zeta, \bar{\zeta})$. Thus an arbitrary congruence of light rays in flat space-time possessing a dilation $\theta \neq 0$* can be rewritten in the form [50]

$$x^{\mu} = z^{\mu}(u) - L(u, \zeta, \bar{\zeta})\hat{\bar{m}}^{\mu}(\zeta, \bar{\zeta}) - \bar{L}(u, \zeta, \bar{\zeta})\hat{m}^{\mu}(\zeta, \bar{\zeta}) + (r - r_0(u, \zeta, \bar{\zeta}))\frac{\hat{l}^{\mu}(\zeta, \bar{\zeta})}{V(u, \zeta, \bar{\zeta})}, \tag{3.2}$$

where $V(u, \zeta, \bar{\zeta}) = V^{\mu}(u)\hat{l}_{\mu}(\zeta, \bar{\zeta})$.

Proposition 2.3.1. The congruence of light rays (3.2) in Minkowski space is shear-free ($\sigma = 0$) if and only if

$$\eth_0(VL) + L\dot{L} = 0. \tag{3.3}$$

The congruence of light rays (3.2) is normal ($\rho = \bar{\rho}$) if and only if

$$\mathrm{Im}\,[L\bar{\eth}_0 V + V\eth_0\bar{L} + L\dot{\bar{L}}] = 0. \tag{3.4}$$

Proof. In order to prove the proposition it suffices to calculate the spin coefficients $\sigma = l_{\mu;\nu}\hat{m}^{\mu}\hat{m}^{\nu}$ and $\rho = l_{\mu;\nu}\hat{m}^{\mu}\hat{\bar{m}}^{\nu}$, where $l^{\mu} = \partial x^{\mu}/\partial r = \hat{l}^{\mu}/V$ is the tangent vector to a light geodesic in the congruence. The covariant derivative

$$\hat{l}_{\mu;\nu} = \frac{1}{2P_0}(\hat{m}_{\mu}\zeta_{,\nu} + \hat{\bar{m}}_{\mu}\bar{\zeta}_{,\nu}) \tag{3.5}$$

can be calculated after using the relation $\partial\hat{l}^{\mu}/\partial\zeta = (2P_0)^{-1}\hat{\bar{m}}^{\mu}$ and taking into account the fact that in a flat space with rectilinear coordinate system the covariant derivative coincides with the partial derivative. Equation (3.5) allows us to obtain

$$\sigma = -\frac{1}{2P_0V}\hat{m}^{\nu}\bar{\zeta}_{,\nu}, \qquad \rho = -\frac{1}{2P_0V}\hat{\bar{m}}^{\nu}\bar{\zeta}_{,\nu}. \tag{3.6}$$

In order to find the quantity $\bar{\zeta}_{,\nu}$, we proceed as follows. Differentiating both sides of (3.2) with respect to x^{ν}, we obtain

$$\delta^{\mu}_{\nu} = p^{\mu}_1 u_{,\nu} + p^{\mu}_2 r_{,\nu} + p^{\mu}_3\zeta_{,\nu} + p^{\mu}_4\bar{\zeta}_{,\nu}, \tag{3.7}$$

* In the case of a congruence with $\theta = 0$, not just one ray but a whole family of rays of the congruence can lie in the light hyperplane $(u, \zeta, \bar{\zeta})$.

where

$$p_1^\mu = V^\mu - \dot{\bar{L}}\hat{m}^\mu - \dot{L}\hat{\bar{m}}^\mu - \dot{r}_0 \frac{\hat{l}^\mu}{V} - (r - r_0)\frac{\hat{l}^\mu}{V^2}\dot{V},$$

$$p_3^\mu = \frac{1}{2P_0}\left[-\eth_0\bar{L}\hat{m}^\mu - \eth_0 L\hat{\bar{m}}^\mu - L\eth_0\hat{\bar{m}}^\mu - \eth_0 r_0\frac{\hat{l}^\mu}{V} + (r - r_0)\frac{\hat{m}^\mu}{V} - (r - r_0)\frac{\hat{l}^\mu}{V^2}\eth_0 V\right],$$

$$p_2^\mu = \hat{l}^\mu/V, \qquad p_4^\mu = \bar{p}_3^\mu.$$

It can be seen that the vector $R_\mu = A\hat{l}_\mu + B\hat{m}_\mu + C\hat{\bar{m}}_\mu$, where

$$A = -V^{-1}[B(\dot{L} + \eth_0 V) + C(\dot{\bar{L}} + \bar{\eth}_0 V)],$$
$$B = -V\eth_0\bar{L} + r - r_0 - \dot{\bar{L}}L - L\bar{\eth}_0 V,$$
$$C = L\dot{L} + V\eth_0 L + L\eth_0 V,$$

satisfies the condition $R_\mu p_1^\mu = R_\mu p_2^\mu = R_\mu p_3^\mu = 0$. We therefore obtain from Eq. (3.7)

$$\bar{\zeta}_{,\nu} = R_\nu/(R_\mu p_4^\mu).$$

A congruence of rays will be shear-free if σ vanishes, i.e., the condition $\hat{m}^\nu R_\nu = 0$ holds, or what is the same, $C = 0$, and Eq. (3.3) is thereby proved.

We obtain analogously, after checking that $R_\mu p_4^\mu$ is a real quantity, that the normality condition $\mathrm{Im}\,\rho = 0$ for the congruence leads to the equality $\mathrm{Im}\,B = 0$, which is equivalent to Eq. (3.4). Proposition 2.3.1 is thereby completely proved.

2.4. Congruences of Light Rays Associated with a World Line. Newman – Unti Coordinates

The simplest special case of a congruence of light rays in Minkowski space is the family of rays described by Eq. (3.2) for $L = 0$. According to Proposition 2.3.1, such a congruence is normal and shear-free. The light rays of this congruence are the generators of light cones whose vertices lie on the world line $x^\mu = z^\mu(u)$. It turns out that even in the most general case, a shear-free normal diverging congruence of light rays in flat space-time is based on some world line. Indeed, we have the following proposition.

Proposition 2.4.1. Let a lightlike vector field l^μ satisfying the following conditions be given in Minkowski space: a) the field l^μ is tangent to a congruence $\Gamma(\mathbf{l})$ of null geodesics; b) the congruence $\Gamma(\mathbf{l})$ is shear-free, i.e., $\sigma = 0$; c) the congruence $\Gamma(\mathbf{l})$ is normal, i.e., $\mathrm{Im}\,\rho = 0$; d) the divergence of $\Gamma(\mathbf{l})$ is different from zero, $\rho \neq 0$. In this case the vectors l^μ coincide with the generators of light cones whose vertieces lie on some world line $z^\mu(u)$.

Proof. It was proved in Chapter 1 that a normal congruence of null geodesics determines a one-parameter family of null hypersurfaces Σ_c, defined by the equation $u(x^\mu) = c =$ const and such that the light rays of the congruence lie completely on these hypersurfaces and $l_\mu = Au_{,\mu}$.

Consider the intersection S of one such null hypersurface $u(x^\mu) = c$ with the hyperplane Π: $x^0 = t =$ const (Fig. 2.3). The two-dimensional surface S is described by the system of equations $x^0 = t$, $u(t, x^i) = c$. Let M be a point on the surface S and l^μ a light vector in the congruence at this point. Let H denote the hyperplane tangent to Σ_c along l^μ and λ the intersection of H with Π. In the two-dimensional space λ choose a pair of orthonormal vectors a^μ and b^μ and form the complex lightlike vectors $m^\mu = (a^\mu + ib^\mu)/\sqrt{2}$ and $m^\mu = (a^\mu - ib^\mu)/\sqrt{2}$ orthogonal to l^μ. We have from the definition of the spin coefficients (1.5.7) that $\rho = l_{\mu;\nu}m^\mu\bar{m}^\nu$ and $\sigma = l_{\mu;\nu}m^\mu m^\nu$,

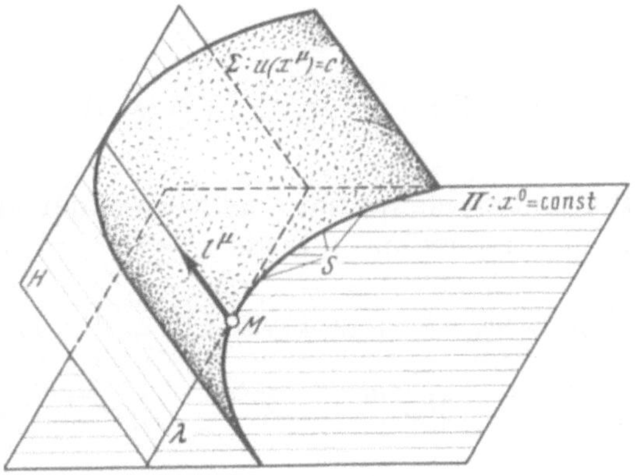

Fig. 2.3

As is easily seen, the conditions $\rho \neq 0$ and $\sigma = 0$ mean in the present case that

$$u_{,ij}\,(a^i a^j + b^i b^j) \neq 0, \tag{4.1}$$

$$u_{,ij}\,(a^i a^j - b^i b^j) = 0, \qquad u_{,ij} a^i b^j = 0. \tag{4.2}$$

These relations show that the paraboloid tangent to the surface S at the point M is circular, and hence the point M of S is a spherical point. By a well-known theorem [64], if S is a two-dimensional surface such that every point is shperical, then S is a sphere. Thus the surface S is a sphere and hence the surface $u(x^\mu) = c$ is a cone. The one-parameter family of light cones $u = c$ defines a world line formed by the vertices of the cones as c varies. If for different c the surfaces $u = c$ do not intersect, the corresponding world line is timelike.

Thus the general form of a shear-free normal divergent congruence of light rays in flat space-time is given by the formula

$$x^\mu = z^\mu(u) + r\,\frac{\hat{l}^\mu(\zeta,\bar{\zeta})}{V(u,\zeta,\bar{\zeta})}, \qquad V(u,\zeta,\bar{\zeta}) = V^\mu(u)\,\hat{l}_\mu(\zeta,\bar{\zeta}), \tag{4.3}$$

where $V^\mu = \partial z^\mu/\partial u$.

Relation (4.3) may also be considered as a formula describing the transition from Cartesian coordinates x^μ to curvilinear coordinates $u, r, \zeta, \bar{\zeta}$. If the coordinates u, r vary over the region $-\infty < u < \infty$, $0 \leq r < \infty$ and $\zeta, \bar{\zeta}$ run over the extended complex plane, then all of Minkowski space is described. In these coordinates, $r = 0$ corresponds to a timelike curve $z^\mu(u)$. Such coordinates were considered by Newman and Unti in [65]. In what follows it turns out to be convenient to obtain an expression for the line element of flat space in these coordinates. Observe that from Eq. (4.3) we have

$$dx^\mu = \left(V^\mu - r\,\frac{\hat{l}^\mu}{V^2}\,\dot{V}\right) du + \frac{\hat{l}^\mu}{V}\,dr + \frac{r}{2P_0 V}\left(\hat{m}^\mu - \frac{\eth_0 V}{V}\,\hat{l}^\mu\right) d\zeta + \frac{r}{2P_0 V}\left(\hat{\bar{m}}^\mu - \frac{\bar{\eth}_0 V}{V}\,\hat{l}^\mu\right) d\bar{\zeta}. \tag{4.4}$$

Hence the line element $ds^2 = \eta_{\mu\nu} dx^\mu dx^\nu$ in the Newman–Unti coordinates has the form

$$ds^2 = \left(1 - 2\,\frac{\dot{V}}{V}\,r\right) du^2 + 2\,du\,dr - \frac{r^2}{2P_0^2 V^2}\,d\zeta\, d\bar{\zeta}, \tag{4.5}$$

where

$$V=\frac{1}{\sqrt{2}(1+\zeta\bar{\zeta})}[(V^0-V^3)+(V^1-iV^2)\zeta+(V^1+iV^2)\bar{\zeta}+(V^0+V^3)\zeta\bar{\zeta}]. \tag{4.6}$$

Expression (4.5) for the line element allows us to consider along with the standard complex null tetrads $\hat{z}^{\mu}_{m}$ the field of complex null tetrads $z^{\mu}_{m}=(l^{\mu}, n^{\mu}, m^{\mu}, \bar{m}^{\mu})$ such that

$$\begin{gathered} l_\mu dx^\mu = du, \qquad n_\mu dx^\mu = dr + \left(\frac{1}{2}-\frac{\dot{V}}{V}r\right)du, \\ m_\mu dx^\mu = -\frac{r}{2P_0V}d\zeta, \qquad \bar{m}_\mu dx^\mu = -\frac{r}{2P_0V}d\bar{\zeta}. \end{gathered} \tag{4.7}$$

In contrast to $\hat{z}^{\mu}_{m}$, the relations

$$l_\mu V^\mu = 1, \qquad m_\mu V^\mu = \bar{m}_\mu V^\mu = 0 \tag{4.8}$$

are satisfied for the vectors of the tetrad z^{μ}_{m} and, as is easily seen, the relation between these two tetrads has the form

$$l^\mu = \frac{\hat{l}^\mu}{V}, \qquad n^\mu = V^\mu - \frac{\hat{l}^\mu}{2V}, \qquad m^\mu = \hat{m}^\mu - \frac{\eth_0 V}{V}\hat{l}^\mu. \tag{4.9}$$

Relation (4.4) permits us to obtain the following expression for the partial derivatives of the new coordinates u, r, ζ, $\bar{\zeta}$ with respect to the Cartesian coordinates:

$$\begin{aligned} u_{,\mu} &= l_\mu = \frac{\hat{l}_\mu}{V}, \\ r_{,\mu} &= n_\mu - \left(\frac{1}{2}-\frac{\dot{V}}{V}r\right)l_\mu = V_\mu - \left(1-\frac{\dot{V}}{V}r\right)\frac{\hat{l}_\mu}{V}, \\ \bar{\zeta}_{,\mu} &= -\frac{2P_0V}{r}m_\mu = -\frac{2P_0V}{r}\left[\hat{m}_\mu - \frac{\eth_0 V}{V}\hat{l}_\mu\right]. \end{aligned} \tag{4.10}$$

It is substantially more complicated to give an explicit description of congruences of light rays in Minkowski space in the case where the congruence is not normal ($\operatorname{Im}\rho \neq 0$, i.e., the congruence has rotation). An elegant method for describing all such congruences, i.e., congruences satisfying conditions (a), (b), and (d) of Proposition 2.4.1, is obtained in [50]. It is shown in that paper that a shear-free congruence of light rays with rotation in real Minkowski space can be obtained from a shear-free normal congruence in the complex extension of Minkowski space. The connection of congruences of light rays with twistors is considered in [66, 67].

2.5. The Electromagnetic Field of a Moving Charge

As a very simple example of the use of Newman–Unti coordinates, we consider in this section the radiation from an electric charge moving along the world line $z^{\mu}(u)$. In the Lorentz gauge the potential $A_{\mu}(x)$ satisfies the equation

$$\Box A_\mu(x) = 4\pi j_\mu, \tag{5.1}$$

where

$$j^\mu(x) = e\int_{-\infty}^{\infty} du V^\mu(u)\,\delta^4(x^\mu - z^\mu(u)) \tag{5.2}$$

and $V^{\mu} = dz^{\mu}/du$. Therefore

$$A_{\mu}(x) = 4\pi e \int d^4x' D^{\text{ret}}(x - x')\, j_{\mu}(x') = 2e \int d^4x' \int_{-\infty}^{\infty} du\, \theta(x^0 - x^{0'})\, \delta((x - x')^2) V_{\mu}(u)\, \delta(x'^{\mu} - z^{\mu}(u)). \tag{5.3}$$

The appearance of $\delta(x'^{\mu} - z^{\mu}(u))$ in the integrand shows that x'^{μ} lies on the line of motion of the charge. On the other hand, the factor $\delta((x - x')^2)$ means that only those points of the trajectory $z^{\mu}(u)$ for which $\Omega \equiv (x^{\mu} - z^{\mu}(u))(x_{\mu} - z_{\mu}(u)) = 0$ give a contribution to the integral. Since

$$d\Omega/du = -2(x^{\mu} - z^{\mu}(u)) V_{\mu} = -2r,$$

we obtain

$$A^{\mu}(x) = eV^{\mu}(u)/r \tag{5.4}$$

the well-known expression for the Liénard–Wiechert potentials.

If we use Eqs. (4.10), the expression for the intensity $F_{\mu\nu} = \partial_{\mu} A_{\nu} - \partial_{\nu} A_{\mu}$ of the field produced by a point charge can be written in the form

$$F_{\mu\nu} = -2e\left(\frac{1}{Vr}\dot{V}_{[\mu}\hat{l}_{\nu]} + \frac{(1 - r\dot{V}/V)}{Vr^2} V_{[\mu}\hat{l}_{\nu]}\right).$$

It is easy to obtain the following expressions:

$$\Phi_0 = 0, \qquad \Phi_1 = -\frac{e}{2r^2}, \qquad \Phi_2 = -\frac{eV}{r}\partial_0\left(\frac{\dot{V}}{V}\right). \tag{5.5}$$

for the tetrad components of $F_{\mu\nu}$ in the complex null tetrad z^{μ}_{m}*

$$\Phi_0 = F_{\mu\nu} l^{\mu} m^{\nu}, \qquad \Phi_1 = \frac{1}{2} F_{\mu\nu}(l^{\mu} n^{\nu} + \bar{m}^{\mu} m^{\nu}), \qquad \Phi_2 = F_{\mu\nu} \bar{m}^{\mu} n^{\nu} \tag{5.6}$$

The quantity Φ_1 describes the Coulomb component of the field, while the component Φ_2 arises from the electric dipole radiation of an accelerated charge. If acceleration is absent ($\dot{V} = 0$) then $\Phi_2 = 0$.

We now calculate the radiation flux of the electromagnetic field from an accelerated moving charged particle. In order to do this we first of all remark that

$$F_{\alpha\beta} F^{\alpha\beta} = -2e^2/r^4$$

and therefore in the wave zone the second term in the expression for the energy-momentum tensor of the electromagnetic field

$$T_{\mu\nu} = \frac{1}{4\pi}\left(-F_{\mu\lambda} F^{\lambda}_{\nu} + \frac{1}{4}\eta_{\mu\nu} F_{\alpha\beta} F^{\alpha\beta}\right) \tag{5.7}$$

can be neglected. If we let $t_{\mu\nu}$ denote the part of $T_{\mu\nu}$ having order r^{-2} in the wave zone, we

---

*We remark that the three complex functions Φ_0, Φ_1, and Φ_2 uniquely determine six real components of the field $F_{\mu\nu}$. The corresponding formula giving an explicit expression for $F_{\mu\nu}$ in terms of the quantities Φ_0, Φ_1, and Φ_2 is given in (A3.12).

obtain

$$t_{\mu\nu} = -\frac{e^2}{4\pi r^2 V^2}[\dot{V}^\lambda \dot{V}_\lambda V^2 + V^\lambda V_\lambda \dot{V}^2]\, l_\mu l_\nu. \tag{5.8}$$

In order to express $t_{\mu\nu}$ in terms of the quantity Φ_2, we observe that if $A^\mu(u)$ and $B^\mu(u)$ are vector fields independent of ζ, $\bar{\zeta}$ and $A \equiv A^\mu \hat{l}_\mu$, $B \equiv B^\mu \hat{l}_\mu$, then

$$A_\mu B^\mu = A^\mu \eta_{\mu\nu} B^\nu = A^\mu (\hat{l}_\mu \hat{n}_\nu + \hat{l}_\nu \hat{n}_\mu - \hat{m}_\mu \hat{\bar{m}}_\nu - \hat{m}_\nu \hat{\bar{m}}_\mu) B^\nu = 2AB - \eth_0 A \bar{\eth}_0 B - \bar{\eth}_0 A \eth_0 B + A \eth_0 \bar{\eth}_0 B + B \eth_0 \bar{\eth}_0 A. \tag{5.9}$$

Using Eq. (5.9) to calculate the expressions $\dot{V}^\lambda \dot{V}_\lambda$ and $V^\lambda V_\lambda$ in Eq. (5.8) and observing that the orthogonality condition for the velocity and the acceleration, $V_\alpha \dot{V}^\alpha = 0$, gives

$$2V\dot{V} + \dot{V}\eth_0\bar{\eth}_0 V + V\eth_0\bar{\eth}_0\dot{V} = \eth_0 V\bar{\eth}_0\dot{V} + \eth_0\dot{V}\bar{\eth}_0 V, \tag{5.10}$$

we obtain finally

$$t_{\mu\nu} = \frac{1}{\pi}\Phi_2\bar{\Phi}_2 l_\mu l_\nu. \tag{5.11}$$

Thus in the wave zone the vector l^μ for the propagation of radiant energy from an accelerated charge coincides with the generator of the corresponding light cone with vertex on the world line of the charge, while the energy flux density is determined by the quantity Φ_2.

In [68] the problem is considered of the radiation of waves with spin s from a point source having multipole moment L and moving in an arbitrary manner; it is shown that the corresponding radiation contains the multipole moments from s to 2L + s.

2.6. Algebraic Properties of the Electromagnetic Field

The following important fact is explained in the study of massless fields (with spin s > 0): Each such field determines from one to 2s congruences of null curves in space. This fact is intimately related to the algebraic classification of massless fields and we defer a complete discussion to the following chapter, where such a classification is carried out in the framework of the spinor formalism. In this section we restrict ourselves to a consideration of the algebraic classification of the electromagnetic field tensor [59].

The algebraic structure of $F_{\mu\nu}$ is closely connected with the eigenvalue problem

$$F_{\mu\nu} l^\nu = \alpha l_\mu.$$

It is easy to see (for $\alpha \neq 0$) that the eigenvectors l^μ (if they exist) are necessarily lightlike. Such "eigenvectors" also arise in a more careful study of the classification problem and form congruences of null curves in a space where the field $F_{\mu\nu}$ is given. Although, as everywhere in this chapter, we consider flat space-time, in view of the purely algebraic character of the question all the assertions obtained in this section carry over directly to the case of an antisymmetric tensor field $F_{\mu\nu}$ in curved space also.

An antisymmetric second-order tensor is alternatively called a bivector. If $\varepsilon_{\alpha\beta\gamma\delta}$ is the totally antisymmetric Levi–Civita symbol, $\varepsilon_{0123} = 1$, then the bivector $\overset{*}{F}_{\alpha\beta}$ adjoint to the given $F_{\alpha\beta}$ is defined by the relation

$$\overset{*}{F}_{\alpha\beta} = \frac{1}{2}\varepsilon_{\alpha\beta\gamma\delta}F^{\gamma\delta}. \tag{6.1}$$

The bivector $F_{\alpha\beta}$ is called simple if there exists a pair of vectors a_α and b_β such that $F_{\alpha\beta} = a_{[\alpha}b_{\beta]}$. In the general case $F_{\alpha\beta}$ is not simple, although it can always be decomposed into a pair of simple bivectors [42, p. 46]

$$F_{\alpha\beta} = a_{[\alpha}b_{\beta]} + c_{[\alpha}d_{\beta]}. \tag{6.2}$$

Proposition 2.6.1. A bivector $F_{\alpha\beta}$ is simple if and only if any of the following conditions holds:

$$\begin{aligned} &\text{a)}\quad F_{\alpha[\beta}F_{\gamma\delta]} = 0,\\ &\text{b)}\quad F_{\alpha\beta}\overset{*}{F}{}^{\alpha\beta} = 0,\\ &\text{c)}\quad \det(F_{\alpha\beta}) = 0. \end{aligned} \tag{6.3}$$

Proof. The proof of necessity and sufficiency of condition (6.3a) can be found, e.g., in [42, p. 43]. The expression $F_{\alpha[\beta}F_{\gamma\delta]}$ is totally antisymmetric and hence it is equal to zero if and only if $\frac{1}{2}\varepsilon^{\alpha\beta\gamma\delta}F_{\alpha\beta}F_{\gamma\delta} = F_{\alpha\beta}\overset{*}{F}{}^{\alpha\beta}$ is zero.

Relation (6.3c) follows from the equality $\det(F_{\alpha\beta}) = (\frac{1}{2}F_{\alpha\beta}\overset{*}{F}{}^{\alpha\beta})^2$.

The transformation

$$F_{\alpha\beta} \to F_{\alpha\beta}(\vartheta) = \cos\vartheta F_{\alpha\beta} + \sin\vartheta \overset{*}{F}_{\alpha\beta} \tag{6.4}$$

is called a dual rotation. It is easy to show that

$$F_{\alpha\beta}(\vartheta)\overset{*}{F}{}^{\alpha\beta}(\vartheta) = F_{\alpha\beta}\overset{*}{F}{}^{\alpha\beta}\cos 2\vartheta - F_{\alpha\beta}F^{\alpha\beta}\sin\vartheta\theta, \tag{6.5}$$

$$F_{\alpha\beta}(\vartheta)F^{\alpha\beta}(\vartheta) = F_{\alpha\beta}F^{\alpha\beta}\cos 2\vartheta + F_{\alpha\beta}\overset{*}{F}{}^{\alpha\beta}\sin 2\vartheta. \tag{6.6}$$

These relations show that there always exists a ϑ_0 such that $F_{\alpha\beta}(\vartheta_0)\cdot\overset{*}{F}{}^{\alpha\beta}(\vartheta_0) = 0$, and hence using a dual rotation we can obtain a simple bivector $F_{\alpha\beta}(\vartheta_0)$.

A simple bivector $F_{\alpha\beta} = a_{[\alpha}b_{\beta]}$ determines the vectors a_α and b_β up to linear transformations; however, the two-dimensional plane $\Pi(\mathbf{a},\mathbf{b})$ spanned by a_α and b_β is uniquely determined by the bivector. The condition that the bivector be simple, $F_{\alpha\beta}$: $F_{\alpha\beta}\overset{*}{F}{}^{\alpha\beta} = 0$, is also at the same time the condition that the dual bivector $\overset{*}{F}{}^{\alpha\beta}$ be simple, and therefore

$$\overset{*}{F}_{\alpha\beta} = c_{[\alpha}d_{\beta]}.$$

Since $\overset{*}{F}_{\alpha\beta} = \frac{1}{2}\varepsilon_{\alpha\beta\gamma\delta}a^{[\gamma}b^{\delta]}$, we have $\overset{*}{F}_{\alpha\beta}a^\beta = \overset{*}{F}_{\alpha\beta}b^\beta = 0$ and hence the plane $\Pi(\mathbf{a},\mathbf{b})$ is orthogonal to the plane $\Pi(\mathbf{c},\mathbf{d})$, defined by the dual bivector.

Proposition 2.6.2. If $F_{\alpha\beta}$ is a simple bivector, there exists a pair of vectors p^α q^β such that $F_{\alpha\beta} = p_{[\alpha}q_{\beta]}$ and $p_\alpha q^\alpha = 0$.

Proof. Consider a simple bivector $F_{\alpha\beta} = a_{[\alpha}b_{\beta]}$. If $a_\alpha a^\alpha \neq 0$, then choosing $p_\alpha = a_\alpha$, $q_\beta = b_\beta - (a_\nu b^\nu)\, a_\beta/(a_\lambda a^\lambda)$, we obtain the desired pair. If $b_\beta b^\beta \neq 0$, we interchange a_α and b_β and again reduce $F_{\alpha\beta}$ to the indicated form. Finally, if $a_\alpha a^\alpha = b_\beta b^\beta = 0$, then we take p_α and q_β to be $p_\alpha = (a_\alpha - b_\alpha)/2$, $q_\beta = (a_\beta + b_\beta)/2$.

We now observe that if a simple bivector is represented in canonical form $F_{\alpha\beta} = p_{[\alpha}q_{\beta]}$, then $F_{\alpha\beta}F^{\alpha\beta} = \frac{1}{2}(p_\alpha p^\alpha)(q_\beta q^\beta)$ and therefore the following cases are possible.

1. $F_{\alpha\beta}F^{\alpha\beta} > 0$. Then both vectors p_α and q_β are spacelike and $\Pi(\mathbf{p},\mathbf{q})$ contains no lightlike vectors. In this case the simple bivector $F_{\alpha\beta}$ is said to be spacelike.

2. $F_{\alpha\beta}F^{\alpha\beta} < 0$. Then one of the vectors (e.g., p_α) is timelike and the other (q_β) is spacelike. There exists a pair of vectors l_α and n_α in the plane $\Pi(\mathbf{p}, \mathbf{q})$ such that $F_{\alpha\beta} = 2\,l_{[\alpha}n_{\beta]}$, $l_\alpha l^\alpha = n_\alpha n^\alpha = 0$, $l_\alpha n^\alpha = \tau \neq 0$; $F_{\alpha\beta}$ is said to be timelike.

3. $F_{\alpha\beta}F^{\alpha\beta} = 0$. Then one of the vectors (p_α) is lightlike while the other (q_β) is spacelike; p_α determines a unique lightlike direction in the plane $\Pi(\mathbf{p}, \mathbf{q})$; $F_{\alpha\beta}$ is said to be lightlike.

It is easy to verify that

$$F_{\alpha\beta}F^{\alpha\beta} = -\overset{*}{F}_{\alpha\beta}\overset{*}{F}^{\alpha\beta}. \tag{6.7}$$

This relation shows that if $F_{\alpha\beta}$ is a timelike (spacelike) simple bivector, then $\overset{*}{F}_{\alpha\beta}$ is a spacelike (timelike) simple bivector. If $F_{\alpha\beta}$ is a lightlike simple bivector, then so is $\overset{*}{F}_{\alpha\beta}$, and the two-dimensional spaces which they generate intersect along a common lighlike direction.

We now turn to the eigenvalue problem for an arbitrary bivector $F_{\alpha\beta}$. Using a duality rotation we construct a simple bivector $F_{\alpha\beta}(\vartheta_0)$ corresponding to $F_{\alpha\beta}$. If $F_{\alpha\beta}(\vartheta_0)$ is timelike, then

$$F_{\alpha\beta}(\vartheta_0) = 2l_{[\alpha}n_{\beta]}.$$

An eigenvector k^β (necessarily lightlike) of the bivector $F_{\alpha\beta}(\vartheta_0)$

$$F_{\alpha\beta}(\vartheta_0)\,k^\beta = \lambda k_\alpha \tag{6.8}$$

coincides with either l^β or n^β, and the corresponding eigenvalue $\lambda = \pm\tau$. Since any vector in the plane determined by $F_{\alpha\beta}(\vartheta_0)$ is orthogonal to $\overset{*}{F}_{\alpha\beta}$,

$$\overset{*}{F}_{\alpha\beta}(\vartheta_0)\,k^\beta = 0. \tag{6.9}$$

Equalities (6.8) and (6.9) can be written in the form

$$k_{[\gamma}F_{\alpha]\beta}(\vartheta_0)\,k^\beta = k_{[\gamma}\overset{*}{F}_{\alpha]\beta}(\vartheta_0)\,k^\beta = 0. \tag{6.10}$$

Since $F_{\alpha\beta}$ and $\overset{*}{F}_{\alpha\beta}$ can be expressed linearly in terms of $F_{\alpha\beta}(\vartheta_0)$ and $\overset{*}{F}_{\alpha\beta}(\vartheta_0)$, we obtain that the eigenvalue problem

$$k_{[\gamma}F_{\alpha]\beta}k^\beta = k_{[\gamma}\overset{*}{F}_{\alpha]\beta}k^\beta = 0 \tag{6.11}$$

in the case where $F_{\alpha\beta}(\vartheta_0)$ is a simple timelike bivector has two and only two distinct solutions.

If $F_{\alpha\beta}(\vartheta_0)$ is a simple spacelike bivector, then $\overset{*}{F}_{\alpha\beta}(\vartheta_0)$ is a simple timelike bivector and the analogous assertion again holds.

If $F_{\alpha\beta}(\vartheta_0)$ is a simple lightlike bivector then

$$F_{\alpha\beta}(\vartheta_0) = l_{[\alpha}a_{\beta]}, \quad \overset{*}{F}_{\alpha\beta}(\vartheta_0) = l_{[\beta}b_{\beta]}, \tag{6.12}$$

$$a_\alpha l^\alpha = b_\alpha l^\alpha = a_\alpha b^\alpha = l_\alpha l^\alpha = 0. \tag{6.13}$$

The vector $F_{\alpha\beta}(\vartheta_0)k^\beta$ lies in the plane $\Pi(\mathbf{l}, \mathbf{a})$ and hence the solution of the eigenvalue problem

$$F_{\alpha\beta}(\vartheta_0)\,k^\beta = \lambda k_\alpha \tag{6.14}$$

has the form $k^\beta = Al^\beta + Ba^\beta$. Substituting this expression into Eq. (6.14), we see that $\lambda = 0$ and $k^\beta \sim l^\beta$. We can see in a completely analogous way that l^β is the unique (up to a factor) solution of the equation

$$\overset{*}{F}_{\alpha\beta}(\vartheta_0)k^\beta = \overset{*}{\lambda}k_\alpha$$

and the corresponding eigenvalue is equal to 0. We remark that in order for the simple bivector $F_{\alpha\beta}(\vartheta_0)$ to be lightlike, it is necessary and sufficient that both the field invariants

$$I = F_{\alpha\beta}F^{\alpha\beta}, \quad J = F_{\alpha\beta}\overset{*}{F}{}^{\alpha\beta}$$

vanish.

By what has been proved above, we can formulate the following proposition.

<u>Proposition 2.6.3.</u> The eigenvalue problem

$$k_{[\gamma}F_{\alpha]\beta}k^\beta = 0, \quad k_{[\gamma}\overset{*}{F}_{\alpha]\beta}k^\beta = 0 \tag{6.15}$$

has two and only two solutions (lightlike directions k^β) if the complex invariant $K = {}^1/_2\,(F_{\alpha\beta}F^{\alpha\beta} + iF_{\alpha\beta}\overset{*}{F}{}^{\alpha\beta})$ is different from zero. If K = 0 then there exists only one solution of problem (6.15) and the corresponding vector k^β is lightlike. The lightlike eigenvectors k^β are called the principal null vectors of the bivector $F_{\alpha\beta}$.

In the case where K = 0 the field $F_{\alpha\beta}$ is said to be degenerate or algebraically special.

We now consider the problem of interpreting the principal null directions. In the reference frame of an observer having 4-velocity V^μ, the electric and magnetic field vectors are determined by the relations $E_\mu = F_{\mu\nu}V^\nu$ and $H_\mu = -\overset{*}{F}_{\mu\nu}V^\nu$. These vectors have the components $E^\mu = (0, \mathbf{E})$, $H^\mu = (0, \mathbf{H})$. The complex invariant K is equal to

$$K = (|\mathbf{H}|^2 - |\mathbf{E}|^2) - 2i\,(\mathbf{E}\mathbf{H}).$$

If $K \neq 0$, there exists a reference frame V^μ in which **E** is parallel to **H** (one of the vectors may be zero). If light rays are directed in this reference frame along the common direction of **E** and **H** and in the opposite direction, the corresponding rays determine two null vectors l^α and n^α. These uniquely determined null vectors are two independent solutions of problem (6.15).

In the algebraically special case, both the invariants I and J vanish and we are dealing with a plane electromagnetic wave. The only principal light vector which exists in this case coincides with the light vector of the propagation of the wave.

2.7. The Mariot – Robinson Theorem. The Property of Sequential Degeneracy

The algebraic type of a bivector is determined by whether its eigen (principal) light directions coincide or not. A bivector of general type is denoted by the symbol [1, 1] when its principal light directions do not coincide. If the bivector is algebraically degenerate, we say that it belongs to type [2]. The geometric characteristics (in particular, the optical scalars) of congruences of principal null directions generated in space-time by the field of the bivector make it possible to carry out the classification of such fields in more detail. This classification takes into account not only the algebraic but also the differential characteristics of the fields. In such a classification, the bivector fields are divided into classes, a single class containing fields of the same algebraic type which are characterized by the vanishing or nonvanishing of certain optical scalars.

In the general case of an arbitrary tensor field $F_{\alpha\beta}$, the possible classes of this type are, generally speaking, nonempty. However, in the case where $F_{\alpha\beta}$ describes an electromagnetic field and is subject to the system of Maxwell equations, additional restrictions appear on the structure of the principal light congruences of the field. In particular we have the following important proposition.

Proposition 2.7.1. (The Mariot – Robinson theorem [69, 70]). A principal lightlike congruence of an algebraically special (type [2]) solution of the free Maxwell equations is geodesic ($k = 0$) and shear-free ($\sigma = 0$).

Proof. We remark first of all that for an algebraically special field $F_{\alpha\beta}$, the complex invariant $K = {}^1/_2\,(F_{\alpha\beta}F^{\alpha\beta} + i\overset{*}{F}{}^{\alpha\beta}F_{\alpha\beta}) = 0$ and hence by Proposition 2.6.1 the bivectors $F_{\alpha\beta}$ and $\overset{*}{F}_{\alpha\beta}$ are simple and lightlike, i.e., we may write

$$F^{+}_{\alpha\beta} \equiv \frac{1}{2}(F_{\alpha\beta} + i\overset{*}{F}_{\alpha\beta}) = \Phi l_{[\alpha}m_{\beta]}, \tag{7.1}$$

where l^{α} is the principal lightlike direction, $m_{\alpha}m^{\alpha} = \bar{m}_{\alpha}\bar{m}^{\alpha} = 0$, $m_{\alpha}\bar{m}^{\alpha} = -1$, $l^{\alpha}m_{\alpha} = l^{\alpha}\bar{m}_{\alpha} = 0$, and $\Phi \neq 0$ is a complex number. Therefore,

$$F^{+\alpha\beta}l_{\alpha} = F^{+\alpha\beta}m_{\alpha} = 0.$$

Differentiating these equalities with respect to x^{β} and taking into account the Maxwell equations in empty space

$$F^{+\alpha\beta}{}_{;\beta} = 0,$$

we obtain

$$F^{+\alpha\beta}l_{[\alpha;\beta]} = F^{+\alpha\beta}m_{[\alpha;\beta]} = 0. \tag{7.2}$$

Using Eqs. (A1.6) and (A1.8), it can be seen that

$$l^{\alpha}m^{\beta}l_{[\alpha;\beta]} = -k/2, \qquad l^{\alpha}m^{\beta}m_{[\alpha;\beta]} = -\sigma/2. \tag{7.3}$$

Substituting expression (7.1) for $F^{+\alpha\beta}$ into Eq. (7.2), we therefore obtain $k = \sigma = 0$.

Remark. The above theorem is also completely valid in the case of twisted space-time since the proof given did not use the Euclidean property of space-time anywhere.

The Mariot–Robinson theorem shows that if there exist type [2] solutions of the free Maxwell equations, then the principal lightlike congruences for these solutions are geodesic and shear-free. However, this theorem does not answer the question of whether there exist such solutions in general. This question can be formulated in a somewhat different way. Assume given an arbitrary bivector field $F_{\alpha\beta}$ of type [2]. We can then write [cf. Eq. (7.1)]

$$F^{+}_{\alpha\beta} = \Phi l_{[\alpha}m_{\beta]}, \tag{7.4}$$

where Φ is an arbitrary complex function, m_{β} a complex lightlike vector, and l_{α} the principal null direction of the bivector $F_{\alpha\beta}$. In which case is the field (7.4) a solution of the free Maxwell equations? The following proposition gives an answer to this question.

Proposition 2.7.2. (Robinson's Theorem [70]). In order for there to exist a solution of the free Maxwell equations of the form (7.4), it is necessary and sufficient that the congruence $\Gamma(\mathbf{l})$ be geodesic and shear-free.

As with any other classification,* the algebraic classification is simultaneously a means for finding exact solutions of the field equations. Such exact solutions having one set of algebraic properties or another can be sought on the basis of a description of the principal null congruences corresponding to them and given in Sections 2.3 and 2.4. The following proposition, for example, was proved in [55] by such a method.

Proposition 2.7.3. A solution of the Maxwell equations in flat space which is regular at infinity is a Liénard–Wiechert field if and only if one of the principal null congruences of the field is geodesic ($\kappa = 0$), shear-free ($\sigma = 0$), normal ($\mathrm{Im}\,\rho = 0$), and has dilation ($\mathrm{Re}\,\rho \neq 0$).

In [50] a complete description is given of the solutions of the free Maxwell equations satisfying all the conditions stated in the formulation of Proposition 2.7.3 except for the condition $\mathrm{Im}\,\rho = 0$ (absence of rotation).

The algebraic properties of the electromagnetic field tensor also turn out to be essential in studying the behavior of a field at large distances from the sources. It is well known [29] that in the presence of electromagnetic radiation the field in the wave zone has order r^{-1} and is reminiscent of the field of a plane wave. A plane wave is characterized by the vanishing of both field invariants $F_{\alpha\beta}F^{\alpha\beta} = \overset{*}{F}_{\alpha\beta}F^{\alpha\beta} = 0$ and in this case the tensor $F_{\alpha\beta}$ is algebraically special. The following theorem proved in [71] holds in complete accord with this fact.

Proposition 2.7.4. (Goldberg – Kerr Theorem). The retarded electromagnetic field from an isolated extended source has the following asymptotic behavior as $r \to \infty$ (the property of sequential degeneration):

$$F_{\mu\nu} = \frac{F^{(1)}_{\mu\nu}}{r} + \frac{F^{(2)}_{\mu\nu}}{r^2} + O(r^{-3}), \tag{7.5}$$

where r is an affine parameter along a principal null congruence of the field $F_{\mu\nu}$ and the fields $F^{(1)}_{\mu\nu}$ and $F^{(2)}_{\mu\nu}$ are independent of r and of type [2] and [1, 1], respectively.

We remark that the theorems presented in this section concerning the properties of principal null congruences, solutions of the free Maxwell equations, and the asymptotic behavior of these solutions admit a direct generalization to the case of free solutions of other massless fields and to the case of solutions of the Einstein equations. However, when the usual tensor methods are used the proof of such results requires rather tedious calculations. It turns out that the use of the spinor formalism permits an essential simplification of the necessary calculations. Therefore, before passing to the study of the solutions of the Einstein equations it is useful to give an exposition of the necessary results concerning spinor analysis.

CHAPTER 3

THE SPINOR FORMALISM. THE NEWMAN–PENROSE EQUATIONS

3.1. Introduction

The concept of spinor introduced by Cartan [75, 76] is fundamental in the description of the representations of the rotation group and the Lorentz group [63, 77]. Every tensor of general degree n turns out to be associated with a spinor having 2n indices. The relation between the spinors and tensors is given by means of the objects $\sigma_\mu^{AA'}$ introduced by Infeld and van

* For instance, the classification of solutions in terms of symmetry groups or holonomy groups.

der Waerden [78]. Spinor fields in the theory of general relativity have been considered by numerous authors* beginning with the paper [79] of Fock and Ivanenko.

Starting in 1960 after the appearance of Penrose's article [12], a new stage was set in the use of spinor methods for the study of the Einstein equations themselves. In this chapter we will mainly follow the papers [1, 5, 12] in the exposition of the spinor formalism and the description of its use in the classification of a gravitational field.

Passage to the spinor description usually gives the impression of a marked complication owing to the doubling of the number of indices for the objects studied. However, this first impression is deceptive and there is in fact an essential simplification achieved in the spinor description of various tensor quantities (and especially in the description of quantities characterizing massless fields). This is most explicitly evident in the study of the algebraic classification (cf. Section 3.5). Another interesting fact is also clarified in the study of the spinor formalism. It turns out that there is a unique correspondence between an arbitrary basis in spinor space and a complex lightlike basis in Minkowski space. This leads to the existence of a rather close connection between the spinor methods and the complex null tetrad formalism considered in the first chapter. In particular, the spin coefficients introduced there coincide with the quantities Γ_{abcd}, which are the components of the spinor connection.

The system of equations obtained by Newman and Penrose [1] which related the spin components of the curvature tensor to the spin coefficients and their derivatives turns out to be extremely convenient both for finding exact solutions of the Einstein equations of one or another algebraic class, and in the study of the general properties of a gravitational field (in particular the asymptotic properties).

3.2. Spinors. Algebraic Properties

The fundamental fact on which the spinor formalism is based is the local isomorphism remarked in the previous chapter between the proper Lorentz group and the group SL(2, **C**) of unimodular complex 2 × 2 matrices (i.e., matrices with determinant equal to one). Speaking more precisely, the group SL(2, **C**) is a twofold covering group of the Lorentz group. Unimodular matrices may also be viewed as linear operators in the two-dimensional complex linear space $\mathbf{C}^2$. We may therefore assume that an arbitrary Lorentz transformation is associated with a certain transformation in $\mathbf{C}^2$ (more precisely, two transformations). The corresponding representation of the Lorentz group is called a spin representation, and the elements (complex vectors) of the linear space $\mathbf{C}^2$ (the spinor space) are called spinors.

Let us consider spinor space in more detail. As in any linear space we can choose a basis (a pair of linearly independent spinors), and then an arbitrary spinor in spinor space is described by a pair of complex coordinates. There exists a natural difference between a spinor as a complete geometric entity and the set of coordinates determined by it in some basis. In order to reflect this difference we agree to denote a spinor, as geometric object, by Greek letters in the formulas, and as indices we agree to use capital Latin letters, e.g., ξ^A, η^B, etc. These indices characterize the geometric properties of these objects and do not take any numerical values. The usual linear operations on spinors are written in the form

$$\lambda^A = \alpha\xi^A + \beta\eta^A.$$

In what follows we agree to write S^0 for spinor space.

Let a pair ζ_a^A (a = 0, 1) of linearly independent spinors be chosen as basis of S^0. Then an arbitrary spinor ξ^A is determined by its coordinates ξ^a in this basis in the form $\xi^A = \xi^a\zeta_a^A$.

* On this subject cf. the papers [80-82] and the references given there.

In the same way as for the indices of the spinor coordinates, the indices of the basis vectors will be denoted by lower case Latin letters $a, b, \ldots$, taking the values 0, 1. As usual, the linear operations on spinors lead to corresponding linear operations on their coordinates.

In the theory of real linear spaces, in addition to the initial space an important role is played by the dual space of linear forms. The elements of this dual space are linear mappings of the vector space into the number line. Since S^0 is a complex linear space, two linear spaces S_0 and S_0' dual to it are defined in a natural way. The elements of S_0 are the linear forms on S^0, i.e., linear mappings of S^0 into **C**. These mappings (elements of S_0) are denoted by Greek letters with capital Latin letters as lower indices. For instance, ω^A: $S^0 \to \mathbf{C}$,

$$\omega_A(\alpha\xi^A + \beta\eta^A) = \alpha\omega_A(\xi^A) + \beta\omega_A(\eta^A).$$

The elements of the space S_0' are denoted by Greek letters with primed capital Latin letters as lower indices: $\omega_{A'}$, $\rho_{B'}$, ... The elements of the space S_0' are antilinear mappings of S^0 into **C**:

$$\omega_{A'}(\alpha\xi^A + \beta\eta^A) = \bar{\alpha}\omega_{A'}(\xi^A) + \bar{\beta}\omega_{A'}(\eta^A).$$

It is easy to define addition of two elements in either of the dual spaces, as well as multiplication by a complex number. S_0 and S_0' are thereby turned into two-dimensional complex linear spaces.

It turns out that in real linear spaces the space dual to the space of linear forms is canonically isomorphic to the original space. This isomorphism is easily established by remarking that the mapping $\omega_A(\xi^A)$ for a constant vector ξ^A and variable form ω_A can be regarded as a linear mapping of the space S_0 into the number line determined by the vector ξ^A. In the case of a complex linear space S^0 we can consider linear and antilinear mappings of each of the dual spaces S_0 and S_0' into **C**. As in the case of a real linear space it is easy to see that the space of linear forms on S_0 and the space of antilinear forms on S_0' are canonically isomorphic to S^0 and are therefore mutually isomorphic. It can be seen analogously that the space of antilinear forms on S_0 is isomorphic to the space of linear forms on S_0'. We denote this new two-dimensional complex linear space by $S^{0'}$ and its elements by Greek letters with primed capital Latin letters as upper indices, e.g., $\xi^{A'}$, $\eta^{B'}$, ... The relation among the various spinor spaces is shown schematically in Fig. 3.1, in which $\lambda(\lambda')$ indicates passage to a space of linear (antilinear) forms, and i' is an antiautomorphism.

The four types of spinors introduced above (the elements of the corresponding spinor spaces) can be used to define spin tensors in exactly the same way as vectors and forms are

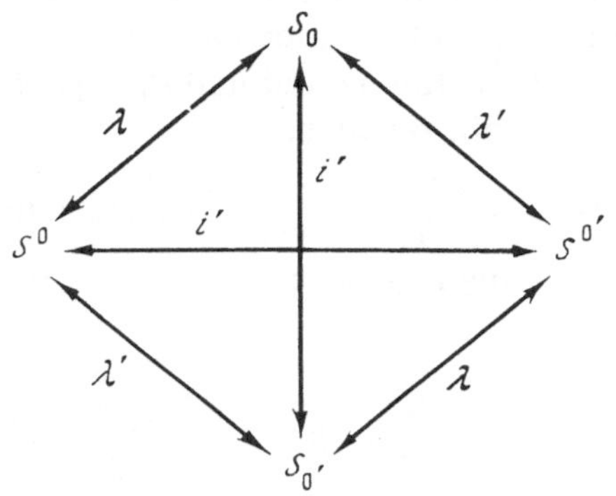

Fig. 3.1

used to define ordinary tensors in the theory of real linear spaces. We have the following definition in complete analogy with the usual coordinate-free definition of tensor [83-84].

Definition. A spin tensor $T^{A_1\cdots A_n B'_1\cdots B'_m}_{C_1\cdots C_l D'_1\cdots D'_p}$ of weight (n, l; m', p') is a multilinear (i.e., linear in each argument) mapping

$$T^{A_1\cdots A_n B'_1\cdots B'_m}_{C_1\cdots C_l D'_1\cdots D'_p}: \underbrace{S_0\times\cdots\times S_0}_{n}\times\underbrace{S^0\times\cdots\times S^0}_{l}\times\underbrace{S_{0'}\times\cdots\times S_{0'}}_{m}\times\underbrace{S^{0'}\times\cdots\times S^{0'}}_{p}\to \mathbf{C}. \tag{2.1}$$

We illustrate this definition using the example of a spin tensor $T^A_{D'}$, of type (1, 0; 0, 1). This spin tensor is uniquely determined by giving its values on an arbitrary pair of spinors ξ_A, $\eta^{D'}$

$$T^A_{D'}(\xi_A,\ \eta^{D'})\equiv T^A_{D'}\xi_A\eta^{D'}\in\mathbf{C}. \tag{2.2}$$

Moreover, the multilinearity property means that

$$T^A_{D'}(\alpha\xi_A+\beta\zeta_A,\ \eta^{D'})=\alpha T^A_{D'}\xi_A\eta^{D'}+\beta T^A_{D'}\zeta_A\eta^{D'},$$
$$T^A_{D'}(\xi_A,\ \alpha\eta^{D'}+\beta\zeta^{D'})=\alpha T^A_{D'}\xi_A\eta^{D'}+\beta T^A_{D'}\xi_A\zeta^{D'}.$$

The number, position, and type of the indices of the spin tensor define its type, and this is their only function. These indices do not take numerical values. The notation $T^A_{D'}\xi_A\eta^{D'}$, which is formally reminiscent of contraction, is conveniently written as $T^A_{D'}(\xi_A,\ \eta^{D'})$ and means that ξ_A is the first and $\eta^{D'}$ the second argument of the multilinear form $T^A_{D'}$. In what follows we will make use of this convention.

The usual rules of tensor algebra can be extended naturally to spin tensors. Spin tensors having the same weight can be added, and they form a complex linear space. The product of a spin tensor of weight (n_1, l_1; m'_1, p'_1) and a spin tensor of weight (n_2; l_2; m'_2, p'_2) is defined in the usual way as a new spin tensor of weight (n_1+n_2, l_1+l_2; $m'_1+m'_2$, $p'_1+p'_2$). The properties of symmetry and antisymmetry of a spin tensor with respect to a pair of abstract indices of the same type are defined as the properties of symmetry and antisymmetry of the corresponding multilinear mapping when the arguments associated with these indices are interchanged.

The existence of a complex structure permits us to introduce a new operation of complex conjugation (new as compared to the case of real linear spaces).

We recall that the elements of the spaces S_0 and S'_0 are defined as linear and antilinear forms on S^0. Let $\omega_A\in S_0$. Then for arbitrary $\xi^A\in S^0$, $\omega_A(\xi^A)\in\mathbf{C}$. Now consider the mapping which associates to ξ^A the value $\overline{\omega_A(\xi^A)}\in\mathbf{C}$. It is easy to see that this mapping is antilinear. If we denote it by $\bar\omega_{A'}\in S_{0'}$, we see that the operation of complex conjugation establishes a canonical isomorphism between S_0 and $S_{0'}$: $i': S_0\to S_{0'}$ such that $i':\omega_A\to\bar\omega_{A'}$. Complex conjugation leads completely analogously to a canonical isomorphism between S^0 and $S^{0'}$, $i': S^0\to S^{0'}$, under which $i':\xi^A\to\bar\xi^{A'}$. Having defined the operation of complex conjugation on spinors, it is easy to extend it to spin tensors.

Definition. A spin tensor $\bar T^{B_1\cdots B_m A'_1\cdots A'_n}_{D_1\cdots D_p C'_1\cdots C'_l}$ of weight (m, p; n', l') complex conjugate to the spin tensor $T^{A_1\cdots A_n B'_1\cdots B'_m}_{C_1\cdots C_l D'_1\cdots D'_p}$ of weight (n, l; m', p') is the multilinear mapping defined as follows:

$$\begin{aligned}&\bar T^{B_1\cdots B_m A'_1\cdots A'_n}_{D_1\cdots D_p C'_1\cdots C'_l}(\underset{1}{\xi}_{B_1},\ \dots,\ \underset{m}{\xi}_{B_m},\ \underset{1}{\eta}^{D_1},\dots,\ \underset{p}{\eta}^{D_p},\ \underset{1}{\bar\lambda}_{A'_1},\dots,\ \underset{n}{\bar\lambda}_{A'_n},\ \underset{1}{\bar\mu}^{C'_1},\dots,\ \underset{l}{\bar\mu}^{C'_l})=\\&=\overline{T^{A_1\cdots A_n B'_1\cdots B'_m}_{C_1\cdots C_l D'_1\cdots D'_p}(\underset{1}{\lambda}_{A_1},\dots,\ \underset{n}{\lambda}_{A_n},\underset{1}{\mu}^{C_1},\dots,\ \underset{l}{\mu}^{C_l},\underset{1}{\bar\xi}_{B'_1},\dots,\ \underset{m}{\bar\xi}_{B'_m},\underset{1}{\bar\eta}^{D'_1},\dots,\underset{p}{\bar\eta}^{D'_p})}.\end{aligned} \tag{2.3}$$

The discussion thus far has been concerned with the construction of the spinor algebra without using coordinates. In practical calculations one usually works with spinors and spin tensors in a definite choice of basis. Choose some basis ζ_a^A in S_A^0. Then the spinors $\bar{\zeta}_{a'}^{A'}$ complex conjugate to ζ_a^A form a dual basis in the space $S^{0'}$. We define a basis ζ_A^a in S_0 by the condition $\zeta_A^a \zeta_b^A = \delta_b^a$, and as basis in $S_{0'}$ we choose $\bar{\zeta}_{A'}^{a'}$. Thus the choice of a basis in S^0 uniquely determines bases in the remaining three spinor spaces which are associated with it.

The components of a spin tensor $T^{A_1 \ldots A_n B'_1 \ldots B'_m}_{C_1 \ldots C_l D'_1 \ldots D'_p}$ in the basis ζ_a^A chosen are defined as follows:

$$T^{a_1 \ldots a_n b'_1 \ldots b'_m}_{c_1 \ldots c_l d'_1 \ldots d'_p} = T^{A_1 \ldots A_n B'_1 \ldots B'_m}_{C_1 \ldots C_l D'_1 \ldots D'_p} \zeta_{A_1}^{a_1} \ldots \zeta_{A_n}^{a_n} \zeta_{c_1}^{C_1} \ldots \zeta_{c_l}^{C_l} \bar{\zeta}_{B'_1}^{b'_1} \ldots \bar{\zeta}_{B'_m}^{b'_m} \bar{\zeta}_{d'_1}^{D'_1} \ldots \bar{\zeta}_{d'_p}^{D'_p}. \tag{2.4}$$

The algebraic operations on spin tensors considered above (addition, multiplication by a number, tensor product, symmetrization, antisymmetrization) lead in the usual way to the corresponding operations on the components of a spin tensor in the chosen basis. The operation of complex conjugation takes a spin tensor T::: with coordinates in the basis ζ_a^A given by $T^{a_1 \ldots a_n b'_1 \ldots b'_m}_{c_1 \ldots c_l d'_1 \ldots d'_p}$, into the new spin tensor $\bar{T}$::: which in the basis ζ_a^A has the coordinates $\bar{T}^{b_1 \ldots b_m a'_1 \ldots a'_n}_{d_1 \ldots d_p c'_1 \ldots c'_l} = \left(\overline{T^{a_1 \ldots a_n b'_1 \ldots b'_m}_{c_1 \ldots c_l d'_1 \ldots d'_p}}\right)$. Let, for example, $Q^{AB'}$ have the coordinates $Q^{00'} = \alpha$, $Q^{01'} = \beta$, $Q^{10'} = \gamma$ and $Q^{11'} = \delta$ in the basis ζ_a^A. Then the complex spin tensor $P^{BA'} = \overline{Q}^{BA'}$ has the following coordinates in the same basis: $P^{00'} = \overline{\alpha}$, $P^{01'} = \overline{\gamma}$, $P^{10'} = \beta$, $P^{11'} = \delta$.

A spin tensor $T^{A_1 \ldots A_n B'_1 \ldots B'_m}_{C_1 \ldots C_l D'_1 \ldots D'_p}$ of weight (n, l; m', p') is said to be Hermitian if n = m, l = p, and

$$\bar{T}^{A_1 \ldots A_n B'_1 \ldots B'_m}_{C_1 \ldots C_l D'_1 \ldots D'_p} = T^{A_1 \ldots A_n B'_1 \ldots B'_m}_{C_1 \ldots C_l D'_1 \ldots D'_p}.$$

For example, the tensor $Q^{AB'}$ considered above is Hermitian if $\alpha = \overline{\alpha}$, $\delta = \overline{\delta}$, and $\beta = \overline{\gamma}$.

The operation of contraction for spin tensors is introduced as follows. Along with a spin tensor $T^{A_1 \ldots A_n B'_1 \ldots B'_m}_{C_1 \ldots C_l D'_1 \ldots D'_p}$, having components $T^{a_1 \ldots a_n b'_1 \ldots b'_m}_{c_1 \ldots c_l d'_1 \ldots d'_p}$, in the basis ζ_a^A, consider a new spin tensor $S^{A_2 \ldots A_n B'_1 \ldots B'_m}_{C_2 \ldots C_l D'_1 \ldots D'_p}$, whose components in the basis ζ_a^A are $S^{aa_2 \ldots a_n b'_1 \ldots b'_m}_{ac_2 \ldots c_l d'_1 \ldots d'_p}$. It is easy to verify that the spin tensor S::: is uniquely determined by this condition and does not depend on the choice of basis ζ_a^A. The spin tensor $S^{A_2 \ldots A_n B'_1 \ldots B'_m}_{C_2 \ldots C_l D'_1 \ldots D'_p}$ is called the contraction of $T^{A_1 \ldots A_n B'_1 \ldots B'_m}_{C_1 \ldots C_l D'_1 \ldots D'_p}$ with respect to the indices A_1 and C_1,

$$S^{A_2 \ldots A_n B'_1 \ldots B'_m}_{C_2 \ldots C_l D'_1 \ldots D'_p} = T^{AA_2 \ldots A_n B'_1 \ldots B'_m}_{AC_2 \ldots C_n D'_1 \ldots D'_p}.$$

Contraction with respect to a pair of primed indices is defined analogously.

We now clarify the way in which spin tensors transform under a transformation in spinor space induced by the action of the group SL(2, **C**). We first of all remark that as usual in the theory of linear spaces, two types of transformations (passive and active) are possible which arise in spinor space from the action of SL(2, **C**). In the first case we consider a basis transformation of the form

$$\zeta_a^A \to \eta_a^A = l_a^b \zeta_b^A, \tag{2.5}$$

defined by a unimodular matric l_a^b. The unimodularity condition (determinant of l_a^b equal to one) can be written as

$$\varepsilon_{ab} = \varepsilon_{cd} l_a^c l_b^d. \tag{2.6}$$

If along with a new basis η_a^A we also consider the bases dual to it in the spaces S_0, $S^{0'}$, $S_{0'}$: η_A^a, $\bar\eta_{a'}^{A'}$, $\bar\eta_{A'}^{a'}$, we then have the following relations:

$$\eta_A^a = L_b^a \zeta_A^b, \qquad \bar\eta_{a'}^{A'} = \bar l^{b'}{}_{a'} \bar\zeta_{b'}^{A'}, \qquad \bar\eta_{A'}^{a'} = \bar L_{b'}^{a'} \bar\zeta_{A'}^{b'}, \tag{2.7}$$

where

$$l_a^b L_b^c = \delta_a^c, \qquad \bar l^{b'}{}_{a'} = \overline{l_a^b}, \qquad \bar L_{b'}^{a'} = \overline{L_b^a}. \tag{2.8}$$

An arbitrary spin tensor $T^{A_1\dots A_n B'_1\dots B'_m}_{C_1\dots C_l D'_1\dots D'_p}$ having coordinates $T^{a_1\dots a_n b'_1\dots b'_m}_{c_1\dots c_l d'_1\dots d'_p}$ in the basis ζ_a^A has coordinates

$$L^{a_1}_{i_1}\dots L^{a_n}_{i_n} l^{f_1}_{c_1}\dots l^{f_l}_{c_l} \bar L^{b'_1}_{g'_1}\dots \bar L^{b'_m}_{g'_m} \bar l^{h'_1}_{d'_1}\dots \bar l^{h'_p}_{d'_p} T^{i_1\dots i_n g'_1\dots g'_m}_{f_1\dots f_l h'_1\dots h'_p}. \tag{2.9}$$

in the new basis $\eta_a^A = l_a^b \zeta_b^A$. Thus, every basis transformation (2.5) leads to a transformation (2.9) of coordinates of the spin tensor. Such a representation of the group SL(2, C) is called passive.

Along with the "passive" way of viewing the action of SL(2, C) there is an "active" way which consists in the following. We assume that the basis ζ_a^A remains unchanged. Then expression (2.9) defines a certain spin tensor $S^{A_1\dots A_n B'_1\dots B'_m}_{C_1\dots C_l D'_1\dots D'_p}$ in the basis ζ_a^A whose coordinates are given by relation (2.9). Every matrix l_a^b in SL(2, C) thus generates an active transformation in the space of spin tensors. Under this transformation each spinor ξ^A goes into a new spinor η^A. It is convenient to write this transformation in the form

$$\eta^A = L_B^A \xi^B, \tag{2.10}$$

where the L_B^A are linear operators in spinor space realizing a representation of SL(2, C). If we contract both sides of (2.10) with a spinor ω_A, we obtain that $L_B^A \omega_A \xi^B \in \mathbf{C}$ and therefore L_B^A is a spin tensor of type (1, 1; 0', 0').

The transformation $l_a^b \in SL(2, \mathbf{C})$ defines active transformations

$$\eta_A = l_A{}^B \xi_B, \qquad \eta^{A'} = \bar L^{A'}{}_{B'} \xi^{B'}, \qquad \eta_{A'} = \bar l_{A'}^{B'} \xi_{B'}, \tag{2.11}$$

acting respectively in the spin spaces S_0, $S^{0'}$, and $S_{0'}$, with

$$l_A{}^B L_B{}^C = \delta_A^C \tag{2.12}$$

and $\bar L_{B'}^{A'}$, $\bar l_{A'}^{B'}$ complex conjugate to the spin tensors L_B^A and l_A^B. (The spin tensor δ_A^B is defined as the spin tensor having the components δ_a^b in an arbitrary basis ζ_a^A.)

A general "active" transformation of the spin tensor $T^{A_1\dots A_n B'_1\dots B'_m}_{C_1\dots C_l D'_1\dots D'_p}$ having the form (2.9) in the basis ζ_a^A can be written as follows in coordinate-free form:

$$L^{A_1}_{I_1}\dots L^{A_n}_{I_n} l^{F_1}_{C_1}\dots l^{F_l}_{C_l} \bar L^{B'_1}_{G'_1}\dots \bar L^{B'_m}_{G'_m} \bar l^{H'_1}_{D'_1}\dots \bar l^{H'_p}_{D'_p} T^{I_1\dots I_n G'_1\dots G'_m}_{F_1\dots F_l H'_1\dots H'_p}. \tag{2.13}$$

If we define the spin tensor ε_{AB} in the basis ζ_a^A by the relation

$$\varepsilon_{ab} = \varepsilon_{AB}\zeta_a^A\zeta_b^B, \tag{2.14}$$

where ε_{ab} is totally antisymmetric, then Eq. (2.6) shows that the components of ε_{AB} do not change under a transformation $l_a^b \in SL(2, C)$ of the basis vectors. On the other hand, if we view l_a^b as an active transformation, Eq. (2.6) shows that the spin tensor ε_{AB} remains invariant under active transformations, i.e.,

$$\varepsilon_{AB} = l_A^C l_B^D \varepsilon_{CD}. \tag{2.15}$$

The invariant spin tensor ε_{AB} in spin space plays a role analogous to the tensor $\eta_{\mu\nu} = \mathrm{diag}(1, -1, -1, -1)$ in Minkowski space. In addition to ε_{AB}, invariant spin tensors ε^{AB}, $\varepsilon_{A'B'}$, and $\varepsilon^{A'B'}$ are defined completely analogously which satisfy the conditions

$$\begin{gathered}\varepsilon^{AB} = L_C^A L_D^B \varepsilon^{CD}, \quad \varepsilon_{A'B'} = \bar{l}_{A'}^{C'} \bar{l}_{B'}^{D'} \varepsilon_{C'D'}, \\ \varepsilon^{A'B'} = \bar{L}_{C'}^{A'} \bar{L}_{D'}^{B'} \varepsilon^{C'D'}, \quad \varepsilon^{AB}\varepsilon_{BC} = \delta_C^A \quad \varepsilon^{A'B'}\varepsilon_{B'C'} = \delta_{C'}^{A'}.\end{gathered} \tag{2.16}$$

The invariant spin tensors ε_{AB} and ε^{AB} ($\varepsilon_{A'B'}$, and $\varepsilon^{A'B'}$) can be used to establish a canonical isomorphism between S_0 and S^0 (between $S_{0'}$ and $S^{0'}$). In other words, these spin tensors can be used to raise and lower abstract indices and hence in what follows, instead of studying general spin tensors of type (n, l; m', p') we may restrict ourselves to considering spin tensors of type (0, n + l; 0, m' + p'), i.e., spin tensors with lower indices only. For simplicity in what follows we will speak of such spin tensors as being spinors of type (n, q) if the spin tensor has n unprimed and q primed indices.

The operations of raising and lowering indices are given by the relations

$$\xi_A = \xi^B \varepsilon_{BA}, \quad \xi^B = \varepsilon^{BA}\xi_A, \quad \xi_{A'} = \xi^{B'}\varepsilon_{B'A'}, \quad \xi^{B'} = \varepsilon^{B'A'}\xi_{A'}. \tag{2.17}$$

The skew symmetry of the spinor ε_{AB} requires a certain amount of care in calculations. In particular, it should be remembered that

$$\xi_A\eta^A = \xi^B\varepsilon_{BA}\eta^A = -\eta^A\varepsilon_{AB}\xi^B = -\eta_A\xi^A. \tag{2.18}$$

Since any completely antisymmetric object with a number of indices greater than the dimension of the space is necessarily equal to zero, we have the relation

$$3\varepsilon_{A[B}\,\varepsilon_{CD]} \equiv \varepsilon_{AB}\,\varepsilon_{CD} + \varepsilon_{AC}\varepsilon_{DB} + \varepsilon_{AD}\varepsilon_{BC} = 0. \tag{2.19}$$

Raising the indices C and D, we obtain

$$\varepsilon_{AB}\varepsilon^{CD} = \delta_A^C\delta_B^D - \delta_A^D\delta_B^C. \tag{2.20}$$

If we apply this formula to an arbitrary spinor ξ_{CD} of type (2, 0) we get

$$\xi_{AB} - \xi_{BA} = \varepsilon_{AB}\,\xi_C{}^C, \tag{2.21}$$

so that

$$\xi_{AB} = \xi_{(AB)} + {}^1\!/_2\varepsilon_{AB}\,\xi_C{}^C. \tag{2.22}$$

It is easy to see that by applying Eq. (2.20) the necessary number of times, we can in analogous fashion write an arbitrary spinor with unprimed indices as the sum of a completely symmetric spinor and spinors obtained from symmetric spinors of lower order by multiplication by a certain number of invariant spinors ε_{AB}. It can be proved (cf., for example, [5]) that such a decomposition is unique. In the case of spinors having both primed and unprimed indices, this operation can be carried out separately for the primed and unprimed indices. A spinor which is completely symmetric in the primed indices and completely symmetric in the unprimed indices will be called symmetric, and the decomposition mentioned above will be called the decomposition into symmetric spinors.

It can be shown [63, 77] that an irreducible representation of the group SL(2, **C**) is realized in the space of symmetric spinors of given type, and that every irreducible finite dimensional representation of SL(2, **C**) can be realized in this way. The corresponding decomposition of spinors into symmetric spinors is therefore nothing but the decomposition of the full linear space of spin tensors into subspaces on which the action of SL(2, **C**) is irreducible.

We remark another fact for the sequel. Two spinors χ_A and ζ_A are parallel ($\chi_A = \alpha\zeta_A$) if and only if $\chi_A\zeta^A = 0$. Indeed, if we choose $\xi_{AB} = \zeta_A\chi_B$ and make use of (2.21) we get the required result. Therefore, any pair of spinors χ_A and ζ_A such that $\chi_A\zeta^A \neq 0$ can be taken as basis in spin space. If $\chi_A\zeta^A = 1$ then

$$\varepsilon_{AB} = \chi_A\zeta_B - \chi_B\zeta_A, \quad \delta^A_B = \chi_B\zeta^A - \chi^A\zeta_B. \tag{2.23}$$

A basis such that $\zeta_{aA}\zeta^A_b = \varepsilon_{ab}$ will be called normalized.

3.3. Spinors and Tensors

The Lorentz group is usually defined as the group of matrics acting in real four-dimensional space which preserves the square of the length of a vector u_μ:

$$\eta_{\mu\nu}\, u^\mu u^\nu = (u^0)^2 - (u^1)^2 - (u^2)^2 - (u^3)^2. \tag{3.1}$$

The group SL(2, **C**) is a double covering group for the connected component of the Lorentz group O(1, 3). Hence the irreducible principal (vector) representation of the group O(1, 3) is equivalent to some irreducible representation of SL(2, **C**). This fact indicates that there must exist a natural connection between ordinary vectors of Minkowski space and spinors.

It turns out that an ordinary vector corresponds to a Hermitian spinor of type (1, 1). In order to show how this correspondence arises, consider a Hermitian spinor $u^{AA'}$ in the basis ζ^A_a. The Hermiticity condition allows us to write the coordinates $u^{aa'}$ in the form of a matrix

$$u^{aa'} = \frac{1}{\sqrt{2}} \left\| \begin{matrix} u^0 + u^1 & u^2 + iu^3 \\ u^2 - iu^3 & u^0 - u^1 \end{matrix} \right\|, \tag{3.2}$$

where the index a numbers the rows and the index a' the columns. We remark that twice the determinant of this matrix coincides with expression (3.1) for the square of the length of $u_\mu = (u^0, u^1, u^2, u^3)$. The set of Hermitean spinors of type (1, 1) forms a real four-dimensional space and is hence isomorphic (as a linear space) to Minkowski space. A scalar product of a pair of spinors $u^{aa'}$ and $v^{aa'}$ can be introduced as follows in the space of Hermitian spinors:

$$(u,\ v) = \varepsilon_{ab}\varepsilon_{a'b'}u^{aa'}v^{bb'}. \tag{3.3}$$

This definition can be written in the form

$$(u,\ v) = \varepsilon_{AB}\varepsilon_{A'B'}u^{AA'}v^{BB'} \tag{3.3'}$$

which is independent of the choice of basis. It is easy to verify that if the spinor $v^{bb'}$ is defined by the matrix

$$v^{bb'} = \frac{1}{\sqrt{2}} \begin{Vmatrix} v^0 + v^1 & v^2 + iv^3 \\ v^2 - iv^3 & v^0 - v^1 \end{Vmatrix},$$

then $(u, v) = u^0v^0 - u^1v^1 - u^2v^2 - u^3v^3$. With respect to the scalar product just introduced, the space of Hermitian spinors of type (1, 1) is isomorphic to Minkowski space not only as a linear space but as a Euclidean space.

Up to this point we have used the coordinates of Hermitian spinors in a definite basis ζ_a^A. We now consider a change of basis in spin space

$$\zeta_a^A \to \eta_a^A = l_a^b \zeta_b^A, \quad l_a^b \in SL(2,\mathbf{C}). \tag{3.4}$$

Under this change of basis the coordinates of the spinor $u^{AA'}$ transform as

$$u^{aa'} \to \tilde{u}^{aa'} = L_b^a \bar{L}_{b'}^{a'} u^{bb'} = \frac{1}{\sqrt{2}} \begin{Vmatrix} \tilde{u}^0 + \tilde{u}^1 & \tilde{u}^2 + i\tilde{u}^3 \\ \tilde{u}^2 - i\tilde{u}^3 & \tilde{u}^0 - \tilde{u}^1 \end{Vmatrix}. \tag{3.5}$$

By the unimodularity condition det (L) = det ($\bar{L}$) = 1 we have

$$(u^0)^2 - (u^1)^2 - (u^2)^2 - (u^3)^2 \equiv 2\det(u) = 2\det(\tilde{u}) \equiv (\tilde{u}^0)^2 - (\tilde{u}^1)^2 - (\tilde{u}^2)^2 - (\tilde{u}^3)^2,$$

i.e., the components of the vector $u_\mu = (u^0, u^1, u^2, u^3)$ undergo a transformation preserving the square of the length (a Lorentz transformation). Thus every basis transformation in spin space defined by a unimodular matrix defines a Lorentz transformation of the basis in vector space. It turns out [63, 77] that every proper Lorentz transformation can be obtained in this way, and moreover there are two and only two distinct matrices L_b^a and $-L_b^a$ leading to the same Lorentz transformation.

The relation given in Eq. (3.2) between the components of the vector u_μ and $u^{AA'}$ can be written in the form

$$u^{aa'} = \sigma_m^{aa'} u^m, \tag{3.6}$$

where

$$\sigma_2^{aa'} = \frac{1}{\sqrt{2}} \begin{Vmatrix} 0 & 1 \\ 1 & 0 \end{Vmatrix}, \quad \sigma_3^{aa'} = \frac{1}{\sqrt{2}} \begin{Vmatrix} 0 & i \\ -i & 0 \end{Vmatrix}, \quad \sigma_1^{aa'} = \frac{1}{\sqrt{2}} \begin{Vmatrix} 1 & 0 \\ 0 & -1 \end{Vmatrix},$$

$$\sigma_0^{aa'} = \frac{1}{\sqrt{2}} \begin{Vmatrix} 1 & 0 \\ 0 & 1 \end{Vmatrix}.$$

Under a change of basis in spin space defined by a matrix $l_a^b \in SL(2, \mathbf{C})$ and the corresponding change of basis in Minkowski space, the quantities $\sigma_m^{aa'}$ remain unchanged. In the general case, if a change of basis in spin space $\zeta_a^A \to l_a^b \zeta_b^A$ and a change of basis $e_m^\mu \to \Lambda_m^n e_n^\mu$ in Minkowski space are performed independently of one another, the quantities $\sigma_m^{aa'}$ undergo the transformation

$$\sigma_m^{aa'} \to \Lambda_m^n L_b^a \bar{L}_{b'}^{a'} \sigma_n^{bb'}. \tag{3.7}$$

This relation allows us to consider the $\sigma_m^{aa'}$ as the components of some geometric object $\sigma_\mu^{AA'}$ with two abstract spinor indices A, A' and one abstract vector index μ. Such objects were introduced in [78] and have become known as the van der Waerden–Infeld symbols. Using the $\sigma_\mu^{AA'}$ it is possible to write the relation between Hermitian spinors and vectors in a form independent of the choice of basis.

$$u^{AA'} = \sigma_\mu^{AA'} u^\mu. \tag{3.8}$$

Equality (3.8) shows that the $\sigma_\mu^{AA'}$ can be viewed as an operator realizing an isomorphic mapping of Minkowski space onto the space of Hermitian spinors. By the fact that these two linear spaces are isomorphic, as just remarked, there exists an operator $\sigma^\mu_{AA'}$ inverse to $\sigma_\mu^{AA'}$ and mapping the space of Hermitian spinors onto Minkowski space:

$$u^\mu = \sigma^\mu_{AA'} u^{AA'}, \tag{3.9}$$

where moreover the relations

$$\sigma^\mu_{AA'} \sigma_\mu^{BB'} = \delta_A^B \delta_{A'}^{B'}, \qquad \sigma^\mu_{AA'} \sigma_\nu^{AA'} = \delta_\nu^\mu \tag{3.10}$$

hold.

The symbols $\sigma_\mu^{AA'}$ and $\sigma^\mu_{AA'}$ realize an isomorphism of Minkowski space with the space of Hermitian spinors not only as linear spaces but also as Euclidean spaces. If we recall that the scalar product in these spaces is defined by Eqs. (3.1) and (3.3') respectively, then we have

$$\eta_{\mu\nu} u^\mu v^\nu = \varepsilon_{AB} \varepsilon_{A'B'} u^{AA'} v^{BB'}, \tag{3.11}$$

where $u^\mu = \sigma^\mu_{AA'} u^{AA'}$, $v^\nu = \sigma^\nu_{BB'} v^{BB'}$. Since the quantities $u^{AA'}$ and $v^{BB'}$ are arbitrary, Eq. (3.11) implies

$$\eta_{\mu\nu} \sigma^\mu_{AA'} \sigma^\nu_{BB'} = \varepsilon_{AB} \varepsilon_{A'B'}. \tag{3.12}$$

If we multiply both sides of this relation by $\eta^{\alpha\beta}\sigma_\beta^{BB'}$, we find that

$$\sigma^\alpha_{AA'} = \eta^{\alpha\beta} \varepsilon_{AB} \varepsilon_{A'B'} \sigma_\beta^{BB'}. \tag{3.13}$$

This formula establishes the connection between the symbols $\sigma_\beta^{BB'}$ and $\sigma^\alpha_{AA'}$.

The relation determined above between vectors and spinors permits us to establish a similar correspondence between arbitrary tensors and spinors, each tensor with n indices corresponding to a Hermitian spinor of type (n, n) according to the rule

$$T_{\alpha_1 \ldots \alpha_m}{}^{\alpha_{m+1} \ldots \alpha_n} \leftrightarrow T_{A_1 A_1' \ldots A_m A_m'}{}^{A_{m+1} A_{m+1}' \ldots A_n A_n'} =$$

$$= \sigma^{\alpha_1}_{A_1 A_1'} \ldots \sigma^{\alpha_m}_{A_m A_m'} \sigma_{\alpha_{m+1}}^{A_{m+1} A_{m+1}'} \ldots \sigma_{\alpha_n}^{A_n A_n'} T_{\alpha_1 \ldots \alpha_m}{}^{\alpha_{m+1} \ldots \alpha_n}. \tag{3.14}$$

We will use the symbol $\leftrightarrow$ to indicate the correspondence described by Eq. (3.14) between tensors and spinors. This allows us to write Eq. (3.12), for example, in the form

$$\eta_{\alpha\beta} \leftrightarrow \varepsilon_{AB} \varepsilon_{A'B'}. \tag{3.15}$$

Relation (3.14) of tensors with spinors shows that the usual algebraic operations on tensors correspond to analogous operations on their spinor images. In particular, we have for the contraction.

$$T^{\dots\alpha\dots}_{\dots}S^{\dots}_{\dots\alpha\dots} \leftrightarrow T^{\dots AA'\dots}_{\dots}S^{\dots}_{\dots AA'\dots}. \tag{3.16}$$

3.4. Spinor Expressions for Certain Tensors. The Geometric Interpretation of a Spinor

In this section we consider the question of how various characteristics of vectors and tensors carry over to their spinor analogs.

Proposition 3.4.1. A complex vector ξ^{α} is lightlike if and only if the spinor $\xi^{AA'}$ corresponding to it can be written in the form $\xi^{AA'} = \zeta^{A}\eta^{A'}$. If ξ^{a} is a real lightlike vector, we have the representation $\xi^{AA'} = \pm\, \xi^{A}\bar{\xi}^{A'}$.

Proof. Just as in the case of real vectors, the spinor image of a complex vector is defined by the relation $\xi^{AA'} = \sigma_{\mu}^{AA'}\xi^{\mu}$. If we make use of the equality

$$\xi^{\alpha}\xi_{\alpha} = \xi^{AA'}\xi_{AA'} = \varepsilon_{AB}\varepsilon_{A'B'}\,\xi^{AA'}\xi^{BB'} = 2\det(\xi^{AA'}),$$

we obtain that $\xi^{\alpha}\xi_{\alpha} = 0$ if and only if $\det(\xi^{AA'}) = 0$. This condition means that in an arbitrary basis, the matrix $\xi^{aa'}$ composed of the components of $\xi^{AA'}$ is singular and can be written in the form $\xi^{aa'} = \zeta^{a}\bar{\eta}^{a'}$. This equality represents the relation $\xi^{AA'} = \zeta^{A}\eta^{A'}$ written in a definite basis, as was required to be proved. In the case where ξ^{α} is a real vector, the corresponding matrix $\xi^{aa'}$ is Hermitian and therefore $\zeta^{a}\bar{\eta}^{a'} = \eta^{a}\bar{\zeta}^{a'}$. It follows from this that $\frac{\zeta^{a}}{\eta^{a}} = \left(\frac{\bar{\zeta}^{a'}}{\bar{\eta}^{a'}}\right) = \lambda$ and λ is a real number. Therefore, $\eta^{a} = \lambda\zeta^{a}$, or if we write $|\sqrt{\lambda}|\,\zeta^{a} = \xi^{a}$, we have $\xi^{aa'} = \pm\xi^{a}\bar{\xi}^{a'}$.

Corollary. The spinor ξ^{A} uniquely determines a real lightlike vector $\xi^{\alpha} = \sigma^{\alpha}_{AA'}\xi^{A}\bar{\xi}^{A'}$. Distinct spinors define the same lightlike vector ξ^{α} if and only if they differ by a complex factor with modulus equal to one.

Proposition 3.4.2. Given a basis ζ^{A}_{a} in spin space, there corresponds a unique complex lightlike basis z^{μ}_{n} in Minkowski space defined by the relations

$$z_1^{\mu} \equiv l^{\mu} = \sigma^{\mu}_{AA'}\zeta^{A}_{0}\bar{\zeta}^{A'}_{0'} \equiv \sigma^{\mu}_{00'}, \qquad z_2^{\mu} \equiv n^{\mu} = \sigma^{\mu}_{AA'}\zeta^{A}_{1}\bar{\zeta}^{A'}_{1'} \equiv \sigma^{\mu}_{11'},$$
$$z_3^{\mu} \equiv m^{\mu} = \sigma^{\mu}_{AA'}\zeta^{A}_{0}\bar{\zeta}^{A'}_{1'} \equiv \sigma^{\mu}_{01'}, \qquad z_4^{\mu} \equiv \bar{m}^{\mu} = \sigma^{\mu}_{AA'}\zeta^{A}_{1}\bar{\zeta}^{A'}_{0'} \equiv \sigma^{\mu}_{10'}. \tag{4.1}$$

If the spinor basis ζ^{A}_{a} is normalized, $\zeta_{aA}\zeta^{B}_{b} = \varepsilon_{ab}$, the vectors in the basis z^{μ}_{n} satisfy the normalization conditions $z_1^{\mu}z_{2\mu} = -z_3^{\mu}z_{4\mu} = 1$ and $z^{\mu}_{m}z_{n\mu} = 0$ in the remaining cases.

Proof. Consider the equation $\sigma^{\mu}_{aa'}\xi^{aa'} = 0$, where $\xi^{aa'}$ is a Hermitian spinor. Since $\sigma^{\mu}_{aa'}$ realizes an isomorphism between the space of Hermitian spinors and Minkowski space, the unique solution of this equation is $\xi^{aa'} = 0$. Therefore the vectors $\sigma^{\mu}_{00'}$, $\sigma^{\mu}_{11'}$, $\sigma^{\mu}_{01'}$ and $\sigma^{\mu}_{10'}$ are linearly independent. It follows from Proposition 3.4.1 that z_1^{μ} and z_2^{μ} are real and z_3^{μ} and z_4^{μ} are complex lightlike vectors. If we use Eqs. (3.10), we check directly that the normalization conditions are satisfied.

Proposition 3.4.3. An antisymmetric tensor of rank two (bivector) $F_{\alpha\beta}$ has the following spinor representation:

$$F_{\alpha\beta} \leftrightarrow \varepsilon_{AB}\bar{\Phi}_{A'B'} + \varepsilon_{A'B'}\Phi_{AB}, \tag{4.2}$$

where Φ_{AB} is a symmetric spinor.

Proof. The antisymmetry of the tensor $F_{\alpha\beta}=-F_{\beta\alpha}$ means that the relation $F_{AA'BB'} = -F_{BB'AA'}$ holds for its spinor analog $F_{AA'BB'}$. We can therefore write

$$F_{AA'BB'}=\frac{1}{2}(F_{AA'BB'}-F_{BB'AA'})=\frac{1}{2}(F_{AA'BB'}-F_{BA'AB'}+$$
$$+F_{BA'AB'}-F_{BB'AA'})=\frac{1}{2}(\varepsilon_{AB}F_{CA'}{}^{C}{}_{B'}+\varepsilon_{A'B'}F_{BC'A}{}^{C'}).$$

We write $\Phi_{AB}={}^1/_2F_{BC'A}{}^{C'}$ and $\Psi_{A'B'}={}^1/_2F_{CA'}{}^{C}{}_{B'}$. Using the antisymmetry of $F_{\alpha\beta}$ and contraction rule (2.18) we have

$$\Phi_{AB}=\frac{1}{2}F_{BC'A}{}^{C'}=-\frac{1}{2}F_{B}{}^{C'}{}_{AC'}=\frac{1}{2}F_{AC'B}{}^{C'}=\Phi_{BA}.$$

The reality of $F_{\alpha\beta}$ means that $F_{AA'BB'}=\overline{F}_{AA'BB'}$, so that

$$\Psi_{A'B'}=\frac{1}{2}F_{CA'}{}^{C}{}_{B'}=\frac{1}{2}\overline{F_{C'A}{}^{C'}{}_{B}}=\overline{\Phi_{AB}}=\overline{\Phi}_{A'B'}.$$

Proposition 3.4.4. If $\varepsilon_{\alpha\beta\gamma\delta}$ is the completely antisymmetric object, then it has the following spinor representation:

$$\varepsilon_{\alpha\beta\gamma\delta}\leftrightarrow E_{ABCDA'B'C'D'}=-i(\varepsilon_{AC}\varepsilon_{BD}\varepsilon_{A'B'}\varepsilon_{C'D'}-\varepsilon_{AB}\varepsilon_{CD}\varepsilon_{A'C'}\varepsilon_{B'D'}), \tag{4.3}$$

$$\varepsilon_{\alpha\beta}{}^{\gamma\delta}\leftrightarrow i(\delta^{C}_{A}\delta^{D}_{B}\delta^{D'}_{A'}\delta^{C'}_{B'}-\delta^{D}_{A}\delta^{C}_{B}\delta^{C'}_{A'}\delta^{D'}_{B'}). \tag{4.3'}$$

Proof. The proof is carried out by decomposing the spinor $E_{ABCDA'B'C'D'}$ into symmetric spinors. Using the complete antisymmetry of $\varepsilon_{\alpha\beta\gamma\delta}$ it can be proved that all the symmetric spinors appearing in this decomposition are equal to zero, and therefore only the quantities ε_{AB} take part in the decomposition of $E_{ABCDA'B'C'D'}$. The proof of Eqs. (4.3) and (4.3') is given in [12].

Proposition 3.4.5. If the bivector $F_{\alpha\beta}$ has the spinor representation (4.2) then the following relations are valid:

$$F^{*}_{\alpha\beta}\equiv\frac{1}{2}\varepsilon_{\alpha\beta\gamma\delta}F^{\gamma\delta}\leftrightarrow i(\varepsilon_{AB}\overline{\Phi}_{A'B'}-\varepsilon_{A'B'}\Phi_{AB}), \tag{4.4}$$

$$F^{+}_{\alpha\beta}\equiv\frac{1}{2}(F_{\alpha\beta}+iF^{*}_{\alpha\beta})\leftrightarrow\varepsilon_{A'B'}\Phi_{AB}, \tag{4.5}$$

$$F^{-}_{\alpha\beta}\equiv\frac{1}{2}(F_{\alpha\beta}-iF^{*}_{\alpha\beta})\leftrightarrow\varepsilon_{AB}\overline{\Phi}_{A'B'}, \tag{4.6}$$

$$F_{\alpha\beta}(\vartheta)\equiv F_{\alpha\beta}\cos\vartheta+F^{*}_{\alpha\beta}\sin\vartheta\leftrightarrow\varepsilon_{AB}\overline{\Phi}_{A'B'}e^{i\vartheta}+\varepsilon_{A'B'}\,\Phi_{AB}e^{-i\vartheta}, \tag{4.7}$$

$$K\equiv\frac{1}{2}(F_{\alpha\beta}F^{\alpha\beta}+iF_{\alpha\beta}F^{*\alpha\beta})=F^{+}_{\alpha\beta}F^{+\alpha\beta}=2\Phi_{AB}\Phi^{AB}. \tag{4.8}$$

Proof. Equations (4.4)-(4.8) written out above are consequences of Eqs. (4.2) and (4.3) and are verified directly.

Proposition 3.4.6. The spinor ξ^{A} uniquely determines a bivector $F_{\alpha\beta}$ by the relation

$$F_{\alpha\beta}=\sigma^{AA'}_{\alpha}\sigma^{BB'}_{\beta}(\varepsilon_{AB}\bar{\xi}_{A'}\bar{\xi}_{B'}+\varepsilon_{A'B'}\xi_{A}\xi_{B}). \tag{4.9}$$

This bivector is simple and lightlike* and can be written in the form

$$F_{\alpha\beta} = \xi_{[\alpha} w_{\beta]}, \tag{4.10}$$

where $\xi^\alpha \leftrightarrow \xi^A \bar{\xi}^{A'}$ is a lightlike vector and w^β is a spacelike vector defined up to a transformation $w^\beta \to w^\beta + \lambda \xi_\beta$. The change of the spinor $\xi^A \to e^{i\vartheta} \xi^A$ corresponds to a dual rotation of the bivector $F_{\alpha\beta}$ by the angle 2ϑ.

Proof. Choose an arbitrary spinor η^A such that $\xi_A \eta^A = 1$. Then by (2.21) we obtain $\varepsilon_{AB} = \xi_A \eta_B - \eta_A \xi_B$. We therefore have

$$F_{\alpha\beta} = \sigma_\alpha^{AA'} \sigma_\beta^{BB'} (\xi_A \eta_B \bar{\xi}_{A'} \bar{\xi}_{B'} - \eta_A \xi_B \bar{\xi}_{A'} \bar{\xi}_{B'} + \bar{\xi}_{A'} \bar{\eta}_{B'} \xi_A \xi_B - \bar{\eta}_{A'} \bar{\xi}_{B'} \xi_A \xi_B) = \xi_\alpha \sigma_\beta^{BB'} \eta_B \bar{\xi}_{B'} - \sigma_\alpha^{AA'} \eta_A \bar{\xi}_{A'} \xi_\beta +$$

$$+ \xi_\alpha \sigma_\beta^{BB'} \bar{\eta}_{B'} \xi_B - \xi_\beta \sigma_\alpha^{AA'} \bar{\eta}_{A'} \xi_A = \xi_{[\alpha} w_{\beta]},$$

where $\xi_\alpha = \sigma_\alpha^{AA'} \xi_A \bar{\xi}_{A'}$ and $w_\beta = \sigma_\beta^{BB'} (\eta_B \bar{\xi}_{B'} + \bar{\eta}_{B'} \xi_B)$. Without violating the condition $\xi_A \eta^A = 1$ we may transform the spinor $\eta^A \to \eta^A + \lambda \xi^A$, and therefore the vector w_β is uniquely determined up to a transformation

$$w_\beta \to \tilde{w}_\beta = \sigma_\beta^{BB'} [(\eta_B + \lambda \xi_B) \bar{\xi}_{B'} + (\bar{\eta}_{B'} + \overline{\lambda \xi}_{B'}) \xi_B] = w_\beta + (\lambda + \bar{\lambda}) \xi_\beta.$$

Proposition 3.4.6 just proved allows us to give the following geometric interpretation to spinors. An arbitrary spinor ξ^A is uniquely determined by the lightlike vector $\xi^\alpha \leftrightarrow \xi^A \bar{\xi}^{A'}$ in Minkowski space and by a two-dimensional area spanned by this vector and defined by the bivector $\xi_{\alpha\beta} \leftrightarrow (\varepsilon_{AB} \bar{\xi}_{A'} \bar{\xi}_{B'} + \varepsilon_{A'B'} \xi_A \xi_B)$. We can therefore depict a spinor in the form of a flag (Fig. 3.2).

Proposition 3.4.7. Assume given a tensor $R_{\alpha\beta\gamma\delta}$ of rank four with the symmetry properties

$$R_{\alpha\beta\gamma\delta} = R_{[\alpha\beta][\gamma\delta]} = R_{[\gamma\delta][\alpha\beta]}; \quad R_{\alpha[\beta\gamma\delta]} = 0, \tag{4.11}$$

analogous to the symmetry properties of the curvature tensor. Then this tensor has the follow-

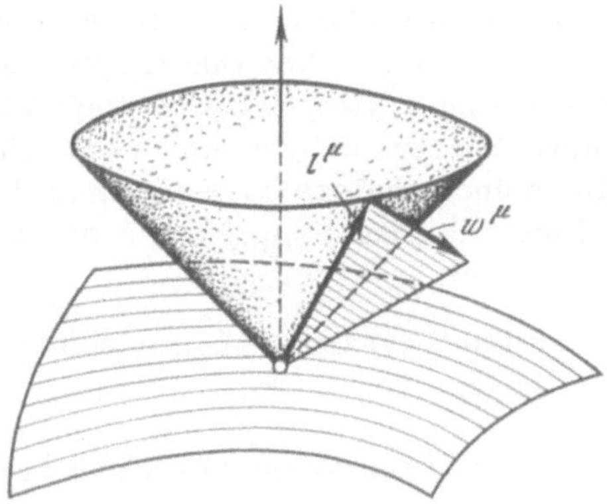

Fig. 3.2

* The definition of a simple lightlike bivector is given in Section 2.6.

ing spinor representation:

$$R_{\alpha\beta\gamma\delta} = \sigma_{\alpha}^{AA'}\sigma_{\beta}^{BB'}\sigma_{\gamma}^{CC'}\sigma_{\delta}^{DD'}[\Psi_{ABCD}\varepsilon_{A'B'}\varepsilon_{C'D'} + \varepsilon_{AB}\varepsilon_{CD}\overline{\Psi}_{A'B'C'D'} +$$
$$+ \Phi_{ABC'D'}\varepsilon_{CD}\varepsilon_{A'B'} + \Phi_{CDA'B'}\varepsilon_{AB}\varepsilon_{C'D'} + 2\Lambda(\varepsilon_{AC}\varepsilon_{BD}\varepsilon_{A'B'}\varepsilon_{C'D'} + \varepsilon_{AB}\varepsilon_{CD}\varepsilon_{A'D'}\varepsilon_{B'C'})], \quad (4.12)$$

where Ψ_{ABCD} is a symmetric spinor and $\Phi_{ABC'D'}$ a Hermitian spinor symmetric with respect to every pair of indices. If the tensor $R_{\alpha\beta\gamma\delta}$ satisfies the additional condition $R_{\alpha\beta}{}^{\alpha}{}_{\delta} = 0$, then the quantities $\Phi_{ABC'D'}$ and Λ in (4.12) vanish.

Proof. The proof is carried out by a method completely analogous to the one used in the proof of Eq. (4.2) and is based on decomposing the spinor $R_{AA'BB'CC'DD'}$ into symmetric spinors (into irreducible components). A detailed proof may be found, for example, in [5].

Proposition 3.4.8. If the curvature tensor $R_{\alpha\beta\gamma\delta}$ has the spinor representation (4.12) then the Ricci tensor $R_{\alpha\beta} = R_{\delta\alpha}{}^{\delta}{}_{\beta}$, scalar curvature $R = R^{\alpha}_{\alpha}$, and Weyl tensor $C_{\alpha\beta\gamma\delta}$ defined by the relation

$$C^{\alpha\beta}{}_{\gamma\delta} = R^{\alpha\beta}{}_{\gamma\delta} - 2\delta^{[\alpha}{}_{[\gamma}R^{\beta]}{}_{\delta]} + \frac{1}{3}\delta^{[\alpha}_{[\gamma}\delta^{\beta]}_{\delta]}R \quad (4.13)$$

have the following representation:

$$R_{\alpha\beta} \leftrightarrow -2\Phi_{ABA'B'} + 6\Lambda\varepsilon_{AB}\varepsilon_{A'B'}, \quad (4.14)$$

$$R = 24\Lambda, \quad (4.15)$$

$$C_{\alpha\beta\gamma\delta} \leftrightarrow \Psi_{ABCD}\varepsilon_{A'B'}\varepsilon_{C'D'} + \varepsilon_{AB}\varepsilon_{CD}\overline{\Psi}_{A'B'C'D'}. \quad (4.16)$$

Proof. The Ricci tensor $R_{\beta\delta}$ is obtained by contracting the curvature tensor $R_{\alpha\beta\gamma\delta}$ with $\eta^{\alpha\delta}$, and hence the spinor analog of the Ricci tensor is obtained from the spinor analog of the curvature tensor by contraction with the quantity $\varepsilon^{AC}\varepsilon^{A'C'}$. Equation (2.20) allows us to write

$$\varepsilon_{AB}\varepsilon^{AC} = \delta^{C}_{B}, \quad \varepsilon_{AB}\varepsilon^{AB} = 2. \quad (4.17)$$

Relation (4.14) is obtained after multiplying the right-hand side of (4.12) by $\varepsilon^{AC}\varepsilon^{A'C'}$ and using Eq. (4.17). Further contraction of (4.14) with $\varepsilon^{AB}\varepsilon^{A'B'}$ leads to Eq. (4.15). Relation (4.16) for the Weyl tensor is easily derived if it is recalled that $C_{\alpha\beta}{}^{\alpha}{}_{\delta} = 0$, and hence there are no terms containing $\Phi_{ABA'B'}$ and Λ in the decomposition (4.12) for the Weyl tensor. Since the Weyl tensor is obtained from the curvature tensor by adding terms linear in the Ricci tensor, the corresponding operation for the spinor analogs does not alter the terms of (4.12) containing Ψ_{ABCD} and $\Psi_{A'B'C'D'}$. This and the fact that $\Phi_{ABA'B'}$ and Λ are absent permit us to obtain Eq. (4.16).

Proposition 3.4.9. If the Weyl tensor $C_{\alpha\beta\gamma\delta}$ has the spinor representation (4.16) then the following relations are valid:

$$^{*}C_{\alpha\beta\gamma\delta} \equiv \frac{1}{2}\varepsilon_{\alpha\beta\rho\sigma}C^{\rho\sigma}{}_{\gamma\delta} \leftrightarrow i[-\Psi_{ABCD}\varepsilon_{A'B'}\varepsilon_{C'D'} + \varepsilon_{AB}\varepsilon_{CD}\overline{\Psi}_{A'B'C'D'}], \quad (4.18)$$

$$C^{*}_{\alpha\beta\gamma\delta} \equiv \frac{1}{2}\varepsilon_{\gamma\delta\rho\sigma}C_{\alpha\beta}{}^{\rho\sigma} \leftrightarrow i[-\Psi_{ABCD}\varepsilon_{A'B'}\varepsilon_{C'D'} + \varepsilon_{AB}\varepsilon_{CD}\overline{\Psi}_{A'B'C'D'}], \quad (4.19)$$

$$^{*}C_{\alpha\beta\gamma\delta} = C^{*}_{\alpha\beta\gamma\delta}, \quad (4.20)$$

$$C^{+}_{\alpha\beta\gamma\delta} \equiv \frac{1}{2}(C_{\alpha\beta\gamma\delta} + iC^{*}_{\alpha\beta\gamma\delta}) \leftrightarrow \Psi_{ABCD}\varepsilon_{A'B'}\varepsilon_{C'D'}, \quad (4.21)$$

$$C^{-}_{\alpha\beta\gamma\delta} \equiv \frac{1}{2}(C_{\alpha\beta\gamma\delta} - iC^{*}_{\alpha\beta\gamma\delta}) \leftrightarrow \overline{\Psi}_{A'B'C'D'}\varepsilon_{AB}\varepsilon_{CD}, \tag{4.22}$$

$$C_{\alpha\beta\gamma\delta}\cos\vartheta + C^{*}_{\alpha\beta\gamma\delta}\sin\vartheta \leftrightarrow e^{-i\vartheta}\Psi_{ABCD}\varepsilon_{A'B'}\varepsilon_{C'D'} + e^{i\vartheta}\overline{\Psi}_{A'B'C'D'}\varepsilon_{AB}\varepsilon_{CD}, \tag{4.23}$$

$$I \equiv C^{+}_{\alpha\beta\gamma\delta}C^{+\alpha\beta\gamma\delta} = \frac{1}{2}[C_{\alpha\beta\gamma\delta}C^{\alpha\beta\gamma\delta} + iC_{\alpha\beta\gamma\delta}C^{*\alpha\beta\gamma\delta}] = 4\Psi_{ABCD}\Psi^{ABCD}, \tag{4.24}$$

$$J \equiv C^{+}_{\alpha\beta}{}^{\gamma\delta}C^{+}_{\gamma\delta}{}^{\lambda\rho}C^{+}_{\lambda\rho}{}^{\alpha\beta} = \frac{1}{2}[C_{\alpha\beta}{}^{\gamma\delta}C_{\gamma\delta}{}^{\lambda\rho}C_{\lambda\rho}{}^{\alpha\beta} + iC^{*}_{\alpha\beta}{}^{\gamma\delta}C_{\gamma\delta}{}^{\lambda\rho}C_{\lambda\rho}{}^{\alpha\beta}] = 8\Psi_{AB}{}^{CD}\Psi_{CD}{}^{LR}\Psi_{LR}{}^{AB}. \tag{4.25}$$

Proof. Equalities (4.18) and (4.19) are easily obtained from expression (4.16) for the spinor analog of the Weyl tensor after using Eq. (4.3') for $\varepsilon_{\alpha\beta}{}^{\gamma\delta}$. Relation (4.20) follows from (4.18) and (4.19). Equations (4.21)-(4.23) are verified by direct calculation. In deriving (4.24) and (4.25) we use the following easily verified relations:

$$\varepsilon^{\alpha\beta\rho\sigma}\varepsilon_{\alpha\beta\pi\mu} = -2(\delta^{\rho}_{\pi}\delta^{\sigma}_{\mu} - \delta^{\rho}_{\mu}\delta^{\sigma}_{\pi}),$$

$$C^{*}_{\alpha\beta\gamma\delta}C^{*\gamma\delta\mu\nu} = -C_{\alpha\beta\gamma\delta}C^{\gamma\delta\mu\nu},$$

$$C_{\alpha\beta\gamma\delta}C^{*\gamma\delta\mu\nu} = C^{*}{}_{\alpha\beta\gamma\delta}C^{\gamma\delta\mu\nu}.$$

3.5. The Canonical Decomposition of Symmetric Spinors. The Petrov – Pirani Classification

The relation given in the preceding section between antisymmetric tensors and the Weyl tensor on the one hand and the symmetric spinors corresponding to them on the other allow us to reduce the problem of the algebraic classification of the electromagnetic and gravitational fields to the problem of classifying symmetric spinors. One of the principal virtues of the spinor formalism is that this classification is extremely simple and is based on the following fact.

Proposition 3.5.1. Every symmetric spinor $\varphi_{A_1A_2\ldots A_n}$ can be written in the form

$$\varphi_{A_1A_2\ldots A_n} = \overset{1}{\alpha}_{(A_1}\overset{2}{\alpha}_{A_2}\ldots\overset{n}{\alpha}_{A_n)}, \tag{5.1}$$

where the spinors $\overset{1}{\alpha}_{A_1}, \overset{2}{\alpha}_{A_2}, \ldots, \overset{n}{\alpha}_{A_n}$ are called the principal spinors. Decomposition (5.1) determines the principal spinors up to complex factors. The lightlike vectors $\xi^{\mu}_{(i)} = \sigma^{\mu}_{AA'}\overset{i}{\alpha}{}^{A}\overset{i}{\bar{\alpha}}{}^{A'}$ corresponding to the principal spinors are called principal lightlike vectors, and the null directions determined by them are called the principal null directions. The principal null directions are uniquely determined by the spinor $\varphi_{A_1A_2\ldots A_n}$.

Proof. By the fundamental theorem of algebra the equation

$$a_n z^n + a_{n-1}z^{n-1} + \ldots + a_0 = 0,\ a_n \neq 0$$

with complex coefficients always has n complex roots $z_1, \ldots, z_n$, or, what is the same,

$$a_n z^n + a_{n-1}z^{n-1} + \ldots + a_0 = a_n(z - z_1)(z - z_2)\ldots(z - z_n).$$

Consider now the symmetric spinor $\varphi_{A_1A_2\ldots A_n}$ in some basis ζ^{A}_{a}, and form the expression $\varphi(\xi) = \varphi_{a_1a_2\ldots a_n}\xi^{a_1}\xi^{a_2}\ldots\xi^{a_n}$. If we take ξ^{a} to be the spinor with components $\xi^{a} = (1, z)$, then

$$\varphi(\xi) = \varphi_{11\cdots1}z^n + n\varphi_{01\cdots1}z^{n-1} + \ldots + \varphi_{00\cdots0} = \sum_{k=0}^{n}\varphi_k C^k_n z^k, \tag{5.2}$$

where $\varphi_{11\ldots10\ldots0} = \varphi_k$ if the index 1 appears k times. Let the first nonvanishing coefficient in polynomial (5.2) be φ_m ($n \geq m \geq 0$). Then we have

$$\varphi(\xi) = \varphi_m (z - z_m) \ . \ . \ . \ (z - z_n).$$

If we write

$$\overset{i}{\alpha}_a = (1, 0), \quad i < m; \qquad \overset{m}{\alpha}_a = (-\varphi_m z_m, \varphi_m);$$
$$\overset{i}{\alpha}_a = (-z_i, 1), \quad m + 1 \leqslant i \leqslant n,$$

we obtain $\varphi(\xi) = (\overset{1}{\alpha}_{a_1}\xi^{a_1})(\overset{2}{\alpha}_{a_2}\xi^{a_2})\ldots(\overset{n}{\alpha}_{a_n}\xi^{a_n})$. This relation can be written in the following form:

$$[\varphi_{A_1A_2\ldots A_n} - \overset{1}{\alpha}_{(A_1}\overset{2}{\alpha}_{A_2} \ . \ . \ . \ \overset{n}{\alpha}_{A_n)}]\, \xi^{A_1}\xi^{A_2} \ . \ . \ . \ \xi^{A_n} = 0,$$

which is independent of the choice of basis, from which Eq. (5.1) follows.

Thus every symmetric spinor $\varphi_{A_1A_2\ldots A_n}$ defines n principal null directions. Two principal null directions coincide if and only if the principal spinors corresponding to them are parallel. The mutual position of the principal null directions is basic to the algebraic classification of symmetric spinors and the tensors defined by them. The following notation is adopted for describing the algebraic type of a symmetric spinor. Let the spinor $\varphi_{A_1A_2\ldots A_n}$ have p distinct principal null directions, each of which is m_i-fold degenerate; we then say that the spinor $\varphi_{A_1A_2\ldots A_n}$ has the type $[m_1, m_2, \ldots, m_p]$. The spinor $\varphi_{A_1A_2\ldots A_n}$ is called algebraically generic if it has the type $[1, 1, \ldots, 1]$. In the contrary case (when at least one principal light direction is degenerate) it is said to be algebraically special.

The Algebraic Classification of the Electromagnetic Field Tensor. The bivector $F_{\alpha\beta}$ is determined by the symmetric spinor Φ_{AB}. By the assertion proved above, we have $\Phi_{AB} = \alpha_{(A}\beta_{B)}$. The following two cases are possible.

1. General type [1, 1]: α_A and β_A are nonparallel and there are two distinct principal null directions. The field invariant K is nonzero (Fig. 3.3).
2. Algebraic special type [2]: α_A and β_A are proportional ($\alpha_A\beta^A = 0$) and there is one twofold degenerate principal null direction, K = 0 (Fig. 3.3).

The principal null directions $l^\alpha = \sigma^\alpha_{AA'}\alpha^A\bar{\alpha}^{A'}$ are a solution of the eigenvalue problem

$$l_{[\alpha}\overset{+}{F}_{\beta]\gamma}l^\gamma = 0.$$

From Table 3.1, which contains the main results of the spinor classification of the electromagnetic field, it is clear that the results reproduce completely the results of Section 2.6, in which the algebraic classification was studied by the tensor method.

The Algebraic Classification of the Weyl Tensor. The Weyl tensor is, by (4.16); uniquely related to the symmetric spinor Ψ_{ABCD}, which can be written in the form

$$\Psi_{ABCD} = \overset{1}{\alpha}_{(A}\overset{2}{\alpha}_B\overset{3}{\alpha}_C\overset{4}{\alpha}_{D)}.$$

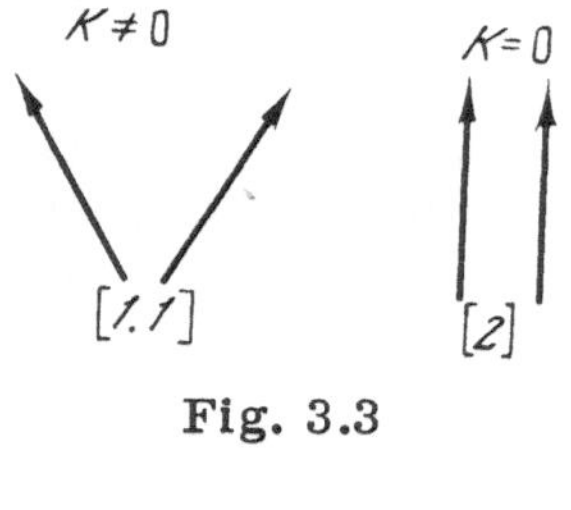

Fig. 3.3

TABLE 3.1

Notation, type	Canonical form Φ_{AB}	Value of the invariant K	Equation for Φ_{AB} ($\lambda \neq 0$)	Equation for $F_{\alpha\beta}$
[1, 1] General	$\Phi_{AB} = \alpha_{(A}\beta_{B)}$	$K \neq 0$	$\Phi_{AB}\xi^A = \lambda\xi_B$	$l_{[\gamma}F^+_{\alpha]\beta}l^\beta = 0$
[2] Special	$\Phi_{AB} = \alpha_{(A}\alpha_{B)}$	$K = 0$	$\Phi_{AB}\xi^A = 0$	$F^+_{\alpha\beta}l^\beta = 0$
[—] Null	$\Phi_{AB} = 0$	$K = 0$	$\Phi_{AB} = 0$	$F^+_{\alpha\beta} = 0$

TABLE 3.2

Notation, Petrov type	Canonical form Ψ_{ABCD}	Value of the invariants I and J	Equation for Ψ_{ABCD} ($\lambda \neq 0$)	Vanishing components	Equations for $C_{\alpha\beta\gamma\delta}$
[1, 1, 1, 1] I	$\alpha_{(A}\beta_B\gamma_C\delta_{D)}$	$I^3 \neq 6J^2$	$\Psi_{ABCD}\alpha^B\alpha^C\alpha^D = \lambda\alpha_A$	—	$l_{[\alpha}C_{\beta]\gamma\delta[\mu}l_{\nu]}l^\gamma l^\delta = 0$
[2, 1, 1] II	$\alpha_{(A}\alpha_B\beta_C\gamma_{D)}$	$I^3 = 6J^2$	$\Psi_{ABCD}\alpha^C\alpha^D = \lambda\alpha_A\alpha_B$	Ψ_0, Ψ_1	$C_{\beta\gamma\delta[\mu}l_{\nu]}l^\gamma l^\delta = 0$
[2, 2] D	$\alpha_{(A}\alpha_B\beta_C\beta_{D)}$	$I^3 = 6J^2$	$\Psi_{ABCD}\alpha^C\alpha^D = \lambda\alpha_A\alpha_B$	Ψ_0, Ψ_1, Ψ_3, Ψ_4	$C_{\beta\gamma\delta[\mu}l_{\nu]}l^\gamma l^\delta = 0$
[3, 1] III	$\alpha_{(A}\alpha_B\alpha_C\beta_{D)}$	$I = J = 0$	$\Psi_{ABCD}\alpha^D = \lambda\alpha_A\alpha_B\alpha_C$	Ψ_0, Ψ_1, Ψ_2	$C_{\beta\gamma\delta[\mu}l_{\nu]}l^\delta = 0$
[4] N	$\alpha_{(A}\alpha_B\alpha_C\alpha_{D)}$	$I = J = 0$	$\Psi_{ABCD}\alpha^D = 0$	Ψ_0, Ψ_1, Ψ_2, Ψ_3	$C_{\beta\gamma\delta\mu}l^\mu = 0$
[—] O	$\Psi_{ABCD} = 0$	$I = J = 0$	$\Psi_{ABCD} = 0$	Ψ_0, Ψ_1 Ψ_2, Ψ_3, Ψ_4	$C_{\beta\gamma\delta\mu} = 0$

It is convenient to denote distinct nonparallel principal spinors by different letters. Table 3.2 enumerates all the possible cases for the mutual position of of the principal null directions of the Weyl tensor (first column). The same column gives the corresponding type in the Petrov classification [3].* The canonical form Ψ_{ABCD} and the corresponding values of the complex invariants I and J defined by Eqs. (4.24) and (4.25) are given in the second and third columns. The fourth column gives the equations which the spinors Ψ_{ABCD} satisfy in each of the cases considered. The fifth column contains information on which components of the Weyl tensor vanish for a choice of complex null tetrads containing the multiple principal light vectors.† The last column presents the corresponding equations to which the Weyl tensor is subject.

In the case where the Weyl tensor belongs to type I we say that the gravitational field is of algebraically generic type. In the remaining cases the gravitational field is said to be algebraically special. An algebraically special gravitational field has at least one multiple principal null direction. The nature of the increase in the algebraic degeneracy of the Weyl tensor when its type is varied is depicted in the Penrose diagram (Fig. 3.4).

It seems appropriate to remark that the clssification of a gravitational field in terms of the algebraic properties of its Weyl tensor can be supplemented by an algebraic classification of the Ricci tensor. (The papers [88, 89], for example, are devoted to this question.)

---

* Concerning the relation between the Petrov classification and the Debever [85]—Penrose [12] algebraic classification based on a study of the principal null directions of the Weyl tensor, of which the spinor variant was treated in this section, see [5, 12, 26, 85-87].

† We use here the notation given in Eq. (A3.2).

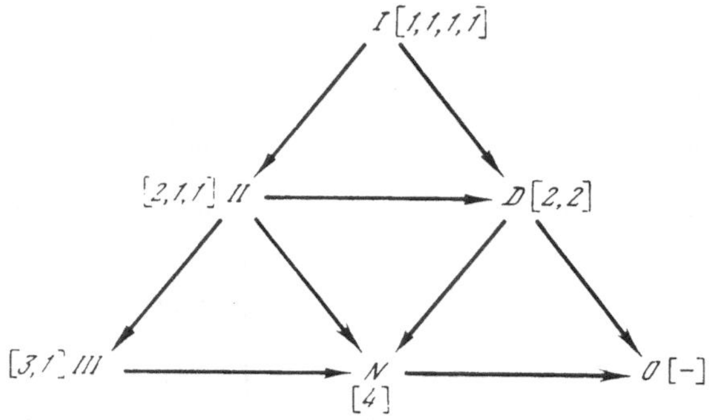

Fig. 3.4

3.6. Spinor Analysis

So far in this chapter we have only considered the algebraic properties of tensors and spinors, and therefore all arguments were carried out relative to a tensor or spinor at a point. In passing to the study of the equations describing the propagation of electromagnetic and gravitational fields in space-time, it is necessary to consider tensor and spinor fields on manifolds. As in the case of tensor analysis, an important role is played by the notion of covariant derivative in the construction of spinor analysis. In this section a definition of covariant derivative of a spinor field is given and its main properties are considered.

If to each point P in some region $U \subset V$ of a space-time manifold V there corresponds a spinor ξ^A, then we say that a spinor field $\xi^A(P)$ is defined in the region U. More precisely, this means the following. Let a smooth field of orthonormalized [with respect to the pseudo-Riemannian metric $g_{\mu\nu}$ of signature (+, −, −, −,) defining the space-time line elements $ds^2 = g_{\mu\nu}dx^\mu dx^\nu$] tetrads e^μ_m be chosen in a region $U \subset V$ on the manifold V. A spinor ξ is defined at the point $P \in U$ if to every orthonormalized tetrad at P there corresponds a pair of complex numbers in such a way that the spinor components (pair of numbers) transforms in the way described in Section 3.2 when we go from one orthonormal frame to another by means of a Lorentz transformation.

The covariant differentiation ∇_X of a spinor field $T^{\cdots}_{\cdots}$ of type (n, m; p', q') along the direction X is defined as a mapping of the spinor field $T^{\cdots}_{\cdots}$ into the spinor field $(\nabla_X T)^{\cdots}_{\cdots}$ of the same type satisfying the following conditions:

1. Linearity:

$$\nabla_X (S^{\cdots}_{\cdots} + T^{\cdots}_{\cdots}) = \nabla_X (S^{\cdots}_{\cdots}) + \nabla_X (T^{\cdots}_{\cdots}), \tag{6.1}$$

$$\nabla_{\alpha X+\beta Y} (T^{\cdots}_{\cdots}) = \alpha \nabla_X (T^{\cdots}_{\cdots}) + \beta \nabla_Y (T^{\cdots}_{\cdots}). \tag{6.2}$$

2. Leibniz rule:

$$\nabla_X (S^{\cdots}_{\cdots} T^{\cdots}_{\cdots}) = (\nabla_X S)^{\cdots}_{\cdots} T^{\cdots}_{\cdots} + S^{\cdots}_{\cdots} (\nabla_X T)^{\cdots}_{\cdots}. \tag{6.3}$$

3. Action on a scalar field [spinor of type (0, 0; 0, 0)]:

$$\nabla_X \varphi = X(\varphi) \equiv X^\mu \partial_\mu \varphi \tag{6.4}$$

4. Reality:

$$\overline{(\nabla_X S)^{\cdots}_{\cdots}} = \nabla_X \overline{(S^{\cdots}_{\cdots})}. \tag{6.5}$$

5. $\nabla_{\mathbf{X}}(\varepsilon_{AB}) = 0, \qquad \nabla_{\mathbf{X}}(\sigma_{\mu}^{AA'}) = 0.$ (6.6)

It follows from relation (6.6) that

$$\nabla_{\mathbf{X}}(g_{\mu\nu}) = 0. \tag{6.7}$$

We remark that conditions (6.1)-(6.4) and (6.7) uniquely determine the Riemannian connection and the operation of covariant differentiation in the case of tensor fields under the additional assumption that the torsion vanishes. In a similar way, axioms (6.1)-(6.6) uniquely determine the operation of covariant differentiation of spinor fields in the torsionless case. Moreover, the covariant derivative introduced in this way, when viewed on tensor fields obtained from Hermitian spinor fields with the aid of $\sigma_{\mu}^{AA'}$, coincides with the usual covariant Riemann derivative.

Proposition 3.6.1. Let $\zeta_a^A(P)$ be a pair of spinor fields on a manifold normalized by the condition

$$\varepsilon_{AB}\zeta_a^A\zeta_b^B = \varepsilon_{ab}. \tag{6.8}$$

Then the covariant derivative of the spinor field $S_{C\cdots D'\cdots}^{A\cdots B'\cdots}$ is uniquely determined by conditions (6.1)-(6.6) and is given in the basis ζ_a^A by the relation

$$(\nabla_{\mathbf{X}}S)_{c\cdots d'\cdots}^{a\cdots b'\cdots} = X^{\mu}(\partial_{\mu}S_{c\cdots d'\cdots}^{a\cdots b'\cdots} + \Gamma_{\mu f}^{\;a}S_{c\cdots d'\cdots}^{f\cdots b'\cdots} + \overline{\Gamma}_{\mu f'}^{\;b'}S_{c\cdots d'\cdots}^{a\cdots f'\cdots} - \Gamma_{\mu c}^{\;f}S_{f\cdots d'\cdots}^{a\cdots b'\cdots} - \overline{\Gamma}_{\mu d'}^{\;f'}S_{c\cdots f'\cdots}^{a\cdots b'\cdots}), \tag{6.9}$$

where

$$\Gamma_{\mu b}^{a} = \frac{1}{2}\sigma_{\lambda}^{ab'}(\sigma_{bb'}^{\nu}\Gamma_{\mu\nu}^{\lambda} + \partial_{\mu}\sigma_{bb'}^{\lambda}), \tag{6.10}$$

$$\overline{\Gamma}_{\mu b'}^{a'} = \frac{1}{2}\sigma_{\lambda}^{ba'}(\sigma_{bb'}^{\nu}\Gamma_{\mu\nu}^{\lambda} + \partial_{\mu}\sigma_{bb'}^{\lambda}) = \overline{\Gamma_{\mu b}^{a}}. \tag{6.11}$$

Proof. First of all, we remark that if $X = X^{\mu}\partial_{\mu}$ then we have using (6.2)

$$(\nabla_{\mathbf{X}}S)_{\cdots}^{\cdots} = X^{\mu}(\nabla_{\mu}S)_{\cdots}^{\cdots}. \tag{6.12}$$

The covariant derivative $\nabla_{\mu}\zeta_a^A$ of a basis spinor field is a spinor field and can therefore be expanded with respect to the basis fields,

$$\nabla_{\mu}\zeta_a^A = \Gamma_{\mu a}^{\;b}\zeta_b^A. \tag{6.13}$$

Using Eqs. (6.3) and (6.4) we can obtain

$$\nabla_{\mu}\bar{\zeta}_{a'}^{A'} = \overline{\Gamma}_{\mu a'}^{\;b'}\bar{\zeta}_{b'}^{A'}, \qquad \nabla_{\mu}\zeta_A^a = -\Gamma_{\mu b}^{\;a}\zeta_A^b, \qquad \nabla_{\mu}\bar{\zeta}_{A'}^{a'} = -\overline{\Gamma}_{\mu b'}^{\;a'}\bar{\zeta}_{A'}^{b'}. \tag{6.14}$$

For simplicity we give the proof of Eq. (6.9) for the case of a spinor field $S_{D'}^{A}$ of type (1, 0; 0', 1'). In order to do this we write

$$(\nabla_{\mu}S)_{d'}^{a} \equiv \zeta_A^a\bar{\zeta}_{d'}^{D'}\nabla_{\mu}S_{D'}^{A} = \zeta_A^a\bar{\zeta}_{d'}^{D'}\nabla_{\mu}(\zeta_b^A\bar{\zeta}_{D'}^{c'}S_{c'}^{b}).$$

Applying the Leibniz rule (6.3) and using (6.13) and (6.14), we have

$$(\nabla_{\mu}S)_{d'}^{a} = \zeta_A^a\bar{\zeta}_{d'}^{D'}[(\partial_{\mu}S_{c'}^{b})\zeta_b^A\bar{\zeta}_{D'}^{c'} + \Gamma_{\mu b}^{\;c}\zeta_c^A\bar{\zeta}_{D'}^{c'}S_{c'}^{b} - \overline{\Gamma}_{\mu f'}^{\;c'}\zeta_b^A\bar{\zeta}_{D'}^{f'}S_{c'}^{b}] = \partial_{\mu}S_{d'}^{a} + \Gamma_{\mu b}^{\;a}S_{d'}^{b} - \overline{\Gamma}_{\mu d'}^{\;c'}S_{c'}^{a}. \tag{6.15}$$

Equation (6.9) for the case of the spinor field $S^{A}_{D'}$ is thereby proved. The proof in the general case proceeds analogously.

We now turn to the derivation of Eqs. (6.10) and (6.11). To do this we show that

$$\Gamma_{\mu a}^{\;a} = 0; \qquad \overline{\Gamma}_{\mu a'}^{\;a'} = 0. \tag{6.16}$$

Indeed, differentiating the relation $\varepsilon_{ab} = \varepsilon_{AB}\zeta_a^A\zeta_b^B$, we have on the basis of (6.4) and (6.6) that

$$0 = \nabla_\mu(\varepsilon_{AB}\zeta_a^A\zeta_b^B) = \varepsilon_{AB}(\Gamma_{\mu a}^{\;c}\zeta_c^A\zeta_b^B + \Gamma_{\mu b}^{\;c}\zeta_a^A\zeta_c^B) = \Gamma_{\mu a}^{\;c}\varepsilon_{cb} + \Gamma_{\mu b}^{\;c}\varepsilon_{ac}. \tag{6.17}$$

The first equality in (6.16) is obtained if we multiply both sides of (6.17) by ε^{ab}. The proof of the second relation in (6.16) is completely analogous.

We now observe that the quantities $\sigma^\lambda_{bb'}$ are a set of vectors and therefore we have

$$\nabla_\mu\sigma^\lambda_{bb'} = \partial_\mu\sigma^\lambda_{bb'} + \Gamma_{\mu\nu}^{\;\lambda}\sigma^\nu_{bb'}. \tag{6.18}$$

On the other hand,

$$\nabla_\mu(\sigma^\lambda_{bb'}) = \nabla_\mu(\sigma^\lambda_{BB'}\zeta_b^B\overline{\zeta}_{b'}^{B'}) = \sigma^\lambda_{BB'}(\Gamma_{\mu b}^{\;c}\zeta_c^B\overline{\zeta}_{b'}^{B'} + \overline{\Gamma}_{\mu b'}^{\;c'}\zeta_b^B\overline{\zeta}_{c'}^{B'}). \tag{6.19}$$

Multiplying (6.19) by $\sigma_\lambda^{ab'} = \sigma_\lambda^{AD'}\zeta_A^a\overline{\zeta}_{D'}^{b'}$, we obtain

$$\sigma_\lambda^{ab'}\nabla_\mu(\sigma^\lambda_{bb'}) = \zeta_A^a\overline{\zeta}_{D'}^{b'}\sigma_\lambda^{AD'}\sigma^\lambda_{BB'}(\Gamma_{\mu b}^{c}\zeta_c^B\overline{\zeta}_{b'}^{B'} + \overline{\Gamma}_{\mu b'}^{c'}\zeta_b^B\overline{\zeta}_{c'}^{B'}) = \zeta_A^a\overline{\zeta}_{B'}^{b'}(\Gamma_{\mu b}^{c}\zeta_c^A\overline{\zeta}_{b'}^{B'} + \overline{\Gamma}_{\mu b'}^{c'}\zeta_b^A\overline{\zeta}_{c'}^{B'}) = 2\Gamma_{\mu b}^{a} + \delta_b^a\overline{\Gamma}_{\mu b'}^{b'} = 2\Gamma_{\mu b}^{a}. \tag{6.20}$$

Comparing (6.18) and (6.20), we finally get the desired relation (6.10). Equation (6.11) is proved analogously.

Using the concept of covariant derivative it is possible to define the parallel transport of a spinor in the usual way. We say that a spinor $S^{\cdots}_{\cdots}$ is parallel transported along the curve γ: $x^\mu = x^\mu(u)$ if

$$\nabla S^{\cdots}_{\cdots} = 0, \qquad \nabla \equiv \frac{dx^\mu}{du}\nabla_\mu. \tag{6.21}$$

If the vector A^μ is parallel transported along the curve γ ($\nabla A^\mu = 0$) then the vectors A^μ define a direction at every point of the curve. The vector field $B^\mu = \lambda(x)A^\mu$, which defines the same field of directions as A^μ, no longer satisfies the equation $\nabla B^\mu = 0$ if $\nabla\lambda \neq 0$. As is easily verified, it instead satisfies the equation $B^{[\mu}\nabla B^{\nu]} = 0$. This equation defines the parallel transport of directions along the curve γ. In complete analogy, if a pair of vectors A^μ and B^μ is given at every point of the curve γ and defines the two-dimensional area $A^{[\mu}B^{\nu]}$, we say that this two-dimensional area is parallel transported along γ if an arbitrary vector of the area again lies in the corresponding two-dimensional area after parallel transport, i.e., the condition

$$A^{[\mu}B^{\nu}\nabla(\alpha A^{\lambda]} + \beta B^{\lambda]}) = 0,$$

holds, or what is equivalent,

$$A^{[\mu}B^{\nu}\nabla A^{\lambda]} = A^{[\mu}B^{\nu}\nabla B^{\lambda]} = 0. \tag{6.22}$$

As shown in Section 3.4, the spinor ξ^A has a flag as geometric image, i.e., a two-dimensional area constructed on the lightlike vector $\xi^\alpha = \sigma^\alpha_{AA'}\xi^A\overline{\xi}^{A'}$. It is easy to see that the following proposition is valid.

Proposition 3.6.2. The spinor ξ^A is parallel transported along a curve γ if and only if both the vector $\xi^a = \sigma^a_{AA'}\xi^A\bar{\xi}^{A'}$ and the two-dimensional area defined by the spinor ξ^A in accordance with (4.10) are parallel transported along the curve γ.

The covariant differentiation operator ∇_μ introduced above takes spinor quantities into geometric objects for which an additional tensor index μ appears along with the spinor indices. It is easy to define instead of this operator a covariant differentiation operator taking spinors into spinors by the following relation:

$$\partial_{AA'} = \sigma^\mu_{AA'}\nabla_\mu, \tag{6.23}$$

or in the basis ζ^A_a

$$\partial_{aa'} = \sigma^\mu_{aa'}\nabla_\mu. \tag{6.24}$$

It was shown in Proposition 3.4.2 that the vectors $\sigma^\mu_{aa'}$ form a complex null tetrad, and therefore the operators $\partial_{aa'}$ are covariant differentiation operators along the lightlike directions defined by the vectors of the tetrad. For what follows it turns out to be convenient to introduce the following standard notation:

$$\begin{aligned} D &= l^\mu\nabla_\mu = \partial_{00'}, \qquad \Delta = n^\mu\nabla_\mu = \partial_{11'},\\ \delta &= m^\mu\nabla_\mu = \partial_{01'}, \qquad \bar{\delta} = \bar{m}^\mu\nabla_\mu = \partial_{10'}. \end{aligned} \tag{6.25}$$

The quantities $\Gamma^b_{\mu a}$ and $\bar{\Gamma}^b_{\mu a'}$ introduced in Eqs. (6.13) and (6.14) are the spinor analogs of the Ricci rotation coefficients. Along with these quantities, it is convenient to introduce the analogous objects possessing only spinor indices,

$$\Gamma_{abcd'} = \sigma^\mu_{cd'}\varepsilon_{fb}\Gamma^f_{\mu a}, \qquad \bar{\Gamma}_{a'b'c'd} = \sigma^\mu_{dc'}\varepsilon_{f'b'}\bar{\Gamma}^{f'}_{\mu a'}. \tag{6.26}$$

Since the basis we use in spin space is normalized, Eq. (6.16) holds, which shows that

$$\Gamma_{abcd'} = \Gamma_{bacd'}, \qquad \bar{\Gamma}_{a'b'c'd} = \bar{\Gamma}_{b'a'c'd}. \tag{6.27}$$

Proposition 3.6.3. The quantities $\Gamma_{abcd'}$ can be expressed directly in terms of derivatives of $\sigma_{\mu ab'}$ as follows:

$$\Gamma_{abcd'} = \frac{1}{2}\varepsilon^{p'q'}\sigma^\mu_{aq'}\sigma^\nu_{cd'}\nabla_\nu\sigma_{\mu bp'} = \frac{1}{2}\varepsilon^{p'q'}\sigma^\mu_{aq'}\partial_{cd'}\sigma_{\mu bp'}, \tag{6.28}$$

or

$$\Gamma_{abcd'} = \frac{1}{2}\varepsilon^{p'q'}\left\{\sigma_{cd'ap'bq'} - \sigma_{cd'bq'ap'} - \sigma_{ap'bq'cd'}\right\}, \tag{6.29}$$

where

$$\sigma_{ab'cd'ef'} = \sigma^{[\mu}_{ab'}\sigma^{\nu]}_{cd'}\sigma_{\mu ef',\nu}. \tag{6.30}$$

Proof. We first prove Eq. (6.28). To do this we remark that

$$\nabla_\nu(\sigma_{\mu bp'}) = \nabla_\nu(\sigma_{\mu AB'}\zeta^A_b\bar{\zeta}^{B'}_{p'}) = \sigma_{\mu AB'}(\nabla_\nu\zeta^A_b\bar{\zeta}^{B'}_{p'} + \zeta^A_b\nabla_\nu\bar{\zeta}^{B'}_{p'})$$

and

$$\sigma^{\mu}_{aq'}\sigma_{\mu AB'} = \sigma^{\mu}_{CD'}\zeta^{C}_{a}\bar{\zeta}^{D'}_{q'}\sigma_{\mu AB'} = \zeta_{Aa}\bar{\zeta}_{B'q'}.$$

We therefore have

$$\frac{1}{2}\varepsilon^{p'q'}\sigma^{\mu}_{aq'}\sigma^{\nu}_{cd'}\nabla_{\nu}\sigma_{\mu bp'} = \frac{1}{2}\varepsilon^{p'q'}\sigma^{\nu}_{cd'}\zeta_{Aa}\bar{\zeta}_{B'q'}\left(\nabla_{\nu}\zeta^{A}_{b}\bar{\zeta}^{B'}_{p'} + \zeta^{A}_{b}\nabla_{\nu}\bar{\zeta}^{B'}_{p'}\right) = \frac{1}{2}\varepsilon^{p'q'}\varepsilon_{p'q'}\sigma^{\nu}_{cd'}\zeta_{aA}\nabla_{\nu}\zeta^{A}_{b} = \Gamma_{bacd'} = \Gamma_{abcd'}.$$

(The second term in the parentheses gives no contribution since $\varepsilon^{p'q'}\bar{\zeta}_{q'B'}\nabla_{\nu}\bar{\zeta}^{B'}_{p'} = 0$.) Equality (6.29) is a consequence of Eq. (6.28) proved above and the identity

$$\Gamma_{abcd'} = \Gamma_{a(bc)d'} + \Gamma_{b(ca)d'} - \Gamma_{c(ab)d'}. \tag{6.31}$$

The relation between the quantities $\Gamma_{abcd'}$ and the spin coefficients introduced in Chapter 1, Eq. (1.5.7), is a straightforward consequence of the assertion just proved. Indeed, the $\sigma_{\mu ab'}$ are vectors and coincide with the vectors $z_{m\mu}$ of the complex null tetrad (4.1). Equations (6.29) and (6.30) show that the $\Gamma_{abcd'}$ are linear combinations of the Ricci rotation coefficients for these tetrads. It can be seen by direct verification that these are precisely the linear combinations which were introduced in the definition of the spin coefficients (1.5.7). The connection between the individual spin coefficients and the quantities $\Gamma_{abcd'}$ is given in Table 3.3.

When applied to a scalar function φ, the covariant differentiation operators ∇_{μ} and ∇_{ν} commute with one another, i.e., $(\nabla_{\mu}\nabla_{\nu} - \nabla_{\nu}\nabla_{\mu})\varphi = 0$. Commutativity no longer holds for the operators $\partial_{aa'}$ introduced in Eq. (6.24). It turns out that the spin coefficients $\Gamma_{abcd'}$ serve as a measure of the lack of commutativity of these operators. More precisely, the following proposition holds.

Proposition 3.6.4. If φ is a scalar function then

$$(\partial_{AB'}\partial_{CD'} - \partial_{CD'}\partial_{AB'})\varphi = 0, \tag{6.32}$$

$$(\partial_{ab'}\partial_{cd'} - \partial_{cd'}\partial_{ab'})\varphi = \{\varepsilon^{fe}(\Gamma_{facd'}\partial_{eb'} - \Gamma_{fcab'}\partial_{ed'}) + \varepsilon^{f'e'}(\bar{\Gamma}_{f'b'd'c}\partial_{ae'} - \bar{\Gamma}_{f'd'b'a}\partial_{ce'})\}\varphi. \tag{6.33}$$

Proof. Equality (6.32) is a direct consequence of the definition (6.23) of the operators $\partial_{AA'}$, Eq. (6.6), and the commutativity of the covariant differentiation operators, $\nabla_{[\mu}\nabla_{\nu]}\varphi = 0$. In order to prove Eq. (6.33), we write its left-hand side in the form

$$\partial_{ab'}\partial_{cd'}\varphi - \partial_{cd'}\partial_{ab'}\varphi = \sigma^{\mu}_{ab'}(\nabla_{\mu}\sigma^{\nu}_{cd'})\nabla_{\nu}\varphi - \sigma^{\nu}_{cd'}(\nabla_{\nu}\sigma^{\mu}_{ab'})\nabla_{\mu}\varphi. \tag{6.34}$$

We further use Eq. (6.19), after writing it in the form

$$\nabla_{\mu}\sigma^{\nu}_{cd'} = \sigma^{\nu}_{CD'}(\Gamma^{e}_{\mu c}\zeta^{C}_{e}\bar{\zeta}^{D'}_{d'} + \bar{\Gamma}^{e'}_{\mu d'}\zeta^{C}_{c}\bar{\zeta}^{D'}_{e'}),$$

TABLE 3.3

$\Gamma_{abcd'} =$	cd'	$ab = 00$	$01, 10$	11
	$0\,0'$	k	ε	π
	$1\,0'$	ρ	α	λ
	$0\,1'$	σ	β	μ
	$1\,1'$	τ	γ	ν

in order to obtain

$$\sigma^{\mu}_{ab'}(\nabla_{\mu}\sigma^{\nu}_{cd'})\nabla_{\nu}\varphi = \Gamma^{e}{}_{cab'}\partial_{ed'}\varphi + \bar{\Gamma}^{e'}{}_{d'b'a}\partial_{ce'}\varphi. \tag{6.35}$$

We have completely analogously that

$$\sigma^{\nu}_{cd'}(\nabla_{\nu}\sigma^{\mu}_{ab'})\nabla_{\mu}\varphi = \Gamma^{e}{}_{acd'}\partial_{eb'}\varphi + \bar{\Gamma}^{e'}{}_{b'd'c}\partial_{ae'}\varphi. \tag{6.36}$$

If we now remember that $\Gamma^{e}_{abd'} = -\varepsilon^{fe}\,\Gamma_{fabd'}$, we obtain Eq. (6.33) by substituting (6.35) and (6.36) into (6.34).

The equality (6.33) just proved permits one to write the commutation relations for the differential operators D, Δ, δ and $\bar{\delta}$. The corresponding equations containing the spin coefficients are presented in explicit form in the Appendix to this chapter [cf. (A3.1)].

3.7. The Newman – Penrose Equations. The Spinor Form of the Bianchi Identities and of the Equations for Massless Fields

In this section we obtain an expression for the spinor components of the curvature tensor in terms of the spin coefficients and their derivatives, which allows us to write the Einstein equations in spinor form. The relation which exists between the spinor formalism and the complex light tetrad formalism makes it possible to get an explicit expression for the components of the curvature tensor in a complex light tetrad by using the results of spinor calculations (which are substantially shorter than the direct calculations in the tetrad formalism).

We begin the derivation of the spinor form of the Einstein equations with the proof of a number of useful relations.

Proposition 3.7.1. We have the following identity:

$$\partial_{AA'}\partial_{BB'} - \partial_{BB'}\partial_{AA'} = \varepsilon_{AB}\partial_{C(A'}\partial^{C}{}_{B')} - \varepsilon_{A'B'}\partial_{(A}{}^{C'}\partial_{B)C'}. \tag{7.1}$$

Proof. We remark first of all that

$$\partial_{AA'}\partial_{BB'} - \partial_{AB'}\partial_{BA'} = \varepsilon_{A'B'}\partial_{AC'}\partial_{B}{}^{C'}. \tag{7.2}$$

Indeed, the left-hand side of (7.2) is antisymmetric in A'B' and is hence proportional to $\varepsilon_{A'B'}$. The factor multiplying $\varepsilon_{A'B'}$ can be found by multiplying the left-hand side by $\varepsilon^{A'B'}$. Using (7.2) we have

$$\partial_{AA'}\partial_{BB'} - \partial_{BB'}\partial_{AA'} = \partial_{AA'}\partial_{BB'} - \partial_{AB'}\partial_{BA'} + \partial_{AB'}\partial_{BA'} - \partial_{BB'}\partial_{AA'} = \varepsilon_{A'B'}\partial_{AC'}\partial_{B}{}^{C'} + \varepsilon_{AB}\partial_{CA'}\partial^{C}{}_{B'}$$

and analogously,

$$\partial_{AA'}\partial_{BB'} - \partial_{BB'}\partial_{AA'} = -(\partial_{BB'}\partial_{AA'} - \partial_{AA'}\partial_{BB'}) = \varepsilon_{A'B'}\partial_{BC'}\partial_{A}{}^{C'} + \varepsilon_{AB}\partial_{CB'}\partial_{A'}^{C}.$$

Combining both of these relations, we see that Eq. (7.1) is correct.

Equation (7.1) just proved is a decomposition of the operator depending on spinor indices on the left-hand side of (7.1), into irreducible parts (operators symmetric with respect to the spinor indices).

Proposition 3.7.2. Let ξ^A be an arbitrary spinor. Then the following equalities hold:

$$\partial_{(A}{}^{A'}\partial_{B)A'}\xi_C = -\Psi_{ABCD}\xi^D + 2\Lambda\xi_{(A}\varepsilon_{B)C}, \tag{7.3}$$

$$\partial_{D(A'}\partial^D{}_{B')}\xi_A = \Phi_{ABA'B'}\xi^B. \tag{7.4}$$

Proof. Equations (7.3) and (7.4) are in essence the spinor analogs of the Ricci identity

$$(\nabla_\mu\nabla_\nu - \nabla_\nu\nabla_\mu)\xi_\lambda = R^\alpha{}_{\lambda\mu\nu}\xi_\alpha, \tag{7.5}$$

which can be viewed as defining the curvature tensor. Instead of finding the spinor analog of Eq. (7.5), it turns out to be more convenient to start with the Ricci identity for a bivector,

$$(\nabla_\alpha\nabla_\beta - \nabla_\beta\nabla_\alpha)\xi_{\gamma\delta} = R^\varepsilon{}_{\gamma\alpha\beta}\xi_{\varepsilon\delta} + R^\varepsilon{}_{\delta\alpha\beta}\xi_{\gamma\varepsilon}, \tag{7.6}$$

where as bivector we take*

$$\xi_{\gamma\delta} \leftrightarrow \xi_{CC'DD'} = \xi_C\xi_D\varepsilon_{C'D'}. \tag{7.7}$$

The spinor analog of Eq. (7.6) has the form

$$(\partial_{AA'}\partial_{BB'} - \partial_{BB'}\partial_{AA'})\xi_{CC'DD'} = R^{EE'}{}_{CC'AA'BB'}\xi_{EE'DD'} + R^{EE'}{}_{DD'AA'BB'}\xi_{CC'EE'}, \tag{7.8}$$

where the spinor analog of the curvature tensor $R_{EE'DD'AA'BB'}$ is given by Eq. (4.12).

Using Eq. (7.1) we rewrite the left-hand side of (7.8) in the form

$$\varepsilon_{C'D'}\,[\varepsilon_{AB}\partial_{C(A'}\partial^C{}_{B')} - \varepsilon_{A'B'}\,\partial_{(A}{}^{C'}\partial_{B)C'}\;(\xi_C\xi_D). \tag{7.9}$$

We may therefore consider separately the parts of Eq. (7.8) which are symmetric and antisymmetric with respect to AB. This is achieved by contracting both sides of (7.8) with $\varepsilon^{A'B'}$ in the first case and ε^{AB} in the second. Contraction with $\varepsilon^{A'B'}$ gives

$$-\partial_{(A}{}^{C'}\partial_{B)C'}(\xi_C\xi_D) = 2\Psi_{ABE(C}\xi_{D)}\xi^E + 4\Lambda\xi_{(C}\varepsilon_{D)\,(A}\xi_{B)}. \tag{7.10}$$

If we make use of the relation

$$\partial_A{}^{C'}\xi_C\partial_{BC'}\xi_D = -\partial_{AC'}\xi_C\partial_B{}^{C'}\xi_D,$$

and apply the Leibniz rule, we can obtain the left-hand side of Eq. (7.10) in the following form:

$$-\xi_C\partial_{(A}{}^{C'}\partial_{B)C'}\xi_D - \xi_D\partial_{(A}{}^{C'}\partial_{B)C'}\xi_C.$$

In order to obtain Eq. (7.3) containing ξ_C linearly, it is necessary to multiply both sides of the above equation by $\eta^C\eta^D$, where η^C is an arbitrary spinor. Then $\xi_C\eta^C$ appears in both sides as a common factor and can be discarded. In the remaining equation the arbitrary spinor η^D appears as a common factor. Since it is arbitrary, we finally obtain the desired equation (7.3). Equality (7.4) is proved analogously, except that in this case we contract Eq. (7.8) with ε^{AB}.

---

* In view of the linearity of Eq. (7.6), it is valid for both real and complex bivectors.

Proposition 3.7.3. The following equality holds:

$$\partial_{fe'}\Gamma_{acdb'} - \partial_{db'}\Gamma_{acfe'} = \varepsilon^{pq}\{\Gamma_{apdb}\Gamma_{qcfe'} + \Gamma_{acpb'}\Gamma_{qdfe'} - \Gamma_{apfe'}\Gamma_{qcdb'} - \Gamma_{acpe'}\Gamma_{qfdb'}\} + \varepsilon^{r's'}\{\Gamma_{acdr'}\overline{\Gamma}_{s'b'e'f} - \Gamma_{acfr'}\overline{\Gamma}_{s'e'b'd}\} + \Psi_{acdf}\varepsilon_{e'b'} + \Lambda\varepsilon_{e'b'}(\varepsilon_{cd}\varepsilon_{af} + \varepsilon_{ad}\varepsilon_{cf}) + \Phi_{acb'e'}\varepsilon_{fd}. \quad (7.11)$$

Proof. For the proof we consider the following expression:

$$\Pi = \zeta_{Ec}[\partial_{fe'}\partial_{db'} - \partial_{db'}\partial_{fe'}]\zeta^E_a. \quad (7.12)$$

Using the equality

$$\partial_{ab'}\zeta^E_c = \varepsilon^{pq}\Gamma_{qcab'}\zeta^E_p$$

and performing the differentiation, we obtain

$$\Pi = \partial_{fe'}\Gamma_{acdb'} - \partial_{db'}\Gamma_{acfe'} + \varepsilon^{pq}\Gamma_{qadb'}\Gamma_{cpfe'} - \varepsilon^{pq}\Gamma_{qafe'}\Gamma_{fpdb'}. \quad (7.13)$$

We now calculate the quantity Π in another way. Using the relations

$$\partial_{ab'} = \zeta^A_a\bar{\zeta}^{B'}_{b'}\partial_{AB'}, \qquad \partial_{AB'}\zeta^C_c = \varepsilon^{CD}\Gamma_{DEAB'}\zeta^E_c, \quad (7.14)$$

we have

$$\Pi = \zeta_{Ec}[\zeta^F_f\bar{\zeta}^{E'}_{e'}\partial_{FE'}(\zeta^D_d\bar{\zeta}^{B'}_{b'})\partial_{DB'}\zeta^E_a - \zeta^D_d\bar{\zeta}^{B'}_{b'}\partial_{DB'}(\zeta^F_f\bar{\zeta}^E_{e'})\partial_{FE'}\zeta^E_a] + \zeta_{Ec}\zeta^F_f\bar{\zeta}^{E'}_{e'}\zeta^D_d\bar{\zeta}^{B'}_{b'}[\partial_{FE'}\partial_{DB'} - \partial_{DB'}\partial_{FE'}]\zeta^F_a.$$

Calculation of the expression in the first set of square brackets gives

$$\Pi_1 = \varepsilon^{r's'}(\overline{\Gamma}_{s'b'e'f}\Gamma_{cadr'} - \overline{\Gamma}_{s'e'b'd}\Gamma_{cafr'}) + \varepsilon^{pq}(\Gamma_{qdfe'}\Gamma_{capb'} - \Gamma_{qfdb'}\Gamma_{cape'}). \quad (7.15)$$

Calculating the expression in the second set of square brackets after using Eqs. (7.1), (7.3), and (7.4), we obtain

$$\Pi_2 = \Psi_{acdf}\varepsilon_{e'b'} + \Lambda\varepsilon_{e'b'}(\varepsilon_{cd}\varepsilon_{af} + \varepsilon_{ad}\varepsilon_{cf}) + \Phi_{acb'e'}\varepsilon_{fd}. \quad (7.16)$$

Comparing the resulting expression $\Pi = \Pi_1 + \Pi_2$, where Π_1 and Π_2 are given by Eqs. (7.15) and (7.16), respectively, with relation (7.13) found previously, we get the desired equality (7.11).

Relations (7.11) were derived in [1] and are called the Newman–Penrose equations (NP equations or NP systems for short). In the Appendix to this chapter we give the explicit form of the NP equations written with the aid of the operators D, Δ, δ and $\bar{\delta}$ [Eq. (A3.4)].

The system of NP equations can be used to find the metric arising due to a known field and matter distribution described by the energy-momentum tensor $T_{\mu\nu}$. The Einstein equations in tensor form

$$R_{\mu\nu} - \frac{1}{2}g_{\mu\nu}R = -\frac{8\pi G}{c^4}T_{\mu\nu} \quad (7.17)$$

are differential equations for the quantities $g_{\mu\nu}$. The principal object in the NP formalism are the complex null tetrads z^μ_m. The metric $g_{\mu\nu}$ is expressed in terms of these quantities in the form

$$g_{\mu\nu} = z_{1\mu}z_{2\nu} + z_{2\mu}z_{1\nu} - z_{3\mu}z_{4\nu} - z_{4\mu}z_{3\nu}.$$

In order to show how the NP equations permit the determination of the unknown quantities z^{μ}_{m} we remark that the spin coefficients are obtained from the $z_{m\mu}$ by means of differentiation operations.* If using (7.17) we express $R_{\mu\nu}$ in terms of $T_{\mu\nu}$ and then substitute Λ and $\Phi_{abc'd'}$ thus determined into the NP equations (7.11), the resulting system of equations makes it possible to find z^{μ}_{m} and Ψ_{abcd}. If in the process of solving the system we are interested only in finding z^{μ}_{m}, it is easy to eliminate Ψ_{abcd} from (7.11). However, it often turns out to be convenient to use Eqs. (7.11) explicitly containing Ψ_{abcd}. The point is that, as was shown in Section 3.5, the algebraic type of a gravitational field is determined by the algebraic properties of the spinor Ψ_{abcd}. In the case where the gravitational field considered is of a definite algebraically degenerate type, using the arbitrariness in the choice of tetrads it is possible to make the spinor component Ψ_{abcd} equal to zero (cf. Table 3.2). For such a choice of tetrads the NP equations (6.11) simplify essentially, and under the above assumptions concerning Ψ_{abcd}, any solution of them will describe a gravitational field of a completely determined type. This question is considered in more detail in the following chapter.

We now derive the spinor form of the Bianchi identities and Maxwell equations.

Proposition 3.7.4. The Bianchi identities

$$\nabla_{[\mu}R_{\alpha\beta]\gamma\delta} = 0 \tag{7.18}$$

in spinor form have the form

$$\partial^{D}{}_{A'}\Psi_{ABCD} = \partial_{(C}{}^{B'}\Phi_{AB)A'B'}, \tag{7.19}$$

$$\partial^{BB'}\Phi_{ABA'B'} = -3\partial_{AA'}\Lambda. \tag{7.20}$$

Proof. If we define the dual curvature tensor by $\overset{*}{R}_{\alpha\beta\gamma\delta} = {}^{1}/_{2}\,\varepsilon_{\gamma\delta}{}^{\mu\nu}R_{\alpha\beta\mu\nu}$, the Bianchi identities can be rewritten as follows:

$$\nabla^{\delta}\overset{*}{R}_{\alpha\beta\gamma\delta} = 0.$$

The spinor analog of the dual curvature tensor $\overset{*}{R}_{\alpha\beta\gamma\delta}$ has the form

$$\begin{aligned}\overset{*}{R}_{\alpha\beta\gamma\delta} \leftrightarrow i\,[\varepsilon_{AB}\varepsilon_{CD}\overline{\Psi}_{A'B'C'D'} - \varepsilon_{A'B'}\varepsilon_{C'D'}\Psi_{ABCD} - \varepsilon_{AB}\varepsilon_{C'D'}\Phi_{CDA'B'} + \\ + \varepsilon_{CD}\varepsilon_{A'B'}\Phi_{ABC'D'} + 2\Lambda\,(\varepsilon_{AC}\varepsilon_{BD}\varepsilon_{A'B'}\varepsilon_{D'C'} + \varepsilon_{AB}\varepsilon_{CD}\varepsilon_{A'C'}\varepsilon_{B'D'})].\end{aligned} \tag{7.21}$$

Letting $\partial^{DD'}$ act on (7.21) and symmetrizing with respect to AB, we obtain

$$\partial^{D}_{C'}\Psi_{ABCD} - \partial^{D'}_{C}\Phi_{ABC'D'} + 2\varepsilon_{C(A}\partial_{B)C'}\Lambda = 0. \tag{7.22}$$

The parts of this equality which are symmetric and antisymmetric with respect to BC give (7.19) and (7.20) respectively. It is straightforward to check that the part of (7.21) antisymmetric with respect to AB gives no new information.

Proposition 3.7.5. In a normalized basis ζ^{A}_{a} the Bianchi identities (7.19), (7.20) have the form

$$\begin{aligned}\partial^{p}_{d'}\Psi_{abcp} - \partial_{(c}{}^{t'}\Phi_{ab)d't'} = \{3\Psi_{pr(ab}\Gamma_{c)}{}^{pr}{}_{d'} + \Psi_{abcp}\Gamma^{p}{}_{r}{}^{r}{}_{d'}\} - 2\Gamma^{p}{}_{(ab}{}^{t'}\Phi_{c)pt'd'} - \\ - \{\overline{\Gamma}_{t'd'v'(a}\Phi_{bc)}{}^{t'v'} + \overline{\Gamma}_{t'}{}^{v'}{}_{v'(a}\Phi_{bc)}{}^{t'}{}_{d'}\},\end{aligned} \tag{7.23}$$

$$\begin{aligned}3\partial_{ab'}\Lambda + \partial^{pt'}\Phi_{apb't'} = \varepsilon^{v'w'}\{\Phi_{apb'}{}^{t'}\overline{\Gamma}_{t'w'v'}{}^{p} + \Phi_{ap}{}^{t'}{}_{w'}\overline{\Gamma}_{b't'v'}{}^{p}\} - \\ - \{\Phi_{prb'}{}^{t'}\Gamma_{a}{}^{pr}{}_{t'} + \Phi_{apb'}{}^{t'}\Gamma^{p}{}_{r}{}^{r}{}_{t'}\}.\end{aligned} \tag{7.24}$$

*We recall that (4.1) shows that z^{μ}_{m} coincides with σ^{μ}_{aa}. Equation (6.28) gives an explicit expression for the spin coefficients in terms of the derivatives of $\sigma^{\mu}_{aa'}$. Another way of finding the spin coefficients consists in applying Eq. (6.33) to the coordinate functions x^{μ}.

Proof. The proof proceeds by direct calculations after substituting $\Psi_{ABCD} = \zeta^a_A\zeta^b_B\zeta^c_C\zeta^d_D \times \Psi_{abcd}$, $\Phi_{ABA'B'} = \zeta^a_A\zeta^b_B\bar{\zeta}^{a'}_{A'}\bar{\zeta}^{b'}_{B'}\cdot\Phi_{aba'b'}$ and $\partial_{AA'} = \zeta^a_A\bar{\zeta}^{a'}_{A'}\partial_{aa'}$ into Eqs. (7.19) and (7.20).

The Bianchi identities in explicit form [1, 5] are given in the Appendix to this chapter [Eq. (A3.7)].

Proposition 3.7.6. The Maxwell equations

$$\nabla_\beta F^{\alpha\beta} = 0, \qquad F_{[\alpha\beta,\,\gamma]} = 0 \tag{7.25}$$

have the spinor form

$$\partial^{BA'}\Phi_{AB} = 0. \tag{7.26}$$

Proof. Applying the operator $\partial^{BB'}$ to the spinor analog $F_{\alpha\beta} \leftrightarrow (\varepsilon_{AB}\overline{\Phi}_{A'B'} + \varepsilon_{A'B'}\Phi_{AB})$ we have

$$\partial_A{}^{B'}\overline{\Phi}_{A'B'} + \partial^B{}_{A'}\Phi_{AB} = 0. \tag{7.27}$$

Remarking that $F_{[\alpha\beta,\gamma]} = 0$, we may now write $\nabla_\beta F^{*\alpha\beta} = 0$. In spinor form this gives

$$\partial_A{}^{B'}\overline{\Phi}_{A'B'} - \partial^B{}_{A'}\Phi_{AB} = 0. \tag{7.28}$$

Equations (7.27) and (7.28) are equivalent to (7.26).

Using the usual arguments, it can be seen that the following proposition is valid.

Proposition 3.7.7. The Maxwell equations (7.26) have the form

$$\partial^{ba'}\Phi_{ab} - \Gamma^{c\ ba'}_{a}\Phi_{cb} - \Gamma^{b\ \ ca'}_{\ c}\Phi_{ab} = 0. \tag{7.29}$$

in the normalized basis ζ^A_a.

Remark. The Maxwell equations (7.26) in spinor form are a special case of an equation describing a massless field of arbitrary spin s > 0 which has the form [90]

$$\partial^{A_1A'}\Phi_{A_1A_2\ldots A_{2s}} = 0. \tag{7.30}$$

In particular, in a normalized basis ζ^A_a the equation describing a massless field of spin 1/2 can be written in the form

$$\partial^{ba'}\Phi_b - \Gamma^b{}_c{}^{ca'}\Phi_b = 0. \tag{7.31}$$

The explicit form of Eqs. (7.29) and (7.31) is given in the Appendix to this chapter [Eqs. (A3.10) and (A3.14)].

In concluding this section we mention a modification of the NP formalism proposed by Geroch, Held, and Penrose [95, 96]. In the variant considered by them, instead of describing geometric quantities using complex null tetrads, use is made of a pair of null vectors at each point together with the two-dimensional surface orthogonal to them. In place of the spin coefficients and operators D, Δ, δ and $\bar{\delta}$, in this approach quantities are used which have definite spin and conformal weights, in particular, the operators ð and $\bar{ð}$ (cf. Sec. 2.2). On the one hand, this permits the description of the NP equation in a more compact form, while on the other hand it makes it possible to simplify the calculations. The application of this formalism to obtain vacuum metrics of type D was considered in [96].

APPENDIX TO CHAPTER 3

1. Commutation Relations for the Operators D, Δ, δ, and $\bar{\delta}$. These relations are obtained from Eqs. (3.6.33) by substituting into the expressions for $\Gamma_{abcd'}$ in terms of the spin coefficients in accordance with Table 3.3, taking into account the definition (3.6.25) of the operators D, Δ, δ, and $\bar{\delta}$:

$$\begin{aligned}
(\Delta D - D\Delta)\,\varphi &= [(\gamma + \bar{\gamma})\,D + (\varepsilon + \bar{\varepsilon})\,\Delta - (\tau + \bar{\pi})\,\bar{\delta} - (\bar{\tau} + \pi)\,\delta]\,\varphi,\\
(\delta D - D\delta)\,\varphi &= [(\bar{\alpha} + \beta - \bar{\pi})\,D + k\Delta - \sigma\bar{\delta} - (\bar{\rho} + \varepsilon - \bar{\varepsilon})\,\delta]\,\varphi,\\
(\delta\Delta - \Delta\delta)\,\varphi &= [-\bar{\nu}D + (\tau - \bar{\alpha} - \beta)\,\Delta + \bar{\lambda}\bar{\delta} + (\mu - \gamma + \bar{\gamma})\,\delta]\,\varphi,\\
(\bar{\delta}\delta - \delta\bar{\delta})\,\varphi &= [(\bar{\mu} - \mu)\,D + (\bar{\rho} - \rho)\,\Delta - (\bar{\alpha} - \beta)\,\bar{\delta} - (\bar{\beta} - \alpha)\,\delta]\,\varphi.
\end{aligned} \tag{A3.1}$$

2. The Newman – Penrose Equations. We introduce the notation

$$\Psi_0 = \Psi_{0000}, \quad \Psi_1 = \Psi_{0001}, \ \Psi_2 = \Psi_{0011}, \ \Psi_3 = \Psi_{0111}, \ \Psi_4 = \Psi_{1111}, \tag{A3.2}$$

$$\begin{aligned}
&\Phi_{00} = \bar{\Phi}_{00} = \Phi_{000'0'}, \quad \Phi_{01} = \bar{\Phi}_{10} = \Phi_{000'1'}, \quad \Phi_{02} = \bar{\Phi}_{20} = \Phi_{001'1'},\\
&\Phi_{10} = \bar{\Phi}_{01} = \Phi_{010'0'},
\end{aligned} \tag{A3.3}$$

$$\begin{aligned}
&\Phi_{11} = \bar{\Phi}_{11} = \Phi_{010'1'}, \quad \Phi_{12} = \bar{\Phi}_{21} = \Phi_{011'1'}, \quad \Phi_{20} = \bar{\Phi}_{02} = \Phi_{110'0'},\\
&\Phi_{21} = \bar{\Phi}_{12} = \Phi_{110'1'}, \ \Phi_{22} = \bar{\Phi}_{22} = \Phi_{111'1'}.
\end{aligned}$$

Equations (3.7.11) can be written in explicit form as follows:

$$\begin{aligned}
&\text{a)}\ D\rho - \bar{\delta}k = (\rho^2 + \sigma\bar{\sigma}) + (\varepsilon + \bar{\varepsilon})\,\rho - \bar{k}\tau - k\,(3\alpha + \bar{\beta} - \pi) + \Phi_{00},\\
&\text{b)}\ D\sigma - \delta k = (\rho + \bar{\rho})\,\sigma + (3\varepsilon - \bar{\varepsilon})\,\sigma - (\tau - \bar{\pi} + \bar{\alpha} + 3\beta)\,k + \Psi_0,\\
&\text{c)}\ D\tau - \Delta k = (\tau + \bar{\pi})\,\rho + (\bar{\tau} + \pi)\,\sigma + (\varepsilon - \bar{\varepsilon})\,\tau - (3\,\gamma + \bar{\gamma})\,k + \Psi_1 + \Phi_{01},\\
&\text{d)}\ D\alpha - \bar{\delta}\varepsilon = (\rho + \bar{\varepsilon} - 2\varepsilon)\,\alpha + \beta\bar{\sigma} - \bar{\beta}\varepsilon - k\lambda - \bar{k}\gamma + (\varepsilon + \rho)\,\pi + \Phi_{10},\\
&\text{e)}\ D\beta - \delta\varepsilon = (\alpha + \pi)\,\sigma + (\bar{\rho} - \bar{\varepsilon})\,\beta - (\mu + \gamma)\,k - (\bar{\alpha} - \bar{\pi})\,\varepsilon + \Psi_1,\\
&\text{f)}\ D\gamma - \Delta\varepsilon = (\tau + \bar{\pi})\,\alpha + (\bar{\tau} + \pi)\,\beta - (\varepsilon + \bar{\varepsilon})\,\gamma - (\gamma + \bar{\gamma})\,\varepsilon + \tau\pi - \nu k + \Psi_2 - \Lambda + \Phi_{11},\\
&\text{g)}\ D\lambda - \bar{\delta}\pi = (\rho\lambda + \bar{\sigma}\mu) + \pi^2 + (\alpha - \bar{\beta})\,\pi - \nu\bar{k} - (3\varepsilon - \bar{\varepsilon})\,\lambda + \Phi_{20},\\
&\text{h)}\ D\mu - \delta\pi = (\bar{\rho}\mu + \sigma\lambda) + \pi\bar{\pi} - (\varepsilon + \bar{\varepsilon})\,\mu - \pi\,(\bar{\alpha} - \beta) - \nu k + \Psi_2 + 2\Lambda,\\
&\text{i)}\ D\nu - \Delta\pi = (\pi + \bar{\tau})\,\mu + (\bar{\pi} + \tau)\,\lambda + (\gamma - \bar{\gamma})\,\pi - (3\varepsilon + \bar{\varepsilon})\,\nu + \Psi_3 + \Phi_{21},\\
&\text{j)}\ \Delta\lambda - \bar{\delta}\nu = -(\mu + \bar{\mu})\,\lambda - (3\gamma - \bar{\gamma})\,\lambda + (3\alpha + \bar{\beta} + \pi - \bar{\tau})\,\nu - \Psi_4,\\
&\text{k)}\ \delta\rho - \bar{\delta}\sigma = \rho\,(\bar{\alpha} + \beta) - \sigma\,(3\alpha - \bar{\beta}) + (\rho - \bar{\rho})\,\tau + (\mu - \bar{\mu})\,k - \Psi_1 + \Phi_{01},\\
&\text{l)}\ \delta\alpha - \bar{\delta}\beta = (\mu\rho - \lambda\sigma) + \alpha\bar{\alpha} + \beta\bar{\beta} - 2\alpha\beta + \gamma\,(\rho - \bar{\rho}) + \varepsilon\,(\mu - \bar{\mu}) - \Psi_2 + \Lambda + \Phi_{11},\\
&\text{m)}\ \delta\lambda - \bar{\delta}\mu = (\rho - \bar{\rho})\,\nu + (\mu - \bar{\mu})\,\pi + \mu\,(\alpha + \bar{\beta}) + \lambda\,(\bar{\alpha} - 3\beta) - \Psi_3 + \Phi_{21},\\
&\text{n)}\ \delta\nu - \Delta\mu = (\mu^2 + \lambda\bar{\lambda}) + (\gamma + \bar{\gamma})\,\mu - \bar{\nu}\pi + (\tau - 3\beta - \bar{\alpha})\,\nu + \Phi_{22},\\
&\text{o)}\ \delta\gamma - \Delta\beta = (\tau - \bar{\alpha} - \beta)\,\gamma + \mu\tau - \sigma\nu - \varepsilon\bar{\nu} - \beta\,(\gamma - \bar{\gamma} - \mu) + \alpha\bar{\lambda} + \Phi_{12},\\
&\text{p)}\ \delta\tau - \Delta\sigma = (\mu\sigma + \bar{\lambda}\rho) + (\tau + \beta - \bar{\alpha})\,\tau - (3\gamma - \bar{\gamma})\,\sigma - k\bar{\nu} + \Phi_{02},\\
&\text{q)}\ \Delta\rho - \bar{\delta}\tau = -(\rho\bar{\mu} + \sigma\lambda) + (\bar{\beta} - \alpha - \bar{\tau})\,\tau + (\gamma + \bar{\gamma})\,\rho + \nu k - \Psi_2 - 2\Lambda,\\
&\text{r)}\ \Delta\alpha - \bar{\delta}\gamma = (\rho + \varepsilon)\,\nu - (\tau + \beta)\,\lambda + (\bar{\gamma} - \bar{\mu})\,\alpha + (\bar{\beta} - \bar{\tau})\,\gamma - \Psi_3.
\end{aligned} \tag{A3.4}$$

3. Expression of the Spin Coefficients in Terms of the Components of Vectors of Light Tetrads. If we take the function φ in Eqs. (A3.1) to be the coordinate function x^μ (μ fixed) and take into account that

$$Dx^\mu = l^\mu, \qquad \Delta x^\mu = n^\mu, \qquad \delta x^\mu = m^\mu, \qquad \bar{\delta}x^\mu = \bar{m}^\mu, \tag{A3.5}$$

we obtain

$$
\begin{aligned}
&\text{a)}\ \Delta l^{\mu} - Dn^{\mu} = (\gamma + \bar{\gamma})\, l^{\mu} + (\varepsilon + \bar{\varepsilon})\, n^{\mu} - (\tau + \bar{\pi})\, \bar{m}^{\mu} - (\bar{\tau} + \pi)\, m^{\mu},\\
&\text{b)}\ \delta l^{\mu} - Dm^{\mu} = (\bar{\alpha} + \beta - \bar{\pi})\, l^{\mu} + kn^{\mu} - \sigma\bar{m}^{\mu} - (\bar{\rho} + \varepsilon - \bar{\varepsilon})\, m^{\mu},\\
&\text{c)}\ \delta n^{\mu} - \Delta m^{\mu} = -\,\bar{\nu} l^{\mu} + (\tau - \bar{\alpha} - \beta)\, n^{\mu} + \bar{\lambda}\bar{m}^{\mu} + (\mu - \gamma + \bar{\gamma})\, m^{\mu},\\
&\text{d)}\ \bar{\delta} m^{\mu} - \delta\bar{m}^{\mu} = (\bar{\mu} - \mu)\, l^{\mu} + (\bar{\rho} - \rho)\, n^{\mu} - (\bar{\alpha} - \beta)\, \bar{m}^{\mu} - (\bar{\beta} - \alpha)\, m^{\mu}.
\end{aligned}
\tag{A3.6}
$$

In these equations z^{μ} denotes the μ-component of a vector and therefore the operators D, Δ, δ, and $\bar{\delta}$ act on z^{μ} as on a simple function.

4. Bianchi Identities. Equations (3.7.23) and (3.7.24) can be written in explicit form as follows:

$$
\begin{aligned}
&\text{a)}\ \bar{\delta}\Psi_0 - D\Psi_1 + D\Phi_{01} - \delta\Phi_{00} = (4\alpha - \pi)\,\Psi_0 - 2\,(2\rho + \varepsilon)\,\Psi_1 + 3k\Psi_2 +\\
&\qquad + (\bar{\pi} - 2\bar{\alpha} - 2\beta)\,\Phi_{00} + 2\,(\varepsilon + \bar{\rho})\,\Phi_{01} + 2\sigma\Phi_{10} - 2k\Phi_{11} - \bar{k}\Phi_{02},\\
&\text{b)}\ \Delta\Psi_0 - \delta\Psi_1 + D\Phi_{02} - \delta\Phi_{01} = (4\gamma - \mu)\,\Psi_0 - 2\,(2\tau + \beta)\,\Psi_1 + 3\sigma\Psi_2 -\\
&\qquad - \bar{\lambda}\Phi_{00} + 2\,(\bar{\pi} - \beta)\,\Phi_{01} + 2\sigma\,\Phi_{11} + (2\varepsilon - 2\bar{\varepsilon} + \bar{\rho})\,\Phi_{02} - 2k\,\Phi_{12},\\
&\text{c)}\ 3\,(\bar{\delta}\Psi_1 - D\Psi_2) + 2(D\Phi_{11} - \delta\Phi_{10}) + \bar{\delta}\Phi_{01} - \Delta\Phi_{00} = 3\lambda\Psi_0 - 9\rho\Psi_2 +\\
&\qquad + 6\,(\alpha - \pi)\,\Psi_1 + 6k\Psi_3 + (\bar{\mu} - 2\mu - 2\gamma - 2\bar{\gamma})\,\Phi_{00} + (2\alpha + 2\pi +\\
&\qquad + 2\bar{\tau})\,\Phi_{01} + 2\,(\tau - 2\bar{\alpha} + \bar{\pi})\,\Phi_{10} + 2\,(2\bar{\rho} - \rho)\,\Phi_{11} + 2\sigma\Phi_{20} - \bar{\sigma}\Phi_{02} - 2\bar{k}\Phi_{12} - 2k\Phi_{21},\\
&\text{d)}\ 3\,(\Delta\Psi_1 - \delta\Psi_2) + 2\,(D\Phi_{12} - \delta\Phi_{11}) + (\bar{\delta}\Phi_{02} - \Delta\Phi_{01}) = 3\nu\Psi_0 + 6\,(\gamma - \mu)\,\Psi_1 -\\
&\qquad - 9\tau\Psi_2 + 6\sigma\Psi_3 - \bar{\nu}\Phi_{00} + 2\,(\bar{\mu} - \mu - \gamma)\,\Phi_{01} - 2\bar{\lambda}\Phi_{10} + 2\,(\tau + 2\bar{\pi})\,\Phi_{11} +\\
&\qquad + (2\alpha + 2\pi + \bar{\tau} - 2\bar{\beta})\,\Phi_{02} + (2\bar{\rho} - 2\rho - 4\bar{\varepsilon})\,\Phi_{12} + 2\sigma\Phi_{21} - 2k\Phi_{22},\\
&\text{e)}\ 3\,(\bar{\delta}\Psi_2 - D\Psi_3) + D\Phi_{21} - \delta\Phi_{20} + 2\,(\bar{\delta}\Phi_{11} - \Delta\Phi_{10}) = 6\lambda\Psi_1 - 9\pi\Psi_2 +\\
&\qquad + 6\,(\varepsilon - \rho)\,\Psi_3 + 3k\Psi_4 - 2\nu\Phi_{00} + 2\lambda\Phi_{01} + 2\,(\bar{\mu} - \mu - 2\bar{\gamma})\,\Phi_{10} +\\
&\qquad + (2\pi + 4\bar{\tau})\,\Phi_{11} + (2\beta + 2\tau + \bar{\pi} - 2\bar{\alpha})\,\Phi_{20} - 2\bar{\sigma}\Phi_{12} + 2\,(\bar{\rho} - \rho - \varepsilon)\,\Phi_{21} - \bar{k}\Phi_{22},\\
&\text{f)}\ 3\,(\Delta\Psi_2 - \delta\Psi_3) + D\Phi_{22} - \delta\Phi_{21} + 2\,(\bar{\delta}\Phi_{12} - \Delta\Phi_{11}) = 6\nu\Psi_1 - 9\mu\Psi_2 +\\
&\qquad + 6\,(\beta - \tau)\,\Psi_3 + 3\sigma\Psi_4 - 2\nu\Phi_{01} - 2\bar{\nu}\Phi_{10} + 2\,(2\bar{\mu} - \mu)\,\Phi_{11} + 2\lambda\Phi_{02} -\\
&\qquad - \bar{\lambda}\Phi_{20} + 2\,(\pi + \bar{\tau} - 2\bar{\beta})\,\Phi_{12} + 2\,(\beta + \tau + \bar{\pi})\,\Phi_{21} + (\bar{\rho} - 2\varepsilon - 2\bar{\varepsilon} - 2\rho)\,\Phi_{22},\\
&\text{g)}\ \bar{\delta}\Psi_3 - D\Psi_4 + \bar{\delta}\Phi_{21} - \Delta\Phi_{20} = 3\lambda\Psi_2 - 2\,(\alpha + 2\pi)\,\Psi_3 + (4\varepsilon - \rho)\,\Psi_4 -\\
&\qquad - 2\nu\Phi_{10} + 2\lambda\Phi_{11} + (2\gamma - 2\bar{\gamma} + \bar{\mu})\,\Phi_{20} + 2\,(\bar{\tau} - \alpha)\,\Phi_{21} - \bar{\sigma}\Phi_{22},\\
&\text{h)}\ \Delta\Psi_3 - \delta\Psi_4 + \bar{\delta}\Phi_{22} - \Delta\Phi_{21} = 3\nu\Psi_2 - 2\,(\gamma + 2\mu)\,\Psi_3 + (4\beta - \tau)\,\Psi_4 -\\
&\qquad - 2\nu\Phi_{11} - \bar{\nu}\Phi_{20} + 2\lambda\Phi_{12} + 2\,(\gamma + \bar{\mu})\,\Phi_{21} + (\bar{\tau} - 2\bar{\beta} - 2\alpha)\,\Phi_{22},
\end{aligned}
\tag{A3.7}
$$

$$
\begin{aligned}
&\text{a)}\ D\Phi_{11} - \delta\Phi_{10} - \bar{\delta}\Phi_{01} + \Delta\Phi_{00} + 3D\Lambda = (2\gamma - \mu + 2\bar{\gamma} - \bar{\mu})\,\Phi_{00} + (\pi - 2\alpha - 2\bar{\tau})\,\Phi_{01} +\\
&\qquad + (\bar{\pi} - 2\bar{\alpha} - 2\tau)\,\Phi_{10} + 2\,(\rho + \bar{\rho})\,\Phi_{11} + \bar{\sigma}\Phi_{02} + \sigma\Phi_{20} - \bar{k}\Phi_{12} - k\Phi_{21},\\
&\text{b)}\ D\Phi_{12} - \delta\Phi_{11} - \bar{\delta}\Phi_{02} + \Delta\Phi_{01} + 3\delta\Lambda = (2\gamma - \mu - 2\bar{\mu})\,\Phi_{01} + \bar{\nu}\Phi_{00} -\\
&\qquad - \bar{\lambda}\Phi_{10} + 2\,(\bar{\pi} - \tau)\,\Phi_{11} + (\pi + 2\bar{\beta} - 2\alpha - \bar{\tau})\,\Phi_{02} + (2\rho + \bar{\rho} - 2\bar{\varepsilon})\,\Phi_{12} + \sigma\Phi_{21} - k\Phi_{22},\\
&\text{c)}\ D\Phi_{22} - \delta\Phi_{21} - \bar{\delta}\Phi_{12} + \Delta\Phi_{11} + 3\Delta\Lambda = \nu\,\Phi_{01} + \bar{\nu}\Phi_{10} - 2\,(\mu + \bar{\mu})\,\Phi_{11} -\\
&\qquad - \lambda\Phi_{02} - \bar{\lambda}\Phi_{20} + (2\pi - \bar{\tau} + 2\bar{\beta})\,\Phi_{12} + (2\beta - \tau + 2\bar{\pi})\,\Phi_{21} + (\rho + \bar{\rho} - 2\varepsilon - 2\bar{\varepsilon})\,\Phi_{22}.
\end{aligned}
\tag{A3.8}
$$

5. The Maxwell Equations. We introduce the notation

$$\Phi_0 = \Phi_{00}, \qquad \Phi_1 = \Phi_{01}, \qquad \Phi_2 = \Phi_{11}. \tag{A3.9}$$

Equations (3.7.26) can be written in explicit form as follows:

$$\begin{aligned} &\text{a)}\quad D\Phi_1 - \bar{\delta}\Phi_0 = (\pi - 2\alpha)\,\Phi_0 + 2\rho\Phi_1 - k\Phi_2, \\ &\text{b)}\quad D\Phi_2 - \bar{\delta}\Phi_1 = -\lambda\Phi_0 + 2\pi\Phi_1 + (\rho - 2\varepsilon)\,\Phi_2, \\ &\text{c)}\quad \delta\Phi_1 - \Delta\Phi_0 = (\mu - 2\gamma)\,\Phi_0 + 2\tau\Phi_1 - \sigma\Phi_2, \\ &\text{d)}\quad \delta\Phi_2 - \Delta\Phi_1 = -\nu\Phi_0 + 2\mu\Phi_1 + (\tau - 2\beta)\,\Phi_2. \end{aligned} \tag{A3.10}$$

Φ_0, Φ_1, and Φ_2 can be described in terms of the electromagnetic field intensity components $F_{\mu\nu}$ in the form

$$\Phi_0 = l^\mu m^\nu F_{\mu\nu}, \qquad \Phi_1 = {}^1\!/_2\,(l^\mu n^\nu + \bar{m}^\mu m^\nu)\,F_{\mu\nu}, \qquad \Phi_2 = \bar{m}^\mu n^\nu F_{\mu\nu}. \tag{A3.11}$$

The following relations permit the expression of $F_{\mu\nu}$ in terms of Φ_0, Φ_1 and Φ_2:

$$\begin{aligned} {}^1\!/_4 F_{\mu\nu} &= -\mathrm{Re}\,(\Phi_1)\, l_{[\mu}\, n_{\nu]} + i\mathrm{Im}\,(\Phi_1)\, m_{[\mu}\bar{m}_{\nu]} + {}^1\!/_2\,(\Phi_2\, l_{[\mu} m_{\nu]} + \\ &\quad + \bar{\Phi}_2\, l_{[\mu}\bar{m}_{\nu]} - \bar{\Phi}_0 n_{[\mu} m_{\nu]} - \Phi_0 n_{[\mu}\bar{m}_{\nu]}), \\ F^{+}_{\mu\nu} &= {}^1\!/_2\,(F_{\mu\nu} + iF^{*}_{\mu\nu}) = 2\,[\Phi_1\,(m_{[\mu}\,\bar{m}_{\nu]} - l_{[\mu}\, n_{\nu]}) + \Phi_2 l_{[\mu} m_{\nu]} - \Phi_0 n_{[\mu}\bar{m}_{\nu]}]. \end{aligned} \tag{A3.12}$$

If we calculate the energy-momentum tensor for an electromagnetic field and use the Einstein equations to determine the quantities $\Phi_{aba'b'}$ and Λ associated with it, we obtain that $(G = c = 1)$

$$\Lambda = 0, \qquad \Phi_{mn} = \Phi_m \bar{\Phi}_n \qquad (m, n = 0, 1, 2), \tag{A3.13}$$

where the Φ_{mn} are defined by Eqs. (A3.3).

6. The Equations for a Two-Component Neutrino. Equations (3.7.31) can be written in explicit form as follows:

$$\begin{aligned} &\text{a)}\quad (D - \rho + \varepsilon)\,\Phi_1 - (\bar{\delta} - \alpha + \pi)\,\Phi_0 = 0, \\ &\text{b)}\quad (\delta + \beta - \tau)\,\Phi_1 - (\Delta + \mu - \gamma)\,\Phi_0 = 0. \end{aligned} \tag{A3.14}$$

CHAPTER 4

ALGEBRAICALLY SPECIAL GRAVITATIONAL FIELDS

4.1. Introduction

The mutual position and multiplicity of the principal null directions (the algebraic type of the field) and the geometric character of the principal null congruences (the vanishing or nonvanishing of particular optical scalars) permit us to carry out a decomposition of gravitational fields into classes. The system of NP equations obtained at the end of the preceding chapter seems rather cumbersome at first sight and not amenable to study. However, the advantage of such a representation of the Einstein equations becomes more apparent when it is used to obtain exact algebraically special solutions belonging to one or another algebraic class. Indeed, the property of belonging to a certain algebraic class means the vanishing of certain spin coefficients and certain spinor components of the Weyl tensor. When these algebraic conditions are used, the NP system simplifies essentially and frequently turns out to be amenable to solution, which permits the description of all or some of the metrics of a given class.

A convenient choice of coordinates plays a large role in solving the NP equations. If one of the principal null congruences $\Gamma(\mathbf{l})$ is geodesic, it is convenient to choose coordinates (x^0, r, x^2, x^3), where r is a canonical affine parameter along $\Gamma(\mathbf{l})$ and x^0, x^2, and x^3 are constant along null geodesics. For such a choice of coordinates the operator D (differentiation along the direction l^μ) has the simple form $D = l^\mu \nabla_\mu = \partial_\gamma$. In this case, for $\Phi_{00} = \Psi_0 = 0$ the solution of Eqs. (A3.4a) and (A3.4b) is easily found. The general solution of these equations was found in Proposition 1.6.6 (Sachs' theorem). Using the expressions found there for the spin coefficients ρ and σ, it is possible in many cases to carry out a complete integration of the radial equations (equations containing the operator D). If such a radial integration is carried out to completion, the remaining previously unused NP equations lead to a system of differential equations, which arise as "constants of integration" in solving the radial equations and are called the transverse equations, for functions of the three variables x^0, x^2, and x^3. Thus, assuming that the metric belongs to one of the algebraically degenerate classes usually permits one to study a system of equations for functions of a smaller number of variables which completely determine the metric. We remark that in this sense the conditions imposed on the spin coefficients play a role analogous to the assumption that the desired solution possesses a certain symmetry. There is, however, an essential difference in that the metrics which result in the algebraic approach often depend on all four variables (although the dependence on the coordinate r is usually rather simple), and in this way solutions can be obtained which have little symmetry or even no symmetry at all [124]. Many of the solutions obtained by this method describe the radiation of gravitational waves.

The standard procedure discussed above for solving the NP equations consists of two qualitatively different parts; the first (solving the radial equations and substituting the solutions found into the equations in order to find the transverse equations) usually reduces to a large number of standard algebraic operations.* The second part of the problem (solution of the transverse equations) is usually more complicated in principle. Indeed, it is the possibility of solving these equations which is decisive in finding the explicit form of the exact solution.

In this chapter the fundamental results are stated concerning the known algebraically special exact solutions of the Einstein equations. In many cases, such exact solutions were first obtained by methods different from the NP formalism. However, the procedure usually employed can be reformulated in the language of the NP equations. Therefore, in our discussion of these solutions we have given references not only to the fundamental papers, but also to papers which consider this problem in the framework of the NP formalism. In our study of the known algebraic degenerate solutions we mainly follow the paper [24].†

4.2. The Goldberg – Sachs Theorem and Its Generalization

An arbitrary tensor field $\Pi_{\alpha\beta\gamma\delta}$ possessing symmetry properties similar to those of the Weyl tensor $C_{\alpha\beta\gamma\delta}$ generates congruences of principal null vectors in space-time (principal null congruences). The Weyl tensor fields $C_{\alpha\beta\gamma\delta}$ are distinguished from other fields $\Pi_{\alpha\beta\gamma\delta}$ in that they satisfy a number of differential relations. In particular, in the case of empty space the

* It is of interest to observe that the existing computer languages which enable transformations to be performed on analytic expressions permit the use of an electronic computer to obtain the transverse equations from the complete system of NP equations. A program in the FORMAC language was used for such calculations in [91], where all the vacuum metrics of type D were obtained. The description of this and other languages used in calculations with analytic expressions, together with a bibliography devoted to this question, can be found in [92, 93]. A concrete program for the analytic calculation of the Einstein tensor is contained in [94].

† The author is grateful to W. Kinnersley for giving him the opportunity to acquaint himself with a preprint of [24].

field $C_{\alpha\beta\gamma\delta}$ satisfies the Bianchi identity

$$\nabla_{[\mu} C_{\alpha\beta]\gamma\delta} = 0, \tag{2.1}$$

or in the spinor language

$$\partial^{D}{}_{A'} \Psi_{ABCD} = 0. \tag{2.2}$$

These differential relations can lead (and in the case of algebraically special fields, they do indeed lead) to the appearance of additional restrictions on the differential invariants (optical scalars) characterizing the geometric properties of the principal null congruences. In complete analogy with the results stated in Section 2.7 for principal null congruences of an electromagnetic field, Goldberg and Sachs [97] proved the following important result.*

<u>Proposition 4.2.1 (The Goldberg – Sachs Theorem).</u> A free gravitational field is algebraically special if and only if it admits a shear-free congruence of null geodesics.

In the presence of a matter distribution, the Bianchi identities (2.2) are replaced by Eq. (3.7.19)

$$\partial^{D}{}_{A'} \Psi_{ABCD} = \partial_{(C}{}^{B'} \Phi_{AB)\,A'B'} \tag{2.3}$$

and in the general case the Goldberg–Sachs theorem becomes false. In order to state a generalization of the Goldberg–Sachs theorem to the case of a gravitational field with sources, we introduce the following notation. Let $\Gamma(l)$ be a principal null congruence, and choose a basis ζ_a^A in spin space such that

$$l^{\mu} = \sigma^{\mu}_{AA'} \zeta_0^A \bar{\zeta}_{0'}^{A'}. \tag{2.4}$$

We agree to say that the sources of the gravitational field have property F_J ($J = 1, 2, 3$) if the equalities

$$\begin{aligned} &\text{a)}\ F_1\colon \zeta_0^A \zeta_0^B \zeta_0^C \partial_{(C}^{B'} \Phi_{AB)\,A'B'} = 0, \\ &\text{b)}\ F_2\colon \zeta_0^B \zeta_0^C \partial_{(C}^{B'} \Phi_{AB)\,A'B'} = 0, \\ &\text{c)}\ F_3\colon \zeta_0^C \partial_{(C}^{B'} \Phi_{AB)\,A'B'} = 0 \end{aligned} \tag{2.5}$$

are satisfied.

It can be shown in exactly the same way as in deriving Eqs. (3.7.23) for the Bianchi identities in the spin basis ζ_a^A that Eqs. (2.5) have the form

$$\partial_{(c}^{t'} \Phi_{ab)\,d't'} = 2\Gamma^{p}{}_{(ab}{}^{t'} \Phi_{c)\,pt'd'} + \bar{\Gamma}_{t'd'v'(a} \Phi_{bc)}{}^{t'v'} + \bar{\Gamma}_{t'}{}^{v'}{}_{v'(a} \Phi_{bc)}{}^{t'}{}_{d'}, \tag{2.6}$$

where moreover Eq. (2.6) must be satisfied for the case

$$\begin{aligned} &F_1:\ \text{for}\ a = b = c = 0; \\ &F_2:\ \text{for}\ b = c = 0; \\ &F_3:\ \text{for}\ c = 0. \end{aligned}$$

The explicit form of these equations can be obtained if Eqs. (A3.7) with $\Psi_0 = \Psi_1 = \Psi_2 = \Psi_3 = \Psi_4 = 0$ are used. The resulting system is said to be truncated. Moreover, condition F_1 holds if

---

* A simpler proof of this theorem is given in [1], which is based on the NP system of equations.

Eqs. (a) and (b) of the truncated system (A3.7) are satisfied. Condition F_2 is equivalent to the validity of Eqs. (a)-(d), and condition F_3 is equivalent to Eqs. (a)-(f) of truncated system (A3.7).

Proposition 4.2.2 (Robinson and Schild [98]; Kundt and Thompson [99]). If a gravitational field admits a shear-free congruence $\Gamma(1)$ of null geodesics and the sources of the field possess property F_1, then the gravitational field is algebraically special and $\Gamma(1)$ is a principal multiple null congruence.

Proposition 4.2.3 (Robinson and Schild [98]; Kundt and Thompson [99]). Let $\Gamma(1)$ be a principal null congruence corresponding to the multiple eigenlight vectors of an algebraically special Weyl tensor. Then the following assertions are valid:

1. If the gravitational field is of type II then $\Gamma(1)$ is geodesic and shear-free if and only if the sources of the gravitational field have property F_1;
2. If the gravitational field is of type III, then $\Gamma(1)$ is geodesic and shear-free if and only if the sources of the gravitational field have property F_2;
3. If the gravitational field is of type N, then $\Gamma(1)$ is geodesic and shear-free if and only if the sources of the gravitational field have property F_3.

The propositions just stated, which together constitute a generalization of the Goldberg–Sachs theorem, were proved in [98] by the tensor method and in [99] by the spinor method. The proof of these theorems can also be carried out using the NP equations by a method analogous to that of [1]. This generalization of the Goldberg–Sachs theorem in particular allows us to conclude the validity of the following proposition.

Proposition 4.2.4. A gravitational field generated by an electromagnetic field possessing a geodesic and shear-free principal null congruence $\Gamma(1)$ is algebraically special and $\Gamma(1)$ is a multiple principal null congruence of the gravitational field.

Proof. We recall that for an electromagnetic field $\Phi_{mn} = \Phi_m\overline{\Phi}_n$. Since $\Gamma(1)$ is a principal null congruence of the electromagnetic field, we have $\Phi_0 = 0$ and therefore $\Phi_{00} = \Phi_{01} = \Phi_{02} = 0$. For shear-free ($\sigma = 0$) and geodesic ($k = 0$) congruences, these equations imply that condition F_1 is satisfied, and the above proposition is a corollary of Proposition 4.2.2.

4.3. The Robinson – Trautman Line Element

Before turning to the description of the known algebraically special exact solutions of the Einstein equations, we obtain a convenient expression for the length element.

Proposition 4.3.1 (Robinson and Trautman [100], Robinson, Robinson, and Zund [101]). Assume that space-time admits a shear-free congruence $\Gamma(1)$ of null geodesics. Then the most general element of length can be written in the form

$$ds^2 = 2\,(l_\alpha dx^\alpha)\,(dr - \operatorname{Re}(\overline{W}d\zeta) - Ul_\beta dx^\beta) - \frac{d\zeta d\overline{\zeta}}{2G\overline{G}}, \tag{3.1}$$

where $l_\alpha dx^\alpha = du - \operatorname{Re}(\overline{Q}d\zeta)$, $G\overline{G}$, W, $\overline{W}$, U are functions of all four coordinates, and Q and $\overline{Q}$ are functions of u, ζ, $\overline{\zeta}$.

Proof. Let $\Gamma(1)$ be a shear-free ($\sigma = 0$) congruence of null geodesics ($k = 0$) and let r be a canonical parameter along it: $l^\mu = dx^\mu/dr$. In the coordinates (x^0, r, x^2, x^3) we have

$$l^\mu = (0,1,0,0), \quad l^\mu r_{,\mu} = 1, \quad g_{11} = g_{\mu\nu}l^\mu l^\nu = 0, \quad l_1 = 0. \tag{3.2}$$

Since r is a canonical parameter, $l^\mu l_{\nu;\mu} = 0$ and hence

$$0 = 2l^\mu l_{[\nu;\mu]} = 2l^\mu l_{[\nu,\mu]} = l^\mu l_{\nu,\mu} - l^\mu l_{\mu,\nu} = l^\mu l_{\nu,\mu}$$

and consequently,

$$\partial l_\nu/\partial r = 0. \tag{3.3}$$

Let $(l^\mu, n^\mu, m^\mu, \bar{m}^\mu)$ be the complex null tetrad associated with the congruence $\Gamma(l)$, and let the coordinates x^0, x^2, x^3 be chosen so that $m^3 \neq 0$. Consider a complex function $\zeta(x)$ satisfying the equations

$$D\zeta = 0, \quad \delta\zeta = 0. \tag{3.4}$$

The first of these equations gives $\zeta = \zeta(x^0, x^2, x^3)$. Commutation relations (A3.1) applied to the function ζ in the case considered ($k = \sigma = \varepsilon = 0$) allows us to write

$$D\delta\zeta = \delta D\zeta - (\bar{\alpha} + \beta - \bar{\pi})\, D\zeta + \bar{\rho}\delta\zeta. \tag{3.5}$$

If the function $\zeta(x^0, x^2, x^3)$ satisfies the condition $\delta\zeta = 0$ on the surface $r = \text{const}$, (3.5) shows that the equation $\delta\zeta = 0$ is satisfied for all r. Therefore, the existence problem for the solution of system (3.4) reduces to whether the equality $\delta\zeta = 0$ holds on the surface $r = \text{const}$. Since $m^3 \neq 0$, this equation can be written as

$$\zeta_{,3} = -(m^0\zeta_{,0} + m^2\zeta_{,2})/m^3$$

and the existence of a solution of this equation follows from the usual theory of differential equations.

Relation (3.4) shows that the function ζ has the properties

$$l^\mu\nabla_\mu\zeta = 0, \quad m^\mu\nabla_\mu\zeta = 0. \tag{3.6}$$

It follows from the equality $l^\mu\nabla_\mu\zeta = 0$ that $\nabla_\mu\zeta = Al_\mu + Bm_\mu + C\bar{m}_\mu$, and the relation $m^\mu\nabla_\mu\zeta = 0$ gives $C = 0$. We therefore have

$$\nabla_\mu\zeta\nabla^\mu\zeta = \nabla_\mu\bar{\zeta}\nabla^\mu\bar{\zeta} = 0. \tag{3.7}$$

If we write $\nabla_\mu\zeta\nabla^\mu\bar{\zeta} = 4G\bar{G}$, where $G = G(x^0, r, \zeta, \bar{\zeta})$, we can obtain

$$m_\mu = -\frac{1}{2G}\nabla_\mu\bar{\zeta}, \quad \bar{m}_\mu = -\frac{1}{2\bar{G}}\nabla_\mu\zeta. \tag{3.8}$$

We use the coordinate transformation $x^0 \rightarrow u = u(x^0, \zeta, \bar{\zeta})$ to reduce $l_\mu dx^\mu$ to the following form:

$$l_\mu dx^\mu = du - {}^1\!/_2\bar{Q}d\zeta - {}^1\!/_2 Qd\bar{\zeta},$$

where $Q = Q(u, \zeta, \bar{\zeta})$.

We write the general expression for the vector n^μ completing the vectors l^μ, m^μ, $\bar{m}^\mu$ to a complex null tetrad z^μ_m in the form

$$n_\mu dx^\mu = Ul_\mu dx^\mu + bdr - {}^1\!/_2\bar{W}d\zeta - {}^1\!/_2 Wd\bar{\zeta}, \tag{3.9}$$

where U, b, W are functions of all four coordinates.

The normalization condition $n_\mu l^\mu = 1$ gives $b = 1$. Using now expression (1.5.5) for the metric $g_{\mu\nu}$ in terms of the components of the complex null tetrad, we arrive at the desired equality (3.1).

It is generally speaking possible to introduce the coordinates $(u, r, \zeta, \bar{\zeta})$ in a neighborhood of an arbitrary point, although it can turn out that for a certain choice of the coordinates $(u, r, \zeta, \bar{\zeta})$, they do not describe all of space-time. The Robinson–Trautman coordinates just introduced are not uniquely determined by the form of the line element (3.1), but admit a transformation of the form

$$\zeta \to \zeta' = f(\zeta), \qquad \bar{\zeta} \to \bar{\zeta}' = \bar{f}(\bar{\zeta}), \tag{3.10}$$

$$u \to u' = h(u, \zeta, \bar{\zeta}), \qquad r \to r' = r/\partial_u h. \tag{3.11}$$

Proposition 4.3.2. If the shear-free congruence $\Gamma(\mathbf{l})$ of null geodesics is normal, then using transformation (3.11) it is possible to make the function Q vanish, i.e., in expression (3.1) for the line element we can put

$$l_\alpha dx^\alpha = du. \tag{3.12}$$

Proof. By the definition of the normal congruence $\Gamma(\mathbf{l})$ we have

$$l_{[\mu} l_{\nu,\lambda]} = 0.$$

This condition means that there exists an integrating factor for the expression $l_\alpha dx^\alpha$, i.e., there exist functions $\tilde{u}$ and $a(\tilde{u}, \zeta, \bar{\zeta})$ such that

$$l_\alpha dx^\alpha = d\tilde{u}/a(\tilde{u}, \zeta, \bar{\zeta}).$$

If we pass to the coordinates $(\tilde{u}, \tilde{r} = r/a, \zeta, \bar{\zeta})$, then in place of (3.1) we obtain the analogous expression

$$ds^2 = 2(l_\alpha dx^\alpha)(d\tilde{r} - \mathrm{Re}(\overline{W} d\zeta) - U l_\beta dx^\beta) - \frac{d\zeta d\bar{\zeta}}{2G\bar{G}},$$

where $l_\alpha dx^\alpha = d\tilde{u}$.

Proposition 4.3.3. Let $\Phi_{00} = \Psi_0 = 0$. If the shear-free congruence $\Gamma(\mathbf{l})$ of null geodesics has a nonzero dilation $\theta = -\mathrm{Re}(\rho) \neq 0$ then the function G in (3.1) has the form

$$G = -P/(r + i\Sigma), \tag{3.13}$$

where P and Σ are real functions of $u, \zeta, \bar{\zeta}$. If $\Gamma(\mathbf{l})$ is a normal congruence, then $\Sigma = 0$ and the line element (3.1) in this case has the form

$$ds^2 = 2du(dr - \mathrm{Re}(\overline{W} d\zeta) - U du) - \frac{r^2}{2P^2} d\zeta d\bar{\zeta}. \tag{3.14}$$

Proof. According to Sachs' theorem proved in the first chapter (Proposition 1.6.6), in the case $\sigma = 0$, $\theta \neq 0$ we have

$$\rho = -1/(r + i\Sigma).$$

We now remark that the line element (3.1) can be written in the form

$$ds^2 = g_{\mu\nu} dx^\mu dx^\nu, \qquad g_{\mu\nu} = 2[l_{(\mu} n_{\nu)} - m_{(\mu} \bar{m}_{\nu)}], \tag{3.15}$$

provided we choose the complex null tetrad as follows:

$$l^\alpha = \delta^\alpha_1; \quad n^\alpha = \delta^\alpha_0 + U\delta^\alpha_1; \quad m^\alpha = \overline{G}\,(Q\delta^\alpha_0 + W\delta^\alpha_1 + \delta^\alpha_2 + i\delta^\alpha_3). \tag{3.16}$$

Since in the present case $k = \sigma = \varepsilon = 0$, Eq. (A3.6b) permits us to obtain

$$DG = \bar{\rho}G. \tag{3.17}$$

Integration of this equation immediately gives expression (3.13). If the congruence $\Gamma(\mathbf{l})$ of null normal, then $\mathrm{Im}\,(\rho) = 0$ and hence $\Sigma = 0$.

Proposition 4.3.4. Let $\Phi_{00} = \Psi_0 = 0$. If the shear-free congruence $\Gamma(\mathbf{l})$ of null geodesics has zero dilation ($\theta = -\mathrm{Re}\,\rho = 0$), then $\Gamma(\mathbf{l})$ is a normal congruence ($\omega = -\mathrm{Im}\,\rho = 0$) and the line element can be written in the following form:

$$ds^2 = 2du\,(dr - \mathrm{Re}\,(\overline{W}d\zeta) - Udu) - P^2 d\zeta\, d\bar{\zeta}, \tag{3.18}$$

where W, $\overline{W}$, U depend on all four coordinates and P is a real function of u, ζ, $\bar{\zeta}$.

Proof. Using Sachs' theorem in the case $\sigma = \theta = 0$, we have $\rho = 0$. Therefore Eq. (3.17) shows that G does not depend on r. If we write $P = |G|^{-1}$ and make use of the result of Proposition 4.3.2, we get Eq. (3.18) for the line element in the present case.

4.4. Metrics without Dilation

Algebraically special solutions of the Einstein equations in a vacuum in the case where the principal light congruence has no dilation were first systematically studied by Kundt [102]. This class of metric contains gravitational fields of all the possible algebraically special types: II, III, D, N, and O. In [102], Kundt gave a complete description of the metrics of types III and N without dilation. These metrics contain the so-called plane gravitational waves.

The algebraically special gravitational fields without dilation possess a shear-free normal principal null congruence $\Gamma(\mathbf{l})$ (Proposition 4.3.4), and consequently the optical scalars ρ and σ vanish:

$$\rho = \sigma = 0. \tag{4.1}$$

If in the general case the spin coefficient $\tau = l_{\mu;\nu} m^\mu n^\nu$ can be made to vanish by using the arbitrariness in the choice of tetrads (1.5.2), then $|\tau|$ does not change under all conformal tetrad transformations when condition (4.1) holds, and hence in this case $|\tau|$ is an additional invariant defined by the congruence $\Gamma(\mathbf{l})$.

The physical meaning of this invariant can be understood by considering the following simple case. Assume given in flat space-time a one-parameter family of plane wave fronts

$$l_\alpha(u)\, x^\alpha = u, \quad l_\alpha l^\alpha = 0. \tag{4.2}$$

A conformal portrait for the wave fronts can be obtained if the point source of the electromagnetic waves is placed at the focus of an ideal lens, where the lens itself can move in such a way that it remains at a constant distance from the source [102]. If the position of the lens does not change with time, the direction of the vector $l_\alpha(u)$ is fixed, i.e.,

$$dl^\alpha/du = b\,(u)\, l^\alpha\,(u), \tag{4.3}$$

and it is therefore possible to select a new parametrization $\tilde{u} = \tilde{u}(u)$ such that (4.2) takes the form

$$\tilde{l}_\alpha x^\alpha = \tilde{u}, \quad d\tilde{l}^\alpha/d\tilde{u} = 0, \quad \tilde{l}_\alpha \tilde{l}^\alpha = 0. \tag{4.4}$$

Proposition 4.4.1. A congruence $\Gamma(l)$ in Minkowski space satisfying condition (4.2) is geodesic, normal, shear-free, and its dilation is equal to zero,

$$k = \rho = \sigma = 0. \tag{4.5}$$

If moreover the invariant $\tau = 0$, the direction of the vector l^α is constant, i.e., (4.3) holds.

Proof. Differentiating (4.2) with respect to x^α, we obtain

$$l_\alpha = au_{,\alpha},$$

where $a = 1 - \dot{l}_\alpha x^\alpha$. Observe that

$$l_{\alpha;\beta} = \frac{dl_\alpha}{du} u_{,\beta} = \frac{1}{a} \dot{l}_\alpha l_\beta.$$

We therefore have

$$\rho = l_{\alpha;\beta} m^\alpha \bar{m}^\beta = \frac{1}{a} (\dot{l}_\alpha m^\alpha)(l_\beta \bar{m}^\beta) = 0,$$

$$\sigma = l_{\alpha;\beta} m^\alpha m^\beta = \frac{1}{a} (\dot{l}_\alpha m^\alpha)(l_\beta m^\beta) = 0,$$

$$\tau = l_{\alpha;\beta} m^\alpha n^\beta = \frac{1}{a} (\dot{l}_\alpha m^\alpha)(l_\beta n^\beta) = \frac{1}{a} \dot{l}_\alpha m^\alpha.$$

If $\tau = 0$ then $\dot{l}_\alpha m^\alpha = 0$, and since $\dot{l}_\alpha l^\alpha = 0$, we get that $\dot{l}^\alpha = bl^\alpha$, so that $|\tau|$ determines the rate of change of direction of the plane wave front.

In the general case, the two-dimensional surface orthogonal to the vectors $\Gamma(l)$ is called the wave surface of the congruence l^μ. Wave surfaces of a congruence $\Gamma(l)$ exist if and only if $\Gamma(l)$ is normal.

In the present case of metrics without dilation the line element has the form (3.18)

$$ds^2 = 2du\,(dr - \mathrm{Re}\,(\overline{W} d\zeta) - U du) - P^2 d\zeta\, d\bar{\zeta}, \tag{4.6}$$

and since $l_\mu dx^\mu = du$, the surfaces r, u = const are wave surfaces. The geometry on these surfaces is defined by the line element

$$dl^2 = P^2 d\zeta\, d\bar{\zeta}. \tag{4.7}$$

Expression (4.6) for the line element fixes the coordinates up to the following transformations:

$$\tilde{u} = f(u), \quad \tilde{r} = \dot{f}^{-1} r, \quad \tilde{\zeta} = \zeta \qquad \text{(change of phase)} \tag{4.8}$$

$$\tilde{u} = u, \quad \tilde{r} = r + g(u, \zeta, \bar{\zeta}), \quad \tilde{\zeta} = \zeta \qquad \text{(change of origin of the canonical parameter)} \tag{4.9}$$

$$\tilde{u} = u, \quad \tilde{r} = r, \quad \tilde{\zeta} = F(u, \zeta) \qquad \text{(conformal transformation on the wave surface).} \tag{4.10}$$

The Gaussian curvature K of the two-dimensional surface (4.7) is defined by

$$K = -\frac{1}{P^2}\Delta(\ln P), \qquad \Delta \equiv \partial_x^2 + \partial_y^2 = 4\partial_\zeta\partial_{\bar\zeta}. \tag{4.11}$$

It was shown by Kundt [102] that a necessary and sufficient condition for a gravitational field without dilation to belong to type III, N, or O is that

$$\Delta(\ln P) = 0.$$

The vanishing of the Gaussian curvature means that P = 1 can be achieved by a transformation (4.10). Such metrics are completely characterized by the property that they possess flat wave surfaces.

In the case of a gravitational field of type N, flat wave surfaces are geodesic and the corresponding solutions are called plane gravitational waves.

If $\tau = 0$ for the congruence $\Gamma(1)$, the corresponding gravitational field belongs to type III, N, or O. For type N metrics the condition $\tau = 0$ means that the direction of propagation l^α of a plane gravitational wave remains constant, and in this case $l_{\alpha;\beta} = 0$, i.e., l^α forms a field of parallel vectors in space. Plane gravitational waves with parallel propagation vectors have been called pp-waves. The metric for pp-waves has the form*

$$ds^2 = -2Udu^2 + 2dudr - d\zeta d\bar\zeta, \tag{4.12}$$

where the function $U = U(u, \zeta, \bar\zeta)$ satisfies the equation

$$\Delta U = 4\partial_\zeta\partial_{\bar\zeta}U = 0. \tag{4.13}$$

The general solution of this equation has the form

$$U = \mathrm{Re}\,[f(u, \zeta)], \tag{4.14}$$

where $f(u, \zeta)$ is an arbitrary complex function analytic with respect to ζ.

The type D metrics without dilation are also completely known. The general solution contains two arbitrary parameters and has the form

$$ds^2 = 2dud\Sigma - f^{-1}dx^2 - fdy^2, \tag{4.15}$$

where

$$f(x) = \frac{2amx + l(a^2 - x^2)}{2a(x^2 + a^2)},$$
$$d\Sigma = \left(\frac{r^2 l}{2a(x^2 + a^2)} - \frac{2r^2a^2 f}{(x^2 + a^2)^2}\right)du + dr - \frac{2rx}{x^2 + a^2}dx + \frac{arf}{x^2 + a^2}dy. \tag{4.16}$$

The parameter a can be made equal to one by an admissible coordinate transformation. This metric was first obtained by Carter [108], and subsequently Kinnersley [91] showed that the class of type D metrics without dilation is completely exhausted by this metric. The three limiting metrics contained in this class and obtained from (4.15) by letting $a, l \to 0$ in various ways are known as the static B-metrics. These metrics are discussed in [105, [109].

---

*Plane gravitational waves were considered in [3, 26, 32, 90, 102-107].

Type II metrics without dilation have been little studied. An explicit solution of this type was obtained by van Stockum [110, 111].

4.5. Metrics without Rotation

The case where the principal null congruence is normal and has dilation was studied by Robinson and Trautman [112]. This solution is easily obtained by using the NP equations, as was shown by Newman and Tamburino [113]. In the present case the metric has the form*

$$ds^2 = H du^2 + 2 du dr - \frac{r^2}{P^2} d\zeta d\bar{\zeta}, \tag{5.1}$$

and the Einstein equations in a vacuum reduce to the following relations:

$$H = K - \frac{2m}{r} - 2\frac{\dot{P}}{P} r, \tag{5.2}$$

$$\partial_u\left(\frac{m}{P^3}\right) = \frac{1}{4P} \Delta K, \tag{5.3}$$

where m = m(u) is the mass of the system and depends on the lag time u; $\Delta = \partial_x^2 + \partial_y^2 = 4\partial_\zeta\partial_{\bar{\zeta}}$; $K = P^2 \Delta \ln P$ is the Gaussian curvature of the two-dimensional surface Σ with line element

$$dl^2 = d\zeta d\bar{\zeta}/P^2. \tag{5.4}$$

The form of the metric (5.1) fixes the choice of coordinates up to a transformation

$$\tilde{u} = g(u), \quad \tilde{r} = \dot{g}^{-1} r, \quad \tilde{\zeta} = f(\zeta). \tag{5.5}$$

For $P = P_0 \equiv {}^1\!/_2(1 + \zeta\bar{\zeta})$, K = 1 and (5.4) gives an expression for the line element on the surface of the unit sphere. In the general case, we can put

$$P = P_0 V, \quad V = V(u, \zeta, \bar{\zeta}) \tag{5.6}$$

and the function (V − 1) describes the deviation of the geometry of the wave surface u, r = const from a spherical geometry.

Using the definition of the operators $\eth_0$ and $\bar{\eth}_0$ (2.2.16), it is possible to rewrite the Robinson–Trautman equation (5.3) in the following form:

$$-\partial_u\left(\frac{m}{V^3}\right) = \frac{1}{V} \eth_0^2 V \bar{\eth}_0^2 V - \bar{\eth}_0^2 \eth_0^2 V, \tag{5.7}$$

and for the Gaussian curvature we have

$$K = V^2 \eth_0 \bar{\eth}_0 (\ln (P_0 V)). \tag{5.8}$$

Solutions of type N are obtained if

$$m = 0, \ \partial_\zeta K = \partial_{\bar{\zeta}} K = 0, \tag{5.9}$$

---

* This metric can be obtained from the general expression (3.14) if we write H = −2U, replace P by $P/\sqrt{2}$, and take into account the fact that the field equations lead to the equality W = 0.

i.e., K = K(u). In a region where K is positive or negative, we can use transformation (5.5) to reduce K to the values +1 or −1 respectively. In order to obtain all the type N solutions, it is necessary to find the general solution of the equation

$$P^2\Delta(\ln P) = K, \quad K = -1,0,1. \tag{5.10}$$

This equation shows that the surface Σ has constant curvature. A particular case of (5.10) is given by

$$P = {}^1\!/_2\,(1 + K\zeta\bar{\zeta}). \tag{5.11}$$

Although this solution leads to the metric of a flat space, it can be used to obtain the most general solution of (5.10) [114]. This is done by introducing a new coordinate $\tilde{\zeta}$ and function $P(u, \tilde{\zeta})$:

$$\tilde{\zeta} = \tilde{\zeta}(u,\zeta), \qquad \tilde{P}^2(u,\tilde{\zeta}) = \frac{\partial\tilde{\zeta}}{\partial\zeta}\frac{\partial\bar{\tilde{\zeta}}}{\partial\bar{\zeta}}P^2. \tag{5.12}$$

For fixed u, (5.12) is a coordinate transformation, although the substitution (5.12) is not a coordinate transformation of four-dimensional space.

If we take $\tilde{\zeta}$ to be

$$\tilde{\zeta} = \frac{a(u)\zeta + b(u)}{c(u)\zeta + d(u)}, \qquad ad - bc = 1, \tag{5.13}$$

transformation (5.12) leads to a flat space metric in the Newman−Unti coordinates (cf. Section 2.4), the vertices of the light cones u = const lying on arbitrary spacelike, lightlike, or timelike world lines for K = −1, 0, or +1, respectively.

In the case K = +1, the wave surfaces u, r = const are spheres. It is known that all the isometric mappings of the sphere onto itself are rigid rotations of the sphere and are given by transformation (5.13), while any other isometric mapping of the sphere into itself has a singularity. Therefore, any nontrivial type N solution of the class considered has an angular singularity. This angular singularity makes it impossible to use such solutions to describe real spherical gravitational waves.

The class of type III metrics is distinguished by the conditions

$$m = 0, \quad \Delta K = 0, \quad \partial_\zeta K \neq 0. \tag{5.14}$$

The solution of Eq. (5.10) given in [112],

$$P = (\zeta + \bar{\zeta})^{3/2}, \quad K = -{}^3\!/_2\,(\zeta + \bar{\zeta}), \tag{5.15}$$

can be used to obtain an infinite set of type III solutions using transformation (5.12). The papers [115, 116] are devoted to type III metrics. The equation ΔK = 0 has no everywhere bounded solutions other than K = const (of corresponding type N), and therefore all the metrics of the class considered have singularities of curvature on the wave surfaces.

The class of type D metrics in the rotation-free case has been described completely. In addition to the Schwarzschild metrics obtained from (5.1) with $P = P_0 = {}^1\!/_2(1 + K\zeta\bar{\zeta})$ and $H = K - 2m/r$, where K = 0, ±1, this class contains the two-parameter generalizations of the

Schwarzschild metrics known by the name of static C-metrics [105, 109],

$$ds^2 = A^{-2}(x + y)^{-2}(Fdt^2 - F^{-1}dy^2 - G^{-1}dx^2 - Gdz^2), \tag{5.16}$$

$$G(x) = 1 - x^2 - 2mAx^3 - e^2A^2x^4, \quad F(y) = -G(-y). \tag{5.17}$$

For e = 0 this metric describes a solution in vacuum, while for e ≠ 0 it is a solution of the Einstein–Maxwell equations. Kinnersley and Walker [117] have shown that this metric describes the gravitational field of a uniformly accelerated moving charged point mass m (with charge e). In the general case, this solution has an angular singularity, although when m = e this singularity is absent. The solution contains both incoming and outgoing electromagnetic and gravitational waves.

For type II metrics (m ≠ 0) we can arrange that m = 1, owing to the arbitrariness in the choice of coordinates (5.5), and the single equation describing such metrics has the form

$$3\dot{P} + {}^1/_4 P^3\Delta\,(P^2\Delta\,(\ln P)) = 0. \tag{5.18}$$

The number of known solutions of this class is extemely small. (On this point see [114, 118].) In [114] the solutions of these equations which are small deviations from the Schwarzschild metric were analyzed.

For regular Robinson–Trautman metrics with a spherical wave surface (i.e., for metrics without an angular singularity) the function V in (5.6) is bounded and nonzero on the sphere. If we integrate Eq. (5.7) over the unit sphere we get

$$-\frac{d}{du}\int\frac{m}{V^3}\,d\Omega = \int\frac{\eth_0^2 V \bar{\eth}_0^2 V}{V}\,d\Omega. \tag{5.19}$$

(Here we have used the fact that if $\eta_{(-1)}$ is a regular function of weight −1 on the sphere, then $\int \eth_0\eta_{(-1)}\,d\Omega = 0$.) It is shown in [16] that the left-hand side of (5.19) describes the rate of change of the total energy of the system. Equality (5.19) is the law of conservation of energy and shows that the change in the energy of the system is connected with the existence of gravitational waves, the intensity of radiation of which is described by the right-hand side of (5.19). If this radiation is absent, the fact that the integrand is positive definite implies that $\eth_0^2 V \equiv 4P_0\partial_\zeta^2 P = 0$, and hence

$$P = \alpha(u) + \beta(u)\,\zeta + \bar{\beta}(u)\,\bar{\zeta} + \gamma(u)\,\zeta\bar{\zeta}.$$

If m ≠ 0 we can arrange that m = 1, owing to the arbitrariness in the choice of coordinates (5.5), and then Eq. (5.7) shows that $\partial_u V = \partial_u P = 0$; P can be reduced to the form $P = P_0 = {}^1/_2(1 + \zeta\bar{\zeta})$ by a coordinate transformation (5.13) with a, b, c, and d not depending on u. We have thus shown that every regular Robinson–Trautman solution in the absence of radiation is a Schwarzschild metric. This result is due to Derry, Isaacson, and Winicour [119].

An interesting physical interpretation of the Robinson–Trautman solution is given in [44].* If a particle of mass m moves in Minkowski space under the influence of a force F^μ, the equation of motion has the form

$$m\dot{v}^\mu = F^\mu.$$

Multiplying both sides by $\hat{l}_\mu(\zeta, \bar{\zeta})$, we obtain ($v = v^\mu\hat{l}_\mu$, $F = F^\mu\hat{l}_\mu$)

$$m\dot{v} = F(u, \zeta, \bar{\zeta}).$$

---

* On this point cf. also [16, 120, 121].

Both sides of this equation have spin weight s = 0 and conformal weight w = 1. It can be seen that the expression

$$W = \alpha v + \beta v^4 \eth_0^2 \bar{\eth}_0^2 v + \gamma v^3 \eth_0^2 v \bar{\eth}_0^2 v$$

has the analogous property. Therefore the equation

$$m\dot{v} = F + W$$

is a nonlinear relativistically invariant equation of motion for a particle with internal degrees of freedom. The Robinson–Trautman equation (5.7) is the particular case where F = 0, $\alpha = 0$, and $\gamma = -\beta = 1/3$.

4.6. Metrics with Rotation

The class of metrics for which the principal light congruence has rotation is the most complicated one to study. Apparently, the only type N solution in this class known at present was obtained by Hauser [122].* This solution admits a single Killing vector field.

Type III metrics with rotation were recently found by Robinson [124, 125] and Held [126].

The type D solutions in vacuum were completely described by Kinnersley [91]. The most general solution of this type, called the C-NUT metric, contains four arbitrary constants corresponding to the mass, NUT parameter ("magnetic" mass), angular momentum, and acceleration. The C-NUT solution is stationary and axially symmetric. Debever [127] and Plebanski and Demianski [128] obtained a seven-parameter family of solutions of type D of the Einstein–Maxwell equations in an extremely simple form. In the coordinates called by Plebanski the Boyer coordinates, the metric has the form

$$ds^2 = (x+y)^{-2}\left[\frac{1+(xy)^2}{G}\,dx^2 + \frac{G}{1+(xy)^2}(dz + y^2 dt)^2 + \frac{1+(xy)^2}{F}\,dy^2 - \frac{F}{1+(xy)^2}(dt - x^2 dz)^2\right]; \tag{6.1}$$

$$\begin{aligned} G &= (-\lambda/6 - g_0^2 + \gamma) + 2nx - \varepsilon x^2 + 2mx^3 + (-\lambda/6 - e_0^2 - \gamma)\,x^4, \\ F &= (-\lambda/6 + g_0^2 - \gamma) + 2ny + \varepsilon y^2 + 2my^3 + (-\lambda/6 + e_0^2 + \gamma)\,y^4, \\ \varepsilon &= -(a^2 - b^2)/[ab\,(a^2 + b^2)], \qquad \gamma = (a^2 + b^2)^{-1}. \end{aligned} \tag{6.2}$$

This solution possesses two commuting Killing vector fields $\partial/\partial t$ and $\partial/\partial z$. The vector field $\partial/\partial t$ is timelike in the domain $F - y^4 G > 0$, and in this domain the gravitational field is stationary. The metric (6.1)-(6.2) describes a gravitational field generated by a system with mass m and angular momentum $L = m a$ moving with acceleration b and having electric and magnetic charge e_0 and g_0 with a λ-term is present. The parameter n is called the "magnetic" mass or the NUT parameter. These parameters can be combined in pairs in a natural way by introducing the complex quantities M = m + in, $E = e_0 + ig_0$, and A = a + ib. The Weyl tensor Ψ_2 is nonzero provided the quantities M and E do not vanish simultaneously. The electromagnetic field is nonzero if $E \neq 0$. The quantities m, n, e_0, g_0, and λ determining the value of the curvature tensor are called the dynamical parameters. The quantities a and b do not appear explicitly in the expression for the curvature tensor, although the properties of the solution depend on them in an essential way: a and b are called the kinematic parameters.

---

*The Hauser solution can be expressed directly in terms of the function f which is a solution of the equation $f'' + {}^3/_{16}(1 + x^2)f = 0$. An explicit expression for f in terms of hypergeometric functions is given in [123].

TABLE 4.1

$m+in,\ a+ib,\ e+ig,\ \lambda$ Debever [127]; Demianski and Plebanski [128]			
$m+in, a, e+ig, \lambda$ Plebanski [129]	$m+in,\ a+ib,\ e+ig$ Kinnersley		$m+in,\ b,\ e+ig,\ \lambda$
$m+in,\ a,\ e,\ \lambda$ Carter [130]	$m+in,\ a,\ e+ig$ Demianski and Newman [131]	$m+in,\ a+ib,\ \lambda$ Carter [130]	$m+in,\ b,\ e,\ \lambda$ Carter [130]
$m+in,\ a,\ \lambda$ Frolov [132]	$m+in,\ e$ Brill [141]	$m+in,\ a+ib$ Kinnersley [91]	$m+in,\ b,\ e$ Levi-Civita [135], Newman and Tamburino [137], Robinson and Trautman [112], Ehlers and Kundt [105]
$m,\ a,\ \lambda$ Demianski [133]	$m,\ a,\ e$ Newman et al. [137] Perjes [138], Zund [155]	$m+in,\ a$ Demianski [134], Kramer and Neugebauer [140], Robinson, Robinson, and Zund [101]	$m+in,\ \lambda$ Demianski [134], Frolov [132]
$m,\ \lambda$ Kottler [142]	$m,\ e$ Reissner [150], Nordström [151] $m,\ a$ Kerr [143]	$m+in$ Newman, Tamburino, and Unti [144], Taub [145]	$e,\ b$ Bertotti [146], Robinson [147]
λ de Sitter [148]	m Schwarzschild [149]		

Plebanski and Demianski [128] have shown that a large number of previously known exact solutions of the Einstein–Maxwell equations can be obtained from the metric (6.1), (6.2) by means of taking suitable limits. The various special cases which arise as a result of these limiting procedures are shown in Table 4.1* which is taken from [128].

It is of interest to observe that all the type D vacuum solutions possess a two-parameter group of motions. A simple explanation of this fact is given in [152].

The general case of type II metrics with rotation was studied by I. Robinson, J. Robinson, and Zund [101]. An analogous study of these metrics in the framework of the NP formalism was carried out by Talbot [153]. The Einstein equations in vacuum enable us to make the form of the metric (3.1) more concrete as follows:

$$ds^2 = 2\,(l_\mu dx^\mu)\,(dr + \mathrm{Re}\,(P^{-1}\,(r + i\Sigma)\,\bar{\omega} d\zeta) - U l_\mu\,(dx^\mu) - {}^1\!/_2 P^{-2}\,(r^2 + \Sigma^2)\,d\zeta d\bar{\zeta}; \tag{6.3}$$

$$l_\mu dx^\mu = du - \mathrm{Re}\,(\bar{Q} d\zeta), \qquad \omega = -\,\omega^0\,(r - i\Sigma)^{-1} + P\partial_u Q,$$
$$U = U^0 + r\partial \ln P/\partial u + \mathrm{Re}\,\{\Psi_2^0\,(r + i\Sigma)^{-1}\}, \tag{6.4}$$

where P, Q, Ψ_2^0, ω^0, U^0 and Σ do not depend on r. The functions P, U^0, and Σ are real whereas Q, Ψ_2^0, and ω^0 are complex; ω^0, U^0, and Σ can be expressed explicitly in terms of P, Q, and their derivatives, and P, Q, and Ψ_2^0 satisfy three nonlinear differential equations.

For $\Sigma = 0$ the congruence $\Gamma(\mathbf{l})$ is normal, and owing to the freedom in choosing coordinate transformations we can arrange that Q is zero. In this case, two of the differential equations mentioned above show that $\Psi_2^0 = \bar{\Psi}_2^0 = -m(u)$, and the last equation reduces to the Robinson–Trautman equation (5.3).

The most general type II solution with rotation known containing three arbitrary holomorphic functions of the complex variable ζ (or what is the same, six real functions of one

* See also the papers [163, 164, 199, 200].

variable) was obtained in [124]. In the general case, this solution does not admit any Killing vector field. Solutions in this class which are less general were obtained by Robinson, Robinson, and Zund [101, 155, 156] (two arbitrary holomorphic functions of ζ), Kerr and Schild [43] (one arbitrary holomorphic function of ζ), Kerr and Debney [154] (three arbitrary parameters), and Kerr [143] (two arbitrary parameters). It is characteristic of all these metrics that the curvature tensor vanishes as r^{-3} at infinity, i.e., there is no gravitational radiation.

A type II gravitational field generated by an electromagnetic field and admitting a shift-free geodesic congruence was studied by Lind [157]. He showed that the metric and the electromagnetic field are completely determined by five functions which do not depend on the affine parameter r and satisfy a system of five nonlinear differential equations. Lind and Newman [50] showed that such solutions can be described as the gravitational and electromagnetic field from a charged source moving along an arbitrary complex timelike curve in complex Riemannian space.

A regular solution of the Einstein–Maxwell equations (i.e., not having angular singularities) for which a principal light congruence is geodesic, shift-free, and has dilation is called a Kerr–Maxwell solution. Lind [161, 162] showed that there do not exist Kerr–Maxwell solutions of types III and N (i.e., such solutions have an angular singularity). The only Kerr–Maxwell solution without radiation is the Kerr–Newman metric [137] belonging to type D.

Trim and Wainwright [158, 159] have studied the general form of metrics admitting a geodesic shear-free congruence with dilation in the case where the gravitational field is generated by a source distribution for which $\Lambda = \Phi_{00} = \Phi_{01} = \Phi_{02} = 0$, and they showed that the general solution of this class for a given source distribution is uniquely determined by three functions P (real), and Q, Ψ_2^0 (complex) which do not depend on r. These functions satisfy three nonlinear differential equations. In [159] an explicit solution is obtained for metrics without radiation. In the case of empty space these solutions go into the solutions of Robinson and Robinson [124]. In the presence of matter, if $\Phi_{11} \neq 0$, then Φ_{11} has the asymptotic behavior $\Phi_{11} = \Phi_{11}^0 r^{-4} + O(r^{-5})$. In this case the general solution $g_{\alpha\beta}^{(0)}$ in vacuum is as follows:

$$g_{\alpha\beta} = g_{\alpha\beta}^{(0)} + 2\Phi_{11}^0 (r^2 + \Sigma^2)^{-1} l_\alpha l_\beta.$$

In [158, 159] expressions are found for type II metrics when an electromagnetic and neutrino field are present.

In concluding this section, we briefly discuss the algebraically generic (type I) gravitational fields. For such metrics, in the case where one of the principal null congruence $\Gamma(\mathbf{l})$ is geodesic a rather complete description can be obtained by solving the NP equations. By the Goldberg–Sachs theorem for type I metrics, $\sigma \neq 0$, and since by Sachs' theorem (Proposition 1.6.6) it can be concluded that when $\sigma \neq 0$ solutions with $\rho = 0$ do not exist, we see that the class of algebraically generic solutions is empty for $\rho = 0$. The case where the congruence $\Gamma(\mathbf{l})$ is normal ($\omega = 0$) was treated completely by Newman and Tamburino [113]. They obtained an explicit expression for the metric both in the case of a spherical $\theta^2 \neq \sigma\bar{\sigma}$ and a cylindrical $\theta^2 = \sigma\bar{\sigma}$ wave surface. Unti and Torrence [160] considered the general case of metrics with rotation and showed that when $\omega \neq 0$ all solutions have cylindrical wave surfaces, i.e., the condition $\rho\bar{\rho} = \sigma\bar{\sigma}$ is satisfied.

4.7. Kerr – Schild Metrics

Many of the physically interesting metrics such as, e.g., the Schwarzschild, Kerr, or pp-wave metrics, can be written in the following form:

$$g_{\mu\nu} = \mathring{g}_{\mu\nu} + 2H l_\mu l_\nu, \quad (7.1)$$

where $l_\mu l^\mu = 0$, $\Gamma(\mathbf{l})$ is a principal null congruence of the field $g_{\mu\nu}$, and $\mathring{g}_{\mu\nu}$ is a flat space metric. The question therefore naturally arises of describing the solutions of the Einstein equations which can be represented in this form. This problem was posed and partly solved by Kerr and Schild [43], and metrics of the form (7.1) are called Kerr–Schild metrics.* In a broader sense one understands Kerr–Schild metrics to be metrics of the form (7.1), where $\mathring{g}_{\mu\nu}$ is any solution of the Einstein equations (the space need not be flat).

We remark first of all that if l^μ is a lightlike vector in the metric $g_{\mu\nu}$ then it is also lightlike relative to the metric $\mathring{g}_{\mu\nu}$. Choose a complex null tetrad $\mathring{z}^m_\mu = (\mathring{n}_\mu, \mathring{l}_\mu, \mathring{\bar{m}}_\mu, \mathring{m}_\mu)$ for the metric $\mathring{g}_{\mu\nu}$ such that $\mathring{l}_\mu = l_\mu$ and we have

$$\mathring{g}_{\mu\nu} = 2[\mathring{n}_{(\mu}\mathring{l}_{\nu)} - \mathring{m}_{(\mu}\mathring{\bar{m}}_{\nu)}]. \tag{7.2}$$

If we put

$$l_\mu = \mathring{l}_\mu, \qquad n_\mu = \mathring{n}_\mu + H\mathring{l}_\mu, \qquad m_\mu = \mathring{m}_\mu, \qquad \bar{m}_\mu = \mathring{\bar{m}}_\mu, \tag{7.3}$$

the vectors l^μ, n^μ, m^μ, $\bar{m}^\mu$ form a complex light tetrad relative to the metric $g_{\mu\nu}$, i.e.,

$$g_{\mu\nu} = 2\,[n_{(\mu}l_{\nu)} - m_{(\mu}\bar{m}_{\nu)}]. \tag{7.4}$$

Proposition 4.7.1. $\Gamma(\mathbf{l})$ is a congruence of light geodesics in the metric $\mathring{g}_{\mu\nu}$ if and only if $\Gamma(\mathbf{l})$ is a congruence of light geodesics in the metric $g_{\mu\nu}$. The optical scalars ρ qnd σ for the congruence $\Gamma(\mathbf{l})$ in both metrics coincide.

Proof. We will denote the spin coefficients calculated for the complex null tetrad $\mathring{z}^\mu_m$ in the metric $\mathring{g}_{\mu\nu}$ in the usual way (cf. 1.5.7) with an upper index $\circ$ added. We remark that $l_{[\mu,\nu]} = \mathring{l}_{[\mu,\nu]}$ and $m_{[\mu,\nu]} = \mathring{m}_{[\mu,\nu]}$. Therefore, using (A1.6), (A1.8), and Eq. (7.3) we can obtain

$$\text{(A1.6)}:\ k = \mathring{k}, \qquad \mathrm{Re}\,\varepsilon = \mathrm{Re}\,\mathring{\varepsilon}, \quad \mathrm{Im}\,\rho = \mathrm{Im}\,\mathring{\rho};$$
$$\text{(A1.8)}:\ \sigma = \mathring{\sigma}, \qquad 2i\,\mathrm{Im}\,\varepsilon - \rho = 2i\,\mathrm{Im}\,\mathring{\varepsilon} - \mathring{\rho}.$$

Consequently, we have

$$k = \mathring{k}, \qquad \varepsilon = \mathring{\varepsilon}, \qquad \sigma = \mathring{\sigma}, \qquad \rho = \mathring{\rho}. \tag{7.5}$$

These relations prove the proposition.

Proposition 4.7.2. $\Gamma(\mathbf{l})$ is a geodesic principal null congruence for the field $\mathring{g}_{\mu\nu}$ if and only if it is a geodesic principal null congruence for the field $g_{\mu\nu}$. If moreover $\Gamma(\mathbf{l})$ is a multiple principal null congruence for the metric $\mathring{g}_{\mu\nu}$ then the same holds for the metric $g_{\mu\nu}$ also.

Proof. If $\Gamma(\mathbf{l})$ is a geodesic principal null congruence for the field $\mathring{g}_{\mu\nu}$ then

$$k = \mathring{\Psi}_0 = 0.$$

Equation (7.5) shows that

$$k = \mathring{k} = 0, \qquad \varepsilon = \mathring{\varepsilon}, \qquad \rho = \mathring{\rho}, \qquad \sigma = \mathring{\sigma}$$

and therefore, comparing Eqs. (A3.4b) written for the metric $\mathring{g}_{\mu\nu}$ and the metric $g_{\mu\nu}$, we

* The survey of Urbantke [165] is devoted to Kerr–Schild metrics.

get that $\Psi_0 = 0$, i.e., $\Gamma(l)$, is a principal null congruence in the metric $g_{\mu\nu}$. If $\Gamma(l)$ is a multiple principal null congruence in the metric $\overset{\circ}{g}_{\mu\nu}$ then

$$\overset{\circ}{k} = \overset{\circ}{\Psi}_0 = \overset{\circ}{\Psi}_1 = 0.$$

By the Goldberg–Sachs theorem, we have $\overset{\circ}{\sigma} = 0$. Therefore Eq. (7.5) shows that

$$k = \sigma = 0.$$

Repeated application of the Goldberg–Sachs theorem permits us to prove that $\Psi_0 = \Psi_1 = 0$, and hence $\Gamma(l)$ is a multiple principal null congruence for the metric $g_{\mu\nu}$.

If in the Kerr–Schild metric $\overset{\circ}{g}_{\mu\nu}$ is the metric of flat space-time, then the Weyl tensor for it vanishes and application of Proposition 4.7.2 permits us to conclude that a gravitational field in vacuum of the form

$$g_{\mu\nu} = \eta_{\mu\nu} + 2Hl_\mu l_\nu, \qquad l_\mu l^\mu = 0 \tag{7.6}$$

is algebraically special.

In order to obtain Kerr–Schild metrics with $\overset{\circ}{g}_{\mu\nu} = \eta_{\mu\nu}$, it is necessary to have a convenient parametric representation for shear-free congruences of null geodesics. To this end the description of such congruences given in Section 2.3 can be used. However, following the paper [43], we proceed in another way. Introduce complex null coordinates u, v, ζ, $\bar{\zeta}$ in Minkowski space such that

$$ds_0^2 = 2\,(du\,dv - d\zeta d\bar{\zeta}). \tag{7.7}$$

The most general expression for the field of null directions l^μ is described by the following equation:

$$l_\mu dx^\mu = du + \bar{Y}d\zeta + Yd\bar{\zeta} + Y\bar{Y}dv, \tag{7.8}$$

where $Y = Y(u, v, \zeta, \bar{\zeta})$. It is convenient to complete the vector field l^μ to a field of complex null tetrads associated with the given congruence as follows:

$$n_\mu dx^\mu = dv, \qquad m_\mu dx^\mu = d\zeta + Ydv, \qquad \bar{m}_\mu dx^\mu = d\bar{\zeta} + \bar{Y}dv. \tag{7.9}$$

For the contravariant quantities we have

$$\begin{aligned} n^\mu\partial_\mu &= \partial_u, \qquad l^\mu\partial_\mu = \partial_v - Y\partial_\zeta - \bar{Y}\partial_{\bar{\zeta}} + Y\bar{Y}\partial_u, \\ m^\mu\partial_\mu &= -\partial_{\bar{\zeta}} + Y\partial_u, \qquad \bar{m}^\mu\partial_\mu = -\partial_\zeta + \bar{Y}\partial_u. \end{aligned} \tag{7.10}$$

Recalling expression (1.5.7) for the spin coefficients k and σ, we have

$$\sigma = l_{\nu,\mu}m^\mu m^\nu, \qquad k = l_{\nu,\mu}l^\mu m^\nu.$$

Hence the condition that the congruence $\Gamma(l)$ is shear-free ($\sigma = 0$) and geodesic ($k = 0$) means that

$$\partial_{\bar{\zeta}}Y - Y\partial_u Y = 0, \qquad \partial_v Y - Y\partial_\zeta Y = 0. \tag{7.11}$$

Moreover, it is easy to obtain that

$$\rho = l_{\nu,\mu}\bar{m}^\mu m^\nu = \partial_\zeta Y - \bar{Y}\partial_u Y; \tag{7.12}$$

$$\tau = l_{\nu,\mu}n^\mu m^\nu = -\partial_u Y. \tag{7.13}$$

The general solution of the quasilinear equations (7.11) can be written in the implicit form

$$F\,(Y,\ \bar{\zeta}Y + u,\ vY + \zeta) = 0, \tag{7.14}$$

where F is an arbitrary analytic function of three complex variables.

The case $\rho \neq 0$ was treated by Kerr and Schild [43]. In this case, the Einstein equations are satisfied if

$$H = mP^{-3}\operatorname{Re}\rho, \qquad P \equiv pY\bar{Y} + qY + \bar{q}\bar{Y} + c,$$

where m, p, c are real and q is a complex constant, while the function F has the form

$$F = \Phi\,(Y) + (qY + c)\,(vY + \zeta) - (pY + \bar{q})\,(\bar{\zeta}Y + u).$$

Here Φ is an arbitrary analytic function. It can be seen [43] that the vector field

$$K = c\partial_u + \bar{q}\partial_\zeta + q\partial_{\bar{\zeta}} + p\partial_v$$

is a Killing vector field for both the metric $\eta_{\mu\nu}$ and metric (7.6).

The case $\rho = 0$ was studied by Debney [74, 166]. As was shown in Section 4.4, when $\rho = \sigma = 0$ the spin coefficient τ is an invariant. If $\tau = 0$ then Eqs. (7.11)-(7.13) show that Y = const, and owing to the arbitrariness in coordinate transformations preserving the form of metric (7.7) it is possible to make Y equal to zero.

The field equations lead to the following relation for H:

$$\frac{\partial^2 H}{\partial\zeta\partial\bar{\zeta}} = 0,$$

i.e., H is the real part of an arbitrary analytic function $F(u, \zeta)$, and the corresponding metric

$$ds^2 = 2du\,dv - 2d\zeta d\bar{\zeta} + 2H du^2$$

describes the gravitational field of a pp-wave.

For $\tau \neq 0$ the function Y is determined by the following implicit equation:

$$F\,(Y, u') = 0, \qquad u' = u + Y\bar{\zeta} + \bar{Y}\,(vY + \zeta).$$

If we pass to the new coordinates

$$u',\ \zeta' = \zeta + Yv, \qquad \bar{\zeta}' = \bar{\zeta} + \bar{Y}v, \qquad v' = v,$$

the Einstein equations are satisfied if H is defined by the relation

$$(1 - \zeta'\partial_{u'}\bar{Y} - \bar{\zeta}'\partial_{u'}Y)\,H = H_1\,(\zeta', \bar{\zeta}', u') + v'H_2\,(u'),$$

where

$$\frac{\partial^2 H_1}{\partial\zeta'\partial\bar{\zeta}'} = 0.$$

Kerr–Schild metrics describing solutions of the Einstein–Maxwell equations were studied by Debney, Kerr, and Schild [18] (the case $\rho \neq 0$) and Debney [74] (the $\rho = 0$).

4.8. Vaidya's Problem

In the presence in space of high-frequency radiation propagating in the direction l^{μ}, the effective energy-momentum tensor is described by the expression

$$T_{\mu\nu} = ql_{\mu}l_{\nu}, \tag{8.1}$$

as was shown in Section 1.3, where q is a function characterizing the energy flux density.

The problem of finding exact solutions of the Einstein equations with right-hand side of the form (8.1) was posed by Vaidya [167-169]. Expression (8.1) is reminiscent of the energy-momentum tensor for an ideal liquid in the absence of pressure (powdered substance) $T_{\mu\nu} = \rho u_{\mu}u_{\nu}$ in the case where the velocity u_{μ} of motion of the liquid tends to the speed of light. Equation (8.1) is therefore sometimes called the energy-momentum tensor of a light liquid.

We recall that in the Eddington–Finkelstein coordinates the Schwarzschild metric has the form

$$ds^2 = \Phi du^2 + 2du\,dr - r^2\,(d\vartheta^2 + \sin^2\vartheta d\varphi^2), \tag{8.2}$$

$$\Phi = 1 - 2m/r. \tag{8.3}$$

Vaidya showed (167-169] that a spherically symmetric solution of the Einstein equation with right-hand side (8.1) has the form (8.2) for

$$\Phi = 1 - 2m\,(u)/r \tag{8.4}$$

and the rate of change of mass m(u) is directly related to $q : q = \dot{m}$.

The solution (8.2), (8.4) obtained by Vaidya belongs to type D and possesses a normal shear-free congruence with $\theta \neq 0$. A more general solution of the same class was obtained by Kinnersley [170]. It contains four arbitrary functions of a single variable (the lag time) and describes the gravitational field of a radiating point particle with accelerated motion due to radiation reaction (three arbitrary functions correspond to a velocity depending arbitrarily on the time), and the mass of the particle decreases with time [the function m(u) is arbitrary]. Frolov and Hlebnikov [171] gave a complete description of type D metrics for systems with energy-momentum tensor (8.1) when a Λ-term is present and the systems admit a normal shear-free congruence with dilation.

Type II Vaidya metrics admitting a normal shear-free congruence with dilation were studied by Frolov [172]. The general solution in this case contains an arbitrary function of three variables corresponding to the value of the radiation intensity and depending on the direction (two coordinates) and the time. An interesting feature of the solution is the existence of gravitational (longwave) radiation caused by the anisotropic and variable nature of the high-frequence radiation described by the energy-momentum tensor (8.1). The condition that long-wave gravitational radiation be absent coincides with the requirement that the solution be of type D.

An analysis of rotation-free metrics ($\omega = 0$) with zero dilation ($\theta = 0$) in the presence of radiation (8.1) was carried out by Kundt [102]. Plebanski [165] gave a complete description of type N metrics.

Vaidya [54] considered solutions with radiation (8.1) which are representable in the form of Kerr–Schild metrics. The Kerr–Vaidya metric is contained among these solutions; it was previously obtained by Kramer [173] and Vaidya and Patel [53] and is a generalization of the Kerr metric to the case where radiation is present. Murenbeeld and Trollope [198] used approximate methods to describe radiation from slowly rotating radiating spheres.

In our study of the generalized Goldberg–Sachs theorem in Section 4.2 we have already addressed our attention to the fact that when matter is present, multiple principal null congruences are not necessarily shear-free (i.e., the ordinary Goldberg–Sachs theorem is not applicable). For an energy-momentum tensor of the type (8.1) we have $\Phi_{22} \neq 0$ while at the same time all the remaining $\Phi_{mn} = 0$. An application of proposition 4.2.3 to the present case shows that for type N solutions a multiple principal null congruence may fail to be geodesic and shear-free. An example of such a solution is given in [165, 202].

In [174] the question was investigated of when a transformation

$$g_{\mu\nu} = \overset{\circ}{g}_{\mu\nu} + Hl_\mu l_\nu$$

of Kerr–Schild type can be used to construct a solution with radiation (8.1) starting from an algebraically special vacuum solution, and metrics with radiation were obtained with radiation which generalize completely the nonsymmetric solution of Robinson and Robinson [124]. Rao [175] obtained a type I solution with radiation without rotation generalizing the Levi–Civita vacuum solution and describing the gravitational field of a radiating homogeneous cylinder of infinite length and finite cross section.

CHAPTER 5

ASYMPTOTIC PROPERTIES OF A GRAVITATIONAL FIELD. QUANTUM THEORY IN AN ASYMPTOTICALLY FLAT SPACE-TIME

5.1. Introduction

From a physical point of view, the most interesting class of gravitational fields are the fields of systems which are bounded in space ("island systems"). In this case it is natural to expect that far from the gravitating bodies the gravitational field is small and the properties of space-time differ little from the properties of empty space. Although the ideas of asymptotic infinity and asymptotically flat space-time are rather simple and natural, certain difficulties arise in trying to give them a mathematically precise meaning. These difficulties are mainly associated with the fact that in twisted space-time the equivalence of all admissible coordinate systems makes it impossible to define infinity by letting some coordinate tend to an infinite value. In order for the notion of asymptotic infinity to be invariant it is necessary to define it in a form not depending on the choice of coordinates.

The earliest studies of the asymptotic properties of a gravitational field were based for the most part on various assumptions concerning the behavior of the metric tensor at spatial infinity. Essential progress was made in the papers of Bondi, van der Burg, and Metzner [13], and Sachs [14, 176], who used characteristic surfaces and were able to derive from simple assumptions asymptotic properties of the metric and curvature tensor. The NP formalism was used by Newman and Penrose [1] and Newman and Unti [177] in studying the asymptotic properties of a gravitational field. In these papers the focus of attention was shifted from studying the behavior of the metric to investigating the asymptotics of the curvature tensor, and under rather natural assumptions they proved the sequential degeneracy property relating the nature of the behavior of the curvature tensor at infinity to its algebraic properties.

Penrose [9] succeeded in formulating a rigorous definition of the concept of asymptotically flat space by using the notion of conformal infinity introduced by him. The main idea of the approach he developed consists in using a conformal transformation $d\hat{s} = \Omega ds$ of the metric with conformal factor Ω decreasing at infinity to turn space-time into a compact manifold, the boundary of which is formed by the points at infinity. In this case, by an asymptotically flat space-time one means a space for which the geometric properties of the boundary coincide with the geometric properties of the boundary of Minkowski space-time. As was shown by Penrose [11], the property of sequential degeneracy of the curvature tensor follows automatically from his definition of an asymptotically flat space. A general approach to the study of ideal (singular or infinite) points of space-time was developed in [178-184].

In the asymptotically flat case, it is natural to expect that in a region where the gravitational field is small there is a certain approximate (asymptotic) symmetry, although the strict symmetry belonging to homogeneous Minkowski space is absent. Such a group of symmetries was originally defined in [13] as a group of transformations preserving the form of a metric characteristic of an asymptotically flat space. The investigations of Sachs [14] showed that the existence of this symmetry group is related to the existence of an asymptotic Killing field. Penrose [11] showed that the group of asymptotic symmetries (now called the Bondi – Metzner – Sachs group, or BMS group for short) can be viewed as a group of conformal transformations at conformal infinity (cf. also [203]). Although the BMS group is infinite dimensional (which causes a number of difficulties in interpreting it physically), it contains the Poincaré group as a subgroup. The structure of the BMS group and a representation of it were considered in [14, 185-190].

The idea of conformal infinity together with the conformal invariance of the equations for massless fields permits the reduction of the scattering problem for massless particles in an asymptotically flat space-time to Cauchy's problem. Penrose [191] showed that instead of studying the asymptotic properties of the fields, it is possible here to consider their local behavior near a lightlike surface at conformal infinity, a regular behavior of the solutions at conformal behavior leading moreover to the property of sequential degeneracy for these fields.

In the exposition of the material in this chapter we have mainly drawn on the papers [10, 11, 14, 182, 191] and the survey [192]. In the last section of this chapter we briefly consider the problems associated with constructing a quantum theory in asymptotically flat spaces. Following the ideas of Penrose [191], Sachs [14], Komar [193, 194] and Hawking [195], we show how the existence of the BMS group of asymptotic symmetries makes it possible to uniquely define the in and out vacuum states, energy and momentum operators, and the operators for other observables in the asymptotic domain.

5.2. Asymptotic Properties of Minkowski Space

In order to give a rigorous meaning to the concept of "point at infinity" we first consider the simplest case of flat space-time. Assume that a gravitational field source undergoes an internal change leading to the radiation of gravitational waves. The leading and trailing radiation fronts propagating with the speed of light form lightlike surfaces N_0 and N_1 (Fig. 5.1). The radiation field is nonzero in the region contained between N_0 and N_1. Let S denote a spacelike hypersurface passing through the source at time t. If we try to define the parameters of the system by going out to spatial infinity along the surface S, we necessarily leave the domain bounded by the intersection $S \cap N_0$. We therefore cannot describe the change in the source using such measurements, since information about such a measurement does not have time to get to the measuring apparatus. In other words, to measure the changes which took place in

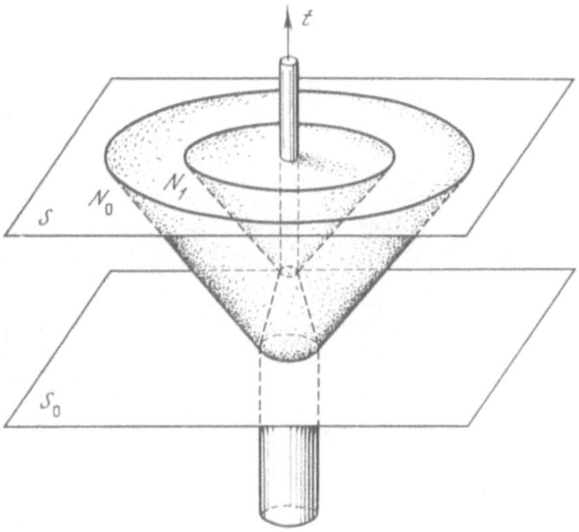

Fig. 5.1

the source as a result of radiation processes, it is necessary to move along a lightlight direction in going away from the source.

Now consider causal (i.e., timelike or lightlike) curves in Minkowski space directed into the future. If such a curve has no endpoint, it defines an "ideal" infinite point in the future space-time. In order to clarify the conditions under which it is natural to assume that a pair of causal curves defines the same "ideal" point, we introduce the concept of the past $I^-[p]$ of the point p; $I^-[p]$ is defined as the set of world points which can be joined to p by a timelike curve directed into the future [10]. In other words, the past of an event p consists of those events which can affect event p by means of signals traveling at speeds less than the speed of light. This definition of $I^-[p]$ is also suitable in the case of twisted space-time. In Minkowski space, the set $I^-[p]$ is the interior of a light cone directed into the past. The future $I^+[p]$ of an event p is defined analogously as the set of world points which can be joined to p by means of timelike curves directed into the past.

An arbitrary point p in Minkowski space is uniquely determined either by its future $I^+[p]$ or by its past $I^-[p]$. For a curve γ, by definition

$$I^\pm[\gamma] = \bigcup_{p\in\gamma} I^\pm[p].$$

If a causal curve directed into the future has an endpoint q, then $I^-[\gamma] = I^-[q]$; in the opposite case the causal curve determines some ideal point in the future at infinity. It is natural to take two ideal points to be the same if the set of events which can affect them are the same, i.e., two curves γ_1 and γ_2 define the same ideal point if $I^-[\gamma_1] = I^-[\gamma_2]$. Curves γ for which $I^-[\gamma]$ coincides with the entire space M determine an ideal point I^+. All timelike lines, for example, belong to this class.

Choosing $I^-[\gamma]$ to be a null line (Fig. 5.2), it can be seen that the situation where γ does not coincide with all of space is possible. If we denote by Π_γ a lightlike hyperplane containing the ray γ, then $I^-[\gamma]$ is comprised of events lying below Π_γ. It can be shown [182] that if $I^-[\gamma]$ does not coincide with all of space-time for a causal curve γ and there is no point p such that $I^-[\gamma] = I^-[p]$, there exists a lightlike hyperplane Π_γ such that the half-space lying below it coincides with $I^-[\gamma]$. If we denote the set of ideal points in the future at infinity by $\mathcal{I}^+$, then there exists a bijective correspondence between points of $\mathcal{I}^+$ and lightlike hyperplanes.

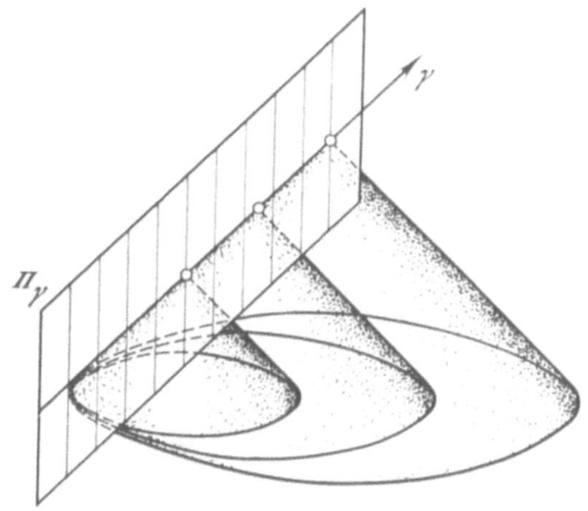

Fig. 5.2

Choose a timelike curve γ_t with parameter u along it equal to the proper time. Every lightlike hyperplane is determined by the value of the time u at which it intersects γ_t together with the direction of the null vector tangent to the hyperplane at the point of intersection. The set of possible directions of light rays is in natural correspondence with the set of points on the two-dimensional sphere S^2, and hence $\mathscr{I}^+$ is a three-dimensional manifold and has the topology of $R \times S^2$.

It is possible to introduce on $\mathscr{I}^+$ not only a differentiable structure but also a metric [10, 11]. In order to do this we observe that if the metric $\eta_{\alpha\beta}$ of flat space-time M is changed by means of a conformal transformation

$$\hat{g}_{\alpha\beta} = \Omega^2 \eta_{\alpha\beta}, \qquad \Omega > 0 \text{ in } M, \tag{2.1}$$

the null geodesics in the metric $\eta_{\alpha\beta}$ go into null geodesics in the metric $\hat{g}_{\alpha\beta}$. The causal structure of the space $(\hat{M}, g_{\alpha\beta})$ coincides with that of Minkowski space $(M, \eta_{\alpha\beta})$, and in particular, the past of a curve γ is the same in both spaces, $I^-[\gamma] = \hat{I}^-[\gamma]$. There is therefore a natural bijective correspondence between the infinite ideal points in both spaces. If r is a canonical parameter along a lightlike geodesic in the metric $\eta_{\alpha\beta}(l^\mu = dx^\mu/dr)$, then the canonical parameter $\hat{r}$ along this lightlike geodesic in the metric $\hat{g}_{\alpha\beta}$ is related to r by the transformation

$$d\hat{r} = \Omega^2 dr, \qquad \hat{l}^\mu = \Omega^{-2} l^\mu. \tag{2.2}$$

Due to the choice of Ω it can be arranged that points at infinity in the space $(M, \eta_{\alpha\beta})$ which correspond to an infinite value of the parameter r will correspond in the space $(\hat{M}, \hat{g}_{\alpha\beta})$ to points with a finite value of the parameter $\hat{r}$. As is easily seen, these points are defined by the equation $\Omega = 0$. Two lightlike lines in the space $(M, \eta_{\alpha\beta})$ define the same point on $\mathscr{I}^+$ if they lie in the same lightlike hyperplane, and in this case they are parallel and the distance between points with the same value of the parameter r does not depend on r. Since $\Omega \to 0$ as $r \to \infty$, these two geodesics become close to one another in the space $(\hat{M}, \hat{g}_{\alpha\beta})$ and converge to the same limit point.

Consider the very simple conformal transformation (2.1). Write the Minkowski metric

$$ds^2 = (dx^0)^2 - (dx^1)^2 - (dx^2)^2 - (dx^3)^2$$

in the coordinates $(u, r, \zeta, \bar{\zeta})$ to get

$$ds^2 = du^2 + 2dudr - r^2\frac{d\zeta d\bar{\zeta}}{P_0^2}, \quad P_0 = \frac{(1+\zeta\bar{\zeta})}{2}, \tag{2.3}$$

where the coordinates $(u, r, \zeta, \bar{\zeta})$ are defined by the relations

$$x^0 = u + r, \quad x^1 + ix^2 = \frac{2r\zeta}{1+\zeta\bar{\zeta}}, \quad x^3 = r\frac{\zeta\bar{\zeta}-1}{1+\zeta\bar{\zeta}}. \tag{2.4}$$

Take Ω to be the quantity r^{-1}. Then by (2.2), $\hat{r} = -r^{-1}$ and we have $(\Omega = r^{-1} = -\hat{r})$

$$ds^2 = \Omega^{-2}d\hat{s}^2, \quad d\hat{s}^2 = \hat{r}^2du^2 - 2dud\hat{r} - \frac{d\zeta d\bar{\zeta}}{P_0^2}. \tag{2.5}$$

Light rays directed outward along the radius are determined by the values of $u, \zeta, \bar{\zeta}$, and these quantities can be used as coordinates on $\mathcal{I}^+$. In these coordinates the metric on the hypersurface of $\mathcal{I}^+$ defined by the equation $\hat{r} = 0$ has the form

$$dl^2 = -d\hat{s}^2|_{\hat{r}=0} = d\zeta d\bar{\zeta}/P_0^2, \tag{2.6}$$

i.e., $\mathcal{I}^+$ is a lightlike hypersurface on which the lightlike curves (generators) are defined by the equation $\zeta, \bar{\zeta} = \text{const}$. A simple calculation gives

$$\frac{\partial\Omega}{\partial\hat{x}^\mu} = (0, 1, 0, 0), \quad \hat{g}^{\mu\nu}\frac{\partial\Omega}{\partial\hat{x}^\mu}\frac{\partial\Omega}{\partial\hat{x}^\nu}\bigg|_{\mathcal{I}^+} = 0, \tag{2.7}$$

i.e., Ω is differentiable on $\hat{M}$ and $\partial\Omega/\partial\hat{x}^\mu$ is a finite lightlike vector on $\mathcal{I}^+$.

Analogously, it is possible to define points in the past at infinity forming the lightlike hypersurface $\mathcal{I}^-$ by taking two causal curves γ_1 and γ_2 directed into the past to define the same point of $\mathcal{I}^-$ if $I^+[\gamma_1]$ and $I^+[\gamma_2]$ coincide. It is easy to show that $\mathcal{I}^-$ and $\mathcal{I}^+$ do not intersect. Summarizing all that has been said above, we have the next proposition.

Proposition 5.2.1. The following statements are valid in the case of flat space-time $(M, \eta_{\alpha\beta})$:

1. There exists a new (nonphysical) space-time $(\hat{M}, \hat{g}_{\alpha\beta})$ with boundary $\mathcal{I}$ such that $\hat{M} \setminus \mathcal{I}$ is diffeomorphic to M and $d\hat{s} = \Omega\, ds$, with $\Omega > 0$ in M;
2. $(\hat{M}, \hat{g}_{\alpha\beta})$ is everywhere smooth, including the boundary;
3. Ω is everywhere smooth and $\Omega = 0$ on $\mathcal{I}$;
4. The metric $\eta_{\alpha\beta}$ satisfies the Einstein equations in vacuum throughout M, and in particular in a neighborhood of $\mathcal{I}$;
5. Every null geodesic in M has two endpoints in $\mathcal{I}$;
6. $\frac{\partial\Omega}{\partial\hat{x}^\mu} \neq 0$ on $\mathcal{I}$ and $\hat{g}^{\mu\nu}\frac{\partial\Omega}{\partial\hat{x}^\mu}\frac{\partial\Omega}{\partial\hat{x}^\nu} = 0$ on $\mathcal{I}$;
7. $\mathcal{I}$ consists of two nonintersecting parts $\mathcal{I} = \mathcal{I}^+ \cup \mathcal{I}^-$, each of the parts having the topology of $R \times S^2$, and the whole of space-time has the topology of R^4.

We now explain how Poincaré transformations acting in Minkowski space act on $\mathcal{I}^+$ and $\mathcal{I}^-$. The x^μ transform as follows under the subgroup of translations:

$$x'^\mu = x^\mu + a^\mu. \tag{2.8}$$

If using (2.4) we calculate the quantities $(u', r', \zeta', \bar{\zeta}')$ for the point x'^{μ}, then we have for large r

$$r' = r + \frac{\mathbf{ra}}{r} + O(r^{-1}),$$
$$u' = u + a^0 - \frac{\mathbf{ra}}{r} + O(r^{-1}), \tag{2.9}$$
$$\zeta' = \zeta + O(r^{-1}), \quad \bar{\zeta}' = \bar{\zeta} + O(r^{-1}),$$

where $\mathbf{ra} = x^1 a^1 + x^2 a^2 + x^3 a^3$. This means that transformation (2.8) induces the transformation

$$u' = u + a_0 + a_1 \frac{\zeta + \bar{\zeta}}{1 + \zeta\bar{\zeta}} + a_2 \frac{\zeta - \bar{\zeta}}{i(1 + \zeta\bar{\zeta})} + a_3 \frac{\zeta\bar{\zeta} - 1}{1 + \zeta\bar{\zeta}}, \quad \zeta' = \zeta, \quad \bar{\zeta}' = \bar{\zeta}. \tag{2.10}$$

on $\mathcal{J}^+$.

It can be shown analogously that the Lorentz rotations $x'^{\mu} = \Lambda^{\mu}_{\nu} x^{\nu}$ lead to the following coordinate transformation on $\mathcal{J}^+$

$$\zeta' = (a\zeta + b)/(c\zeta + d), \quad ad - bc = 1; \tag{2.11}$$

$$u' = Ku, \ K = (1 + \zeta\bar{\zeta})\,[(a\zeta + b)(\bar{a}\bar{\zeta} + \bar{b}) + (c\zeta + d)(\bar{c}\bar{\zeta} + \bar{d})]^{-1}, \tag{2.12}$$

and the line element on $\mathcal{J}^+$ takes the form*

$$dl'^2 = d\zeta' d\bar{\zeta}'/(P_0' K)^2. \tag{2.13}$$

Relations (2.10) and (2.11) allow us to obtain the following proposition.

Proposition 5.2.2. The Poincaré group transformations in Minkowski space induce the following transformations on $\mathcal{J}^+$:

$$\zeta' = \frac{a\zeta + b}{c\zeta + d}, \quad u' = K\left(u + a_0 + a_1 \frac{\zeta + \bar{\zeta}}{1 + \zeta\bar{\zeta}} + a_2 \frac{\zeta - \bar{\zeta}}{i(1 + \zeta\bar{\zeta})} + a_3 \frac{\zeta\bar{\zeta} - 1}{1 + \zeta\bar{\zeta}}\right), \tag{2.14}$$

where $K(\zeta, \bar{\zeta})$ is defined by Eq. (2.12), and we have the analogous transformations on $\mathcal{J}^-$.

5.3. Asymptotically Flat Spaces.

The Bondi – Metzner – Sachs Group of Asymptotic Symmetries

In analogy with the case of Minkowski space considered in the preceding section, it is possible to define boundary points at infinity for an arbitrary space-time [182]. In order to do this one considers causal curves γ on the space-time manifold, and if there does not exist a point p such that $I^-[p] = I^-[\gamma]$, one takes the curve γ to define some boundary ("ideal") point. Causal curves for which the past $I^-[\gamma]$ is the same determine the same boundary point. With this definition of boundary point, in the general case the set of boundary points contains not only points at asymptotic infinity but also various singular points (singularities) of space-time, if they exist. The construction of the boundary points of space-time described above is evidently conformally invariant.

---

* It is helpful to compare this result with Proposition 2.2.1. The evident analogy with the equations given there becomes clear if we take into account the fact that the surface u = 0 goes into itself under a Lorentz transformation and that the result of Proposition 2.2.1 is applicable to the transformation law of the coordinates ζ, $\bar{\zeta}$ defining the lightlike direction.

The central idea in the definition of the notion of asymptotically flat space-time consists in arranging its points at infinity in exactly the same way as for the infinite points of Minkowski.

Definition. The space-time $(M, g_{\alpha\beta})$ is asymptotically flat if:

1. There exists a new (unphysical) space-time $(\hat{M}, \hat{g}_{\alpha\beta})$ with boundary $\mathscr{I}$ such that $\hat{M} \setminus \mathscr{I}$ is diffeomorphic to M and $d\hat{s} = \Omega ds$, with $\Omega > 0$ in M;
2. $(\hat{M}, \hat{g}_{\alpha\beta})$ is everywhere smooth, including the boundary;
3. Ω is everywhere smooth and $\Omega = 0$ on $\mathscr{I}$;
4. The metric $g_{\alpha\beta}$ satisfies the Einstein equations in vacuum in a neighborhood of $\mathscr{I}$;
5. Every maximally extended lightlike geodesic in $\hat{M}$ has two endpoints in $\mathscr{I}$.

It is easy to see that all five requirements of the definition are completely analogous to the first five assertions of Proposition 5.2.1 concerning the structure of infinity in Minkowski space. Moreover, using this definition it is possible to prove the following proposition [10, 191, 192].

Proposition 5.3.1. In an asymptotically flat space-time $(M, g_{\alpha\beta})$:

1. $\frac{\partial\Omega}{\partial\hat{x}^\mu} \neq 0$ on $\mathscr{I}$ and $\hat{g}^{\mu\nu}\frac{\partial\Omega}{\partial\hat{x}^\mu}\frac{\partial\Omega}{\partial\hat{x}^\nu} = 0$ on $\mathscr{I}$, i.e., $\mathscr{I}$ is a lightlike hypersurface;
2. $\mathscr{I}$ consists of two nonintersecting parts, $\mathscr{I} = \mathscr{I}^+ \cup \mathscr{I}^-$, each of the parts having the topology of $R \times S^2$, and the whole space M has the topology of R^4;
3. $\Psi_{ABCD} = 0$ on $\mathscr{I}$, i.e., $(M, g_{\alpha\beta})$ is flat at infinity;
4. The property of sequential degeneracy holds, i.e., the spinor components of the Weyl tensor decay according to the following rule along an arbitrary lightlike geodesic γ going out to $\mathscr{I}^+$:

$$\Psi_0 \sim r^{-5}, \quad \Psi_1 \sim r^{-4}, \quad \Psi_2 \sim r^{-3}, \quad \Psi_3 \sim r^{-2}, \quad \Psi_4 \sim r^{-1},$$

where r is a canonical parameter along γ.

It is convenient to introduce coordinates as follows on $\mathscr{I}^+$ (and analogously on $\mathscr{I}^-$), cf. Fig. 5.3. Since $\mathscr{I}^+$ is a lightlike hypersurface, it is generated by a two-parameter family of null curves (generators). Since the generators of $\mathscr{I}^+$ are in bijective correspondence with the points of S^2 (part 2) of Proposition 5.3.1), they can be parametrized by the complex stereographic coordinates ζ, $\bar{\zeta}$ on S^2. A one-parameter family of nonintersecting spacelike sections (each of which is diffeomorphic to S^2) can be chosen on $\mathscr{I}^+$, and these sections can be

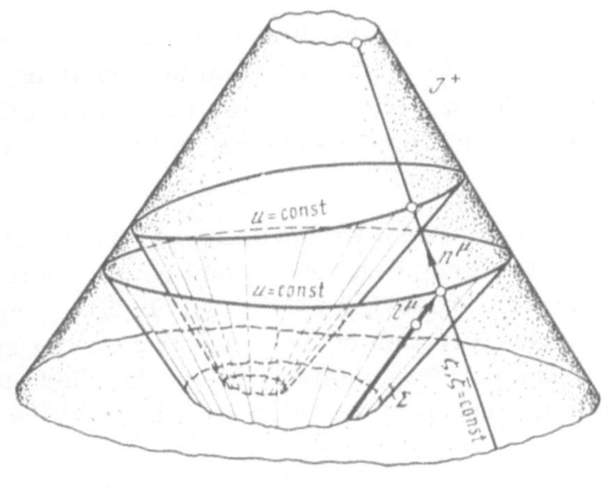

Fig. 5.3

enumerated by a monotonic increasing parameter u in such a way that u = const on every section. In the coordinates $(u, \zeta, \bar{\zeta})$, the metric on $\mathcal{J}^+$ has the form

$$dl^2 = d\zeta d\bar{\zeta}/P^2\ (u, \zeta, \bar{\zeta}), \tag{3.1}$$

and the coordinates $(u, \zeta, \bar{\zeta})$ themselves are determined up to the transformations

$$\begin{aligned} \zeta' &= (a\zeta + b)/(c\zeta + d), \quad ad - bc = 1 \text{ (conformal transformation of the sphere } S^2 \text{ onto itself)}, \\ u' &= G\,(u, \zeta, \bar{\zeta}) \text{ (passage to a new family of sections)}. \end{aligned} \tag{3.2}$$

The choice of a definite coordinate system on $\mathcal{J}^+$ makes it possible to introduce coordinates uniquely in a neighborhood of $\mathcal{J}^+$ in the four-dimensional space $(\hat{M}, \hat{g}_{\alpha\beta})$ by means of the following procedure (cf. Fig. 5.3). Choose an arbitrary section u = const and consider the null geodesics intersecting this section orthogonally and not lying in $\mathcal{J}^+$. As was shown in Section 1.6, these null geodesics form a lightlike hypersurface Σ in space-time, and we choose the coordinate u to be constant on Σ. We choose a canonical affine parameter $\hat{r}$ along the lightlike geodesics, and if we take into account the fact that a lightlike geodesic is determined by its boundary point in $\mathcal{J}^+$ with coordinates $(u, \zeta, \bar{\zeta})$, we obtain in a neighborhood of $\mathcal{J}^+$ a rigid coordinate system $(u, \hat{r}, \zeta, \bar{\zeta})$ defined by the choice of coordinates on $\mathcal{J}^+$. It can be seen that in this coordinate system the metric has the form

$$\hat{g}^{\alpha\beta} = \left\| \begin{matrix} 0 & 1 & 0 & 0 \\ 1 & \hat{g}^{11} & \hat{g}^{12} & \hat{g}^{13} \\ 0 & \hat{g}^{12} & & \\ 0 & \hat{g}^{13} & \multicolumn{2}{c}{\hat{g}^{AB}} \end{matrix} \right\|, \quad A, B = 2, 3; \quad \det(\hat{g}^{\alpha\beta}) = \det(\hat{g}^{AB}). \tag{3.3}$$

Under conformal transformations the lightlike geodesics go into lightlike geodesics, and hence the procedure considered above defines a fixed coordinate system $(u, r, \zeta, \bar{\zeta})$ in physical space-time $(M, g_{\alpha\beta})$. In these coordinates, the metric is given by the expression [201]

$$ds^2 = Wdu^2 + 2Bdudr - r^2 h_{AB}\,(dx^A - U^A du)(dx^B - U^B du), \tag{3.4}$$

where A, B = 2, 3, W, B, U^A, and h_{AB} are functions of x^α and the coordinates x^A are constant along light rays, with

$$\lim_{r\to\infty} (ds^2) = du^2 + 2du\,dr - r^2\,(d\vartheta^2 + \sin^2\vartheta d\varphi^2). \tag{3.5}$$

The requirement that there exist coordinates $u_0 \le u \le u_1$, $r_0 \le r < \infty$, $0 \le \vartheta < \pi$, and $0 \le \varphi < 2\pi$ in which the metric takes the form (3.4) and the functions W, B, U^A, and h_{AB} behave in the region $r \to \infty$ so that condition (3.5) holds, can be taken as the definition of an asymptotically flat space, provided certain additional assumptions are made on the asymptotic behavior of these functions.*

Using the example of flat space-time, we have already seen that the group of motions in $(M, \eta_{\alpha\beta})$ generates a group of transformations on $\mathcal{J}^+$ and on $\mathcal{J}^-$. In the general case, there is no symmetry of any kind in a twisted space-time $(M, g_{\alpha\beta})$. In an asymptotically flat space, far away from the sources $(r \to \infty)$ the deviation of the space from a flat space tends to zero and it is natural to expect that although strict symmetry is absent, there is an "asymptotic" symmetry. In order to describe this concept in more detail we will need the following definition [14, 191].

---

* A more detailed discussion of this question is contained in [8, 14].

Definition. A function $f(u, r, x^2, x^3)$ is called asymptotically smooth of order r^{-k}: $f = O^*(r^{-k})$ provided the following equalities hold in coordinates (3.4):

$$f = O(r^{-k}), \quad \frac{\partial f}{\partial r} = O(r^{-k-1}), \quad \frac{\partial f}{\partial u}, \frac{\partial f}{\partial x^2}, \frac{\partial f}{\partial x^3} = O(r^{-k}),$$
$$\frac{\partial^2 f}{\partial r^2} = O(r^{-k-2}), \quad \frac{\partial^2 f}{\partial r \partial u} = \ldots = O(r^{-k-1}), \tag{3.6}$$
$$\frac{\partial^2 f}{\partial u \partial x^2} = \ldots = O(r^{-k}), \ldots, \text{etc.}$$

If the space-time admits a group of motions, then the vector field ξ^α generating this motion satisfies the Killing equation

$$\xi_{\alpha;\beta} + \xi_{\beta;\alpha} = 0. \tag{3.7}$$

If ξ^α is an arbitrary vector field in space-time, we denote by $\delta g_{\alpha\beta}$ the quantity $\delta g_{\alpha\beta} = -\xi_{\alpha;\beta} - \xi_{\beta;\alpha}$.

Definition. We say that space-time admits a group of asymptotic symmetries [14] provided there exists a nontrivial vector field ξ^α for which, in the coordinates (3.4),

$$\delta g_{11} = 0, \quad \delta g_{1A} = 0, \quad \delta g_{AB} g^{AB} = 0,$$
$$\delta g_{00} = O^*(r^{-1}), \quad \delta g_{0A} = O^*(1), \quad \delta g_{01} = O^*(r^{-2}), \quad \delta g_{AB} = O^*(r). \tag{3.8}$$

Such transformations preserve the form and the asymptotic properties of (3.4). The group generated by the transformations associated with such vector fields ξ^α is called the Bondi–Metzner–Sachs group (or BMS group for short) [13, 14].

Penrose ([11, p. 180]) showed that the BMS group of asymptotic symmetries can be defined as the group of transformations of the hypersurface $\mathcal{J}^+$ (or $\mathcal{J}^-$) preserving the strong conformal geometry on $\mathcal{J}^+$. This geometry is introduced on $\mathcal{J}^+$ in the following way. Using the fact that under transformation (3.3) the quantity P changes in accordance with the law

$$P' = \dot{G}^{-1} P, \tag{3.9}$$

and choosing a new coordinate u', we can arrange that the surfaces u' = const are isometric to the unit sphere, i.e., $P' = P_0 = \frac{1}{2}(1 + \zeta\bar{\zeta})$. In this metric on $\mathcal{J}^+$, the shift and divergence of the generators is equal to zero. In this case we can define parallel transport of vectors on $\mathcal{J}^+$ with the aid of the Christoffel symbols defined on $(\hat{M}, \hat{g}_{\alpha\beta})$.

These rules for transporting vectors allow us to establish an integrable equivalence between isotropic angles which does not depend on the concrete choice of metric $\hat{g}_{\alpha\beta}$.

Coordinates $(u, \zeta, \bar{\zeta})$ on $\mathcal{J}^\pm$ with respect to which $dl^2 = d\zeta d\bar{\zeta}/P_0^2$ are called Bondi coordinates. The transformations preserving the strong conformal geometry coincide with the transformations (3.2), (3.3) which realize a transition from one set of Bondi coordinates to another. It can be seen that the transformations defined by these transformation conditions, the generators of the BMS group, have the form

$$\zeta' = (a\zeta + b)/(c\zeta + d), \quad u' = K(u + \alpha(\zeta, \bar{\zeta})), \tag{3.10}$$

where

$$K = (1 + \zeta\bar{\zeta})\,[(a\zeta + b)/(\bar{a}\bar{\zeta} + \bar{b}) + (c\zeta + d)(\bar{c}\bar{\zeta} + \bar{d})]^{-1} \tag{3.11}$$

and $a(\zeta, \bar{\zeta})$ is an arbitrary real function. The BMS group is essentially larger than the Poincaré symmetry group of flat space-time. If we compare (3.10) with the transformations (2.14) induced on $\mathcal{I}^{\pm}$ by the action of the Poincaré group in $(M, \eta_{\alpha\beta})$, it is clear that for

$$\alpha(\zeta, \bar{\zeta}) = a_0 + a_1 \frac{\zeta + \bar{\zeta}}{1 + \zeta\bar{\zeta}} + a_2 \frac{\zeta - \bar{\zeta}}{i(1 + \zeta\bar{\zeta})} + a_3 \frac{\zeta\bar{\zeta} - 1}{1 + \zeta\bar{\zeta}} \equiv$$
$$\equiv a_0 + a_1 \sin\vartheta \cos\varphi + a_2 \sin\vartheta \sin\varphi + a_3 \cos\vartheta \tag{3.12}$$

we achieve a complete correspondence. The transformations

$$\zeta' = \zeta, \quad u' = u + \alpha(\zeta, \bar{\zeta}), \tag{3.13}$$

where α is an arbitrary function, are called supertranslations. For $\alpha(\zeta, \bar{\zeta})$ defined by Eq. (3.12), the transformations (3.13) form the subgroup of translations. The structure of the BMS group was investigated by Sachs [14], who proved the following proposition.

Proposition 5.3.2. The supertranslations (3.13) form a normal Abelian subgroup of the BMS group such that the corresponding quotient group is isomorphic to the (orthochronous) Lorentz group. The translations form a normal four-dimensional subgroup of the BMS group. The subgroup of translations is the only normal four-dimensional subgroup of the BMS group.

McCarthy [185] showed using the simple example of a two-dimensional space that the existence of a BMS group substantially larger than the Poincaré group as a group of asymptotic symmetries is related to the slow decay of the gravitational field at infinity. Representations of the BMS group were studied in [186-190]. The fact that the four-dimensional translation subgroup is uniquely determined in the whole BMS group permits a natural way of defining the energy-momentum both of a gravitating system itself in an asymptotically flat space and of different physical fields in a given asymptotically flat space.

Difficulties arise mainly in attempts to define the angular momentum of a system and the center of mass. These difficulties are caused by the nonuniqueness in choosing the Lorentz subgroup from the BMS group.

5.4. Massfree Fields in an Asymptotically Flat Space

The notion of conformal infinity $\mathcal{I}$ of space-time turns out to be important in studying the the asymptotic properties of massless fields. This problem was systematically studied by Penrose [10, 191]. The fruitfulness of the idea of conformal infinity is directly related to the conformal invariance of the equations for massless fields.

We recall that in spinor form the equation for a massless field with spin s > 0 has the form (3.7.30)

$$\partial^{A_1 A'} \varphi_{A_1 A_2 \ldots A_{2s}} = 0, \quad s = 1/2,\ 1,\ 3/2, \ldots \tag{4.1}$$

In the definition of conformal infinity for physical space-time $(M, g_{\alpha\beta})$, we introduce a space $(\hat{M}, \hat{g}_{\alpha\beta})$ such that

$$\hat{g}_{\alpha\beta} = \Omega^2 g_{\alpha\beta}, \quad \hat{g}^{\alpha\beta} = \Omega^{-2} g^{\alpha\beta}. \tag{4.2}$$

If we assume that the symbols $\sigma^{\mu}_{AB'}$ and $\sigma_{\mu}^{AB'}$ relating spinors and tensors under the transformation (4.2) do not change, Eq. (3.3.12) allows us to conclude that

$$\hat{\varepsilon}_{AB} = \Omega \varepsilon_{AB}, \quad \hat{\varepsilon}^{AB} = \Omega^{-1} \varepsilon^{AB},$$
$$\hat{\varepsilon}_{A'B'} = \Omega \varepsilon_{A'B'}, \quad \hat{\varepsilon}^{A'B'} = \Omega^{-1} \varepsilon^{A'B'}. \tag{4.3}$$

Let $\Gamma_{AA'}$ denote the quantity $\Omega^{-1}\partial_{AA'}\Omega$, and let $\partial_{AB'}$ be the covariant spinor derivative with respect to the metric $\hat{g}_{\alpha\beta}$. Then it can be seen that

$$\hat{\partial}_{AA'}T_{BB'\cdots}^{DD'\cdots} = \partial_{AA'}T_{BB'\cdots}^{DD'\cdots} - \Gamma_{BA'}T_{AB'\cdots}^{DD'\cdots} - \Gamma_{AB'}T_{BA'\cdots}^{DD'\cdots} + \varepsilon_A{}^D\Gamma_{FA'}T_{BB'\cdots}^{FD'\cdots} + \varepsilon_{A'}{}^{D'}\Gamma_{AF'}T_{BB'\cdots}^{DF'\cdots}. \tag{4.4}$$

[For the proof of this equality it suffices to show that the operator $\hat{\partial}_{AA'}$ thus defined satisfies axioms (3.6.1)–(3.6.7).]

Using (4.4) it is easy to verify that if $\varphi_{A_1\ldots A_{2s}}$ is a solution of Eq. (4.1) then

$$\hat{\varphi}_{A_1\ldots A_{2s}} = \Omega^{-1}\varphi_{A_1\ldots A_{2s}} \tag{4.5}$$

is a solution of the equation

$$\hat{\partial}^{A_1 A'}\hat{\varphi}_{A_1 A_2\ldots A_{2s}} = 0, \tag{4.6}$$

i.e., it is enough to prove the conformal invariance of Eq. (4.1). The conformal invariance of the Maxwell equations, which is a particular case (for s = 1) of the conformal invariance considered above of Eq. (4.1), is easily established if we take into account that $\hat{F}_{\alpha\beta} = F_{\alpha\beta}$ and write the Maxwell equations in the form

$$\partial_\alpha(\sqrt{-g}\, g^{\alpha\beta}g^{\gamma\delta}F_{\beta\delta}) = 0, \qquad \partial_{[\alpha}F_{\beta\gamma]} = 0. \tag{4.7}$$

The scalar massless field described by the equation

$$(\Box - R/6)\,\varphi = 0, \tag{4.8}$$

is also conformally invariant.

The property of conformal invariance of Eq. (4.1) makes it possible to establish a bijective correspondence between massless fields $\varphi_{A_1\ldots A_{2s}}$ in physical space-time $(M, g_{\alpha\beta})$ and the solutions $\hat{\varphi}_{A_1\ldots A_{2s}}$ of the analogous equation (4.6) in the space $(M, g_{\alpha\beta})$.

<u>Definition.</u> A massless field $\varphi_{A_1\ldots A_m}$ is called asymptotically regular if the field $\hat{\varphi}_{A_1\ldots A_m}$ on $\hat{M}$ is related to $\varphi_{A_1\ldots A_m}$ by Eq. (4.5) in the domain $\hat{M}\setminus\mathscr{I}$ and is bounded and continuous on $\mathscr{I}$.

The asymptotic regularity condition distinguishes a class of solutions of Eq. (4.1) having "good" asymptotic properties which play a role analogous to the Sommerfeld radiation condition. In particular, a monochromatic wave of infinite length incident from infinity does not satisfy the asymptotic regularity condition. In order to discuss in more detail the asymptotic properties of massless fields, choose an arbitrary light geodesic $\gamma : x^\mu = x^\mu(r)$, where r is a canonical parameter along γ, and let ζ_0^A be a spinor such that $dx^\mu/dr \equiv l^\mu = \sigma^\mu_{AA'}\bar{\zeta}_0^A\zeta_0^{A'}$ and ζ_0^A is parallel transported along γ. Choose an additional spinor ζ_1^A parallel along γ such that $\zeta_{0A}\zeta_1^A = 1$. We denote by $\varphi_{(i)}$ the quantity $\varphi_{A_1\ldots A_i A_{i+1}\ldots A_m}\zeta_1^{A_1}\ldots\zeta_1^{A_i}\zeta_0^{A_{i+1}}\ldots\zeta_0^{A_m}$. Penrose [191] proved the following proposition.

<u>Proposition 5.4.1.</u> The limit

$$\lim_{r\to\pm\infty}(r^{m+1-i}\varphi_{(i)}) = \Phi^{\pm}_{(i)} \tag{4.9}$$

exists for an arbitrary asymptotically regular field $\varphi_{A_1\ldots A_m}$ in an asymptotically flat space for an arbitrary null geodesic γ. For $\Phi^{+}_{(m)}$ $(\Phi^{-}_{(m)})$ this limit is the same for all null geodesics end-

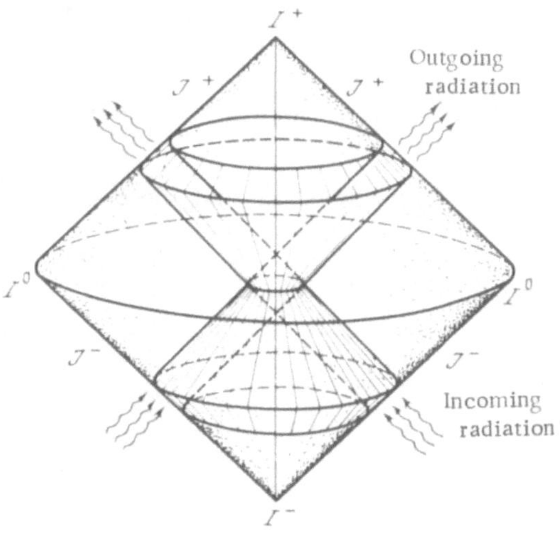

Fig. 5.4

ing in the same point on $\mathscr{I}^+$ ($\mathscr{I}^-$) and is equal to the limiting value at this point of the function $\hat{\varphi}_{(m)} = \hat{\varphi}_{A_1\ldots A_m}\zeta_1^{A_1}\ldots\zeta_1^{A_m}$.

We will call the quantity $\Phi^+ \equiv \Phi^+_{(m)}$ ($\Phi^- \equiv \Phi^-_{(m)}$) the image of the function $\varphi_{A_1\ldots A_m}$ on $\mathscr{I}^+$ (on $\mathscr{I}^-$). Relation (4.9) shows that

$$\varphi_{(i)} = O\,(r^{i-m-1}), \tag{4.10}$$

i.e., the property of sequential degeneracy holds, viz., the vector l^μ is an i-fold principal null vector for the coefficient of r^{i-m-1} in the expansion of the spinor $\varphi_{A_1\ldots A_m}$ in powers of r^{-1}.

It was shown by Penrose that if on an arbitrary lightlike Cauchy surface u = const we choose a spinor field ζ_1^A such that $u_{,\mu} \sim \sigma_{\mu AA'}\zeta_1^A\bar{\zeta}_1^{A'}$, then the complex function $\varphi_{(m)}=\varphi_{A_1\ldots A_m}\zeta_1^{A_1}\times \ldots \zeta_1^{A_m}$ completely determines the solution of Eq. (4.1) in the entire space.* Since in an asymptotically flat space the lightlike surface $\mathscr{I}^+$ ($\mathscr{I}^-$) is a Cauchy surface, the image on $\mathscr{I}^+$ (or $\mathscr{I}^-$) of the function $\varphi_{A_1\ldots A_m}$ uniquely determines a solution in spacetime. Thus the introduction of conformal infinity and the conformal invariance of massless fields make it possible to consider the ordinary Cauchy problem instead of the scattering problem in an asymptotically flat space (Fig. 5.4).

5.5. Quantum Theory in an Asymptotically Flat Space. The Hawking Effect

In constructing a quantum theory in an external gravitational field, the main difficulty is in defining the vacuum state. The physical reason for this is that in the region of a strong gravitational field, where the characteristic space-time curvature has the order l^{-2}, it is difficult (and perhaps impossible in theory) to distinguish a real quantum with wavelength $\lambda \lessgtr l$ from virtual quanta with analogous wavelength. The usual definition of vacuum state as the state of lowest energy is possible only in a stationary gravitational field. The stationarity of the

* This is in agreement with the fact that two arbitrary real functions are necessary to describe a massless field of arbitrary spin s > 0.

metric leads to the existence of a global timelike Killing vector field ξ^{μ}, and according to Noether's theorem, the invariant energy integral has the form

$$E = \int_{\Sigma} T^{\mu\nu}\xi_{\nu}\sqrt{-g}\, d\sigma_{\mu}, \tag{5.1}$$

where Σ is an arbitrary Cauchy surface and $d\sigma_{\mu}$ the volume element on it. For massless fields the analogous expression also determines an invariant integral in the case where ξ^{μ} is a conformal Killing vector field, i.e., satisfies the equation

$$\xi_{\alpha;\beta} + \xi_{\beta;\alpha} = 2f g_{\alpha\beta}. \tag{5.2}$$

In the general case where such a symmetry of space-time does not exist, it is not possible to define the energy integral and nonuniqueness arises in choosing the vacuum state.*

In an asymptotically flat space, massless particles moving at the speed of light sooner or later leave the region of strong gravitational field and enter the asymptotic domain where the deviation of the space-time metric from a flat one is small. Analogously, if we follow the motion of a particle into the past we can see that in the distant past the particle was located far from the sources of the gravitational field and moved almost as if it were free. These considerations permit us to hope that it is possible to uniquely define the scattering S-matrix and the asymptotic vacuum state for massless fields in an asymptotically flat space. In this section we demonstrate using the example of a massless scalar field how the idea of conformal infinity permits the construction of scattering theory in an asymptotically flat space.

The Heisenberg operator φ describing a massless field satisfies the equation of motion

$$(\Box - R/6)\,\varphi = 0. \tag{5.3}$$

The solution of this equation can be written in the form

$$\varphi(x) = \sum_{n} (u_n(x)\,\mathbf{a}_n + \bar{u}_n(x)\,\mathbf{a}_n^{*}), \tag{5.4}$$

where $\{u_n, \bar{u}_n\}$ is a basis, i.e., a complete system of complex-valued asymptotically regular solutions of Eq. (5.3) normalized by the condition

$$\begin{aligned} \langle u_n, u_m\rangle &\equiv i\int_{\Sigma} \bar{u}_n \overleftrightarrow{\partial}_{\mu} u_m g^{\mu\nu}\sqrt{-g}\, d\sigma_{\nu} = \delta_{mn}, \\ \langle \bar{u}_n, u_m\rangle &= \langle u_n, \bar{u}_m\rangle = 0. \end{aligned} \tag{5.5}$$

The integration is performed over an arbitrary Cauchy surface. It can be verified that for

* It seems appropriate to remark that in the coordinates (x^0, x^i) the expression $H[x^0] = \int_{x^0=\text{const}} T^{\nu}{}_{0}\sqrt{-g}\, d^3x$ is the Hamiltonian. As in both the classical and the quantum theories, the Hamiltonian determines the evolution of the system from a surface $x^0 = \text{const}$ to a "neighboring" surface $x^0 + \delta x^0 = \text{const}$. Therefore, the value of the Hamiltonian depends on the choice of the time x^0 on a space-time manifold, and for various choices of x^0 the states minimizing the quantity H will be different. When there exists a global timelike Killing vector field ξ^{α}, we can choose the time t such that $dx^{\alpha}/dt = \xi^{\alpha}$ to be the "standard" physical time. For such a choice the value of the Hamiltonian H[t] coincides with the energy E defined by Eq. (5.1).

any two solutions of Eq. (5.3) the inner product $\langle f, g\rangle$ does not depend on the choice of Σ and has the property

$$\overline{\langle f, g\rangle} = \langle g, f\rangle,$$
$$\langle \bar{f}, \bar{g}\rangle = -\langle g, f\rangle, \qquad \langle \alpha f, \beta g\rangle = \bar{\alpha}\beta\langle f, g\rangle. \tag{5.6}$$

The canonical quantization rules lead to the following commutation relations for the vectors $\mathbf{a}_n$, $\mathbf{a}^*_n$:

$$[\mathbf{a}_n, \mathbf{a}^*_m] = \delta_{nm}, \qquad [\mathbf{a}_n, \mathbf{a}_m] = [\mathbf{a}^*_n, \mathbf{a}^*_m] = 0. \tag{5.7}$$

The completeness of the basis $\{u_n, \bar{u}_n\}$ permits us to write

$$\mathbf{a}_n = \langle u_n, \varphi\rangle, \qquad \mathbf{a}^*_n = -\langle \bar{u}_n, \varphi\rangle. \tag{5.8}$$

When there exists a global timelike Killing vector field, choosing the u_n to be the positive frequency solutions leads to a unique definition of the vacuum by means of the conditions

$$\mathbf{a}_n|0\rangle = 0. \tag{5.9}$$

In the general case the basis is not uniquely determined. Two arbitrary bases $\{u_n, \bar{u}_n\}$ and $\{v_n, \bar{v}_n\}$ are related by the transformation

$$u_n = \sum_m (\alpha_{nm} v_m + \beta_{nm}\bar{v}_m), \tag{5.10}$$

where $\alpha_{nm} = \langle v_m, u_n\rangle$, $\beta_{nm} = -\langle \bar{v}_m, u_n\rangle$. The matrices $\boldsymbol{\alpha}$ and $\boldsymbol{\beta}$ satisfy the condition

$$\begin{Vmatrix} \alpha & \beta \\ \bar{\beta} & \bar{\alpha} \end{Vmatrix} \cdot \begin{Vmatrix} \alpha^+ & -\beta^T \\ \beta^+ & \alpha^T \end{Vmatrix} = \begin{Vmatrix} I & 0 \\ 0 & I \end{Vmatrix}. \tag{5.11}$$

Under this transformation the operators $\mathbf{a}_n$, $\mathbf{a}^*_n$ go into new operators $\mathbf{b}_n$, $\mathbf{b}^*_n$ by means of the Bogolyubov transformation

$$\mathbf{a}_n = \sum_m (\bar{\alpha}_{nm}\mathbf{b}_m - \bar{\beta}_{nm}\mathbf{b}^*_m),$$
$$\mathbf{a}^*_n = \sum_m (\alpha_{nm}\mathbf{b}^*_m - \beta_{nm}\mathbf{b}_m). \tag{5.12}$$

We now consider along with the original real space-time $(M, g_{\alpha\beta})$ the (non-physical) space-time $(\hat{M}, \hat{g}_{\alpha\beta})$. By the conformal invariance of Eq. (5.3), the Heisenberg operator $\hat{\boldsymbol{\varphi}} = \Omega^{-1}\boldsymbol{\varphi}$ satisfies the equation

$$(\hat{\Box} - \hat{R}/6)\hat{\boldsymbol{\varphi}} = 0, \tag{5.13}$$

where here and below the quantities with a caret refer to the space $(\hat{M}, \hat{g}_{\alpha\beta})$. If we take into account that $d\sigma_\mu$ does not change under a conformal transformation, we obtain

$$i\int_\Sigma \bar{\hat{f}}\overleftrightarrow{\partial_\mu}\hat{h}\hat{g}^{\mu\nu}\sqrt{-\hat{g}}\,d\sigma_\nu \equiv \langle \hat{f}, \hat{h}\rangle = \langle f, h\rangle, \tag{5.14}$$

where $\hat{f} = \Omega^{-1}f$ and $\hat{h} = \Omega^{-1}h$. Choosing a coordinate system (3.3) in a neighborhood of $\mathscr{I}^+$ and taking Σ to be the surface $\mathscr{I}^-$, we get

$$\langle f, h\rangle = \langle\!\langle F^+, H^+\rangle\!\rangle_{\mathscr{I}^+}. \tag{5.15}$$

Here $\langle\!\langle F^+, H^+\rangle\!\rangle_{\mathscr{I}^+}$ is the inner product of the images F^+ and H^+ of the functions f and h on $\mathscr{I}^+$ defined by*

$$\langle\!\langle F^+, H^+\rangle\!\rangle_{\mathscr{I}^+} = i\int \bar{F}^+\overleftrightarrow{\partial}_u H^+\, du\, d\Omega, \tag{5.16}$$

where $d\Omega$ is the surface element on the surface with metric (3.1). In the Bondi coordinates, $d\Omega = \sin\vartheta d\vartheta d\varphi$.

We define the operator $\mathbf{\Phi}_{\text{out}}$ by the condition

$$\langle\!\langle U_n^{+'}, \mathbf{\Phi}_{\text{out}}\rangle\!\rangle_{\mathscr{I}^+} = \langle u_n, \boldsymbol{\varphi}\rangle \equiv \mathbf{a}_n. \tag{5.17}$$

Using (5.4) we can write

$$\mathbf{\Phi}_{\text{out}} = \sum_n (U_n^+\mathbf{a}_n + \bar{U}_n^+\mathbf{a}_n^*). \tag{5.18}$$

A transformation (5.10) of basis functions induces a corresponding transformation for the images on $\mathscr{I}^+$.

In an asymptotically flat space the BMS group of asymptotic symmetries acts on $\mathscr{I}^+$. In the Bondi coordinates the generators P_μ of the translation subgroup (uniquely defined in the full BMS group) can be written in the form [cf. Eq. (3.12)].

$$\begin{gathered} P_0 = \frac{\partial}{\partial u}, \quad P_1 = \sin\vartheta\cos\varphi\frac{\partial}{\partial u}, \\ P_2 = \sin\vartheta\sin\varphi\frac{\partial}{\partial u}, \quad P_3 = \cos\vartheta\frac{\partial}{\partial u}. \end{gathered} \tag{5.19}$$

The operators

$$\begin{aligned} \mathbf{H} \equiv \mathbf{P}_0 &= \int du\, d\Omega\, \partial_u\mathbf{\Phi}_{\text{out}}\, \partial_u\mathbf{\Phi}_{\text{out}}, \\ \mathbf{P}_1 &= \int du\, d\Omega \sin\vartheta\cos\varphi\, \partial_u\mathbf{\Phi}_{\text{out}}\, \partial_u\mathbf{\Phi}_{\text{out}}, \\ \mathbf{P}_2 &= \int du\, d\Omega \sin\vartheta\sin\varphi\, \partial_u\mathbf{\Phi}_{\text{out}}\, \partial_u\mathbf{\Phi}_{\text{out}}, \\ \mathbf{P}_3 &= \int du\, d\Omega \cos\vartheta\, \partial_u\mathbf{\Phi}_{\text{out}}\, \partial_u\mathbf{\Phi}_{\text{out}}, \end{aligned} \tag{5.20}$$

which realize a representation of the generators of the translation subgroup have the interpretation of energy-momentum operators of outgoing particles. The fundamental state $|0\rangle_{\mathscr{I}^+}$, characterizing the energy minimum can be defined by the condition

$$\mathbf{a}_n|0\rangle_{\mathscr{I}^+} = 0, \tag{5.21}$$

where the annihilation operators $\mathbf{a}_n$ arise in the expansion of $\mathbf{\Phi}_{\text{out}}$ with respect to the basis $\{U_n^+, \bar{U}_n^+\}$ of positive frequency (relative to the Bondi coordinate u) functions on $\mathscr{I}^+$. If we denote by $N_{\text{out}}[\ ]$ the operation of normal ordering relative to the vacuum $|0\rangle_{\mathscr{I}^+}$, the (finite)

* Here and below, the image on $\mathscr{I}^+$ ($\mathscr{I}^-$) of functions or operators denoted by a lower case letter will be denoted by the corresponding capital letter with upper index plus (or minus).

energy operator on $\mathscr{I}^+$ can be written in the form

$$\mathbf{H}_{\text{out}} = \int du\, d\Omega\, N_{\text{out}}\, [\partial_u \mathbf{\Phi}_{\text{out}}\, \partial_u \mathbf{\Phi}_{\text{out}}]. \tag{5.22}$$

If we let $\{V_n^-, \bar{V}_n^-\}$ denote a basis of positive frequency functions on $\mathscr{I}^-$ (positive with respect to the Bondi coordinate v), the vacuum state on $\mathscr{I}^-$ is analogously defined by the equation

$$\mathbf{b}_n\, |\, 0\rangle_{\mathscr{I}^-} = 0, \tag{5.23}$$

where the $\mathbf{b}_n$ are particle annihilation operators on $\mathscr{I}^-$ and arise from expanding $\mathbf{\Phi}_{\text{in}}$ with respect to the basis $\{V_n^-, \bar{V}_n^-\}$:

$$\mathbf{\Phi}_{\text{in}} = \sum_n (V_n^- \mathbf{b}_n + \bar{V}_n^- \mathbf{b}_n^*). \tag{5.24}$$

In the case where there are no initial particles and the gravitational field creates particles which then go off to infinity, the energy flux of the particles as measured by a distant observer during a certain length of time is given by the expression

$$E\,[f] = \int du\, d\Omega_{\mathscr{I}^-} \langle 0\,|\, N_{\text{out}}\, [\partial_u \mathbf{\Phi}_{\text{out}}\, \partial_u \mathbf{\Phi}_{\text{out}}]\, |\, 0\rangle_{\mathscr{I}^-} f\,(u), \tag{5.25}$$

where the function f(u) for the measuring apparatus is equal to one for u corresponding to the observation interval and vanishes smoothly outside this interval. In order to calculate the quantity E[f] we substitute expansion (5.18) into (5.25):

$$\begin{aligned} E\,[f] = \int du\, d\Omega f\,(u) \sum_{n,\, m} \{ & \partial_u U_n^+ \partial_u U_m^+{}_{\mathscr{I}^-} \langle 0\,|\, \mathbf{a}_n \mathbf{a}_m\, |\, 0\rangle_{\mathscr{I}^-} + \\ & + \partial_u \bar{U}_n^+ \partial_u \bar{U}_m^+{}_{\mathscr{I}^-} \langle 0\,|\, \mathbf{a}_n^* \mathbf{a}_m^*\, |\, 0\rangle_{\mathscr{I}^-} + \partial_u U_n^+ \partial_u \bar{U}_m^+{}_{\mathscr{I}^-} \langle 0\,|\, \mathbf{a}_m^* \mathbf{a}_n\, |\, 0\rangle_{\mathscr{I}^-} + \\ & + \partial_u \bar{U}_n^+ \partial_u U_m^+{}_{\mathscr{I}^-} \langle 0\,|\, \mathbf{a}_n^* \mathbf{a}_m\, |\, 0\rangle_{\mathscr{I}^-} \}. \end{aligned} \tag{5.26}$$

If the observation interval is sufficiently great (i.e., the characteristic energy of the outgoing particles is $\varepsilon \gg h/T$ and the energy flux carried away by particles with energy less than h/T can be neglected), the first two terms in the curly brackets containing the product of two functions with identical frequencies can be omitted. The last two terms in brackets are equal, and therefore we have

$$E\,[f] = 2\int du\, d\Omega f\,(u) \sum_{n,\, m} \partial_u \bar{U}_n^+ \partial_u U_m^+{}_{\mathscr{I}^-} \langle 0\,|\, \mathbf{a}_n^* \mathbf{a}_m\, |\, 0\rangle_{\mathscr{I}^-}. \tag{5.27}$$

We now make use of Eq. (5.12) in order to obtain

$$E\,[f] = 2\int du\, d\Omega f\,(u) \sum_{n,\, m,\, k} \partial_u \bar{U}_n^+ \partial_u U_m^+ \beta_{nk} \bar{\beta}_{mk}. \tag{5.28}$$

Recall that

$$\beta_{nm} = -\langle \bar{v}_m, u_n\rangle = -\langle\!\langle \bar{V}_m^+, U_n^+ \rangle\!\rangle_{\mathscr{I}^+} = -\langle\!\langle \bar{V}_m^-, U_n^- \rangle\!\rangle_{\mathscr{I}^-}, \tag{5.29}$$

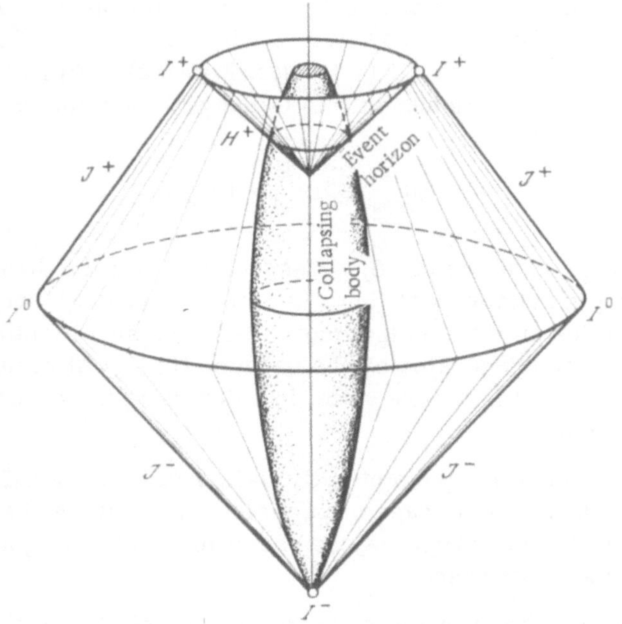

Fig. 5.5

so that finally the problem of calculating the energy flux of particles created by the gravitational field reduces to calculating the coefficients β_{nm}. By (5.29) these coefficients are defined by the expansion of the functions $U_{\underline{n}}^-$, whose images on $\mathcal{J}^+$ are positive frequency functions, in terms of the negative frequency functions $\overline{V}_m^-$ on $\mathcal{J}^-$. Thus, in order to solve the quantum problem of particle creation it is sufficient to study the behavior of complex c-number solutions. This fact is valid for the calculation of any observables (not just the energy) and is, as well known, a consequence of the linearity of the Heisenberg equation (5.3).

The most interesting results of the method discussed above are obtained by applying it to the calculation of the energy flux of particles created in the gravitational field of a black hole. Hawking [15, 195] discovered that the effect of spontaneous creation of particles from a vacuum in the field of a black hole leads to a stationary radiation having a Planck spectrum with characteristic temperature $\Theta = \hbar c^3/8\pi GM$, where M is the mass of a Schwarzschild black hole.* In this section we limit ourselves to a few isolated remarks touching on some of the peculiarities of the problem which are due to the presence of a black hole.

First of all, when a black hole is present (Fig. 5.5), spacetime does not strictly speaking satisfy the definition of asymptotically flat space given in Section 5.3 since there exist light rays (entering the black hole) which do not go off to infinity. On the other hand, the structure of conformal infinity does not change essentially in the presence of a black hole. A modification of the definition [10] makes it possible to consider also spaces containing black holes while retaining the advantages of the notion of conformal infinity and the asymptotic Euclidean property; the modification consists in allowing the existence of other boundary points of space-time

* The survey [196], in which references to the corresponding papers can be found, is devoted to black holes, the Hawking effect, and the astrophysical consequences of the phenomenon of quantum evaporation of black holes. Quantum processes in black holes with variable parameters were considered by Volovich, Zagrebnov, and Frolov [197] using a formalism analogous to the one discussed in this section.

along with the conformal infinity $\mathscr{I}$ constructed as described above. Another important fact is that in the presence of a black hole, $\mathscr{I}^+$ does not by itself form a Cauchy surface. Such a Cauchy surface is given by the union $\mathscr{I}^+ \cup H^+$ (cf. Fig. 5.5), where H^+ is the event horizon (the boundary of the black hole). Hence on this Cauchy surface the vacuum state is represented in the form of a tensor product

$$|0\rangle_{\text{out}} = |0\rangle_{\mathscr{I}^+} \otimes |0\rangle_{H^+}.$$

It turns out that the concrete choice of $|0\rangle_{H^+}$ (associated with the difficulty in defining a particle in a region of strong gravitatational field) is unimportant if we are only interested in the characteristics of particles on $\mathscr{I}^+$. In this case there is natural factorization of the full state space $\mathscr{H}_{\text{out}}$ modulo the space generated by the vacuum $|0\rangle_{H^+}$. Although a finite state is described by a vector in $\mathscr{H}_{\text{out}}$ and is a pure state, in calculating the observables on $\mathscr{I}^+$ it is necessary to use the density matrix.

The quantization method in asymptotically flat spaces discussed in this section can be generalized straightforwardly to the case of particles with spin and used in the most diverse problems; in particular, this approach may be useful in constructing an S-matrix theory for massless fields in ordinary flat space.

In conclusion I would like to express my great appreciation to I. A. Egorova and B. S. Kupinskii for their aid in preparing the manuscript.

LITERATURE CITED

1. E. Newman and R. Penrose, J. Math. Phys., 3:566 (1962).
2. A. Z. Petrov, Uchen. Zap. Kazan. Univ., Vol. 114, No. 8 (1954).
3. A. Z. Petrov, New Methods in the General Theory of Relativity [in Russian], Nauka, Moscow (1966).
4. F. A. E. Pirani, Phys. Rev., 105:1089 (1957).
5. F. A. E. Pirani, Lectures on General Relativity, in: Brandeis Summer Institute in Theoretical Physics, Vol. 1 (1964), p. 249.
6. P. Jordan, J. Ehlers, and R. Sachs, Akad. Wiss. Mainz Abh. Math.-Naturwiss. Kl., No. 1, 2 (1961).
7. R. K. Sachs, Proc. Roy. Soc., A264:309 (1961).
8. R. Sachs, in: Gravitation and Topology, Benjamin, New York (1964).
9. R. Penrose, Phys. Rev. Lett., 10:66 (1963).
10. R. Penrose, The Structure of Space-time [Russian translation], Mir, Moscow (1972).
11. R. Penrose, in: Gravitation and Topology, Benjamin, New York (1964).
12. R. Penrose, Ann. Phys. 10:171 (1960).
13. H. Bondi, M. G. J. van der Burg, and A. W. K. Metzner, Proc. Roy. Soc., A269:21 (1962).
14. R. Sachs, Phys. Rev., 128:2851 (1962).
15. S. W. Hawking, Commun. Math. Phys., 43:199 (1975).
16. R. W. Lind, J. Messmer, and E. T. Newman, J. Math. Phys., 13:1884 (1972).
17. R. K. Sachs, J. Math. Phys., 3:908 (1962).
18. G. Dautcourt, J. Math. Phys., 8:1492 (1967).
19. P. N. Demmie and A. I. Janis, J. Math. Phys., 14:793 (1973).
20. D. W. Pajerski and E. T. Newman, J. Math. Phys., 12:1929 (1971).
21. S. A. Teukolsky, Phys. Rev. Lett., 29:1114 (1972).
22. S. A. Teukolsky, Astrophys. J., 185:635 (1973).
23. W. H. Press and S. A. Teukolsky, Astrophys. J., 185:649 (1973).

24. W. Kinnersley, Recent Progress in Exact Solutions, in: Summaries of Talks of the Seventh International Conference on Gravitation and General Relativity, Tel Aviv (1974).
25. E. T. Newman and A. I. Janis, "Research in general relativity," Pittsburgh Univ. Preprint ARL 70-0136 (1970).
26. V. D. Zakharov, Gravitational Waves and Einstein's Theory of Gravitation [in Russian], Nauka, Moscow (1972).
27. R. Courant, Partial Differential Equations [Russian translation], Mir, Moscow (1964).
28. J. Hadamard, Lectures on Cauchy's Problem in Linear Partial Differential Equations, Dover, New York (1952).
29. L. D. Landau and E. M. Lifshits, The Theory of Fields [in Russian], Nauka, Moscow (1973).
30. M. Born and E. Wolf, Principles of Optics, fourth Edition, Pergamon, New York (1970).
31. M. Kline and I. W. Kay, Electromagnetic Theory and Geometrical Optics, New York (1965).
32. C. W. Misner, K. S. Thorne, and J. A. Wheeler, Gravitation, Freeman, San Francisco (1973).
33. R. A. Isaacson, Phys. Rev., 166:1263 (1968).
34. R. A. Isaacson, Phys. Rev., 166:1272 (1968).
35. G. C. Debney, R. P. Kerr, and A. Schild, J. Math. Phys., 10:1842 (1969).
36. G. C. Debney and J. D. Zund, Tensor, 25:47 (1972).
37. A. Trautman, Bull. Acad. Polon. Sci., Cl. 3, 5:273 (1957).
38. A. Lichnérowicz, Théories Relativistes de la Gravitation et de l'Électromagnetisme, Masson and Cie., Paris (1955).
39. V. A. Fock, Theory of Space, Time, and Gravitation, 2nd edition, Macmillan, New York (1964).
40. R. Zulanke and P. Vintgen, Differential Geometry and Vector Bundles [Russian translation], Mir, Mowcow (1975), p. 127.
41. P. K. Rashevskii, The Geometric Theory of Partial Differential Equations [in Russian], Gostekhizdat, Leningrad (1947), p. 91.
42. J. A. Schouten, Tensor Analysis for Physicists [Russian translation], Nauka, Moscow (1965).
43. R. P. Kerr and A. Schild, Atti del Convegno sulla Relativita Generale, G. Barbera (editor), Firenze (1965), p. 173.
44. A. Held, E. T. Newman, and R. Posadas, J. Math. Phys., 11:3145 (1970).
45. J. N. Goldberg, A. J. Macfarlane, E. T. Newman, F. Rohrlich, and E. C. G. Sudarshan, J. Math. Phys., 8:2155 (1967).
46. R. W. Lind, J. Messmer, and E. T. Newman, J. Math. Phys., 13:1879 (1972).
47. E. T. Newman and R. Penrose, J. Math. Phys., 7:863 (1966).
48. B. Aronson, R. Lind, J. Messmer, and E. Newman, J. Math. Phys., 12:2462 (1971).
49. B. Aronson and E. T. Newman, J. Math. Phys., 13:1847 (1972).
50. R. W. Lind and E. T. Newman, J. Math. Phys., 15:1103 (1974).
51. P. C. Vaidya, Tensor, 24:315 (1972).
52. P. C. Vaidya, Tensor, 27:276 (1973).
53. P. C. Vaidya and L. K. Patel, Phys. Rev., D7:3590 (1973).
54. P. C. Vaidya, Proc. Cambridge Philos. Soc., 75:383 (1974).
55. E. T. Newman, J. Math. Phys., 15:44 (1974).
56. H. S. Ruse, Proc. London Math. Soc., 41:302 (1936).
57. J. L. Synge, Univ. Toronto Stud. Appl. Math. Sec., 1:1 (1935).
58. G. C. Debney and J. D. Zund, Tensor, 22:333 (1971).
59. J. D. Zund and E. Brown, Tensor, 22:180 (1971).
60. G. C. Debney and J. D. Zund, Tensor, 25:53 (1972).
61. J. D. Zund, Tensor, 27:355 (1973).

62. J. D. Zund, Tensor, 28:283 (1974).
63. M. A. Naimark, Linear Representations of the Lorentz Group [in Russian], Fizmatgiz, Moscow (1958).
64. A. V. Pogorelov, Lectures on Differential Geometry [in Russian], Khar'kov State Univ., Khar'kov (1967), p. 135.
65. E. Newman and T. Unti, J. Math. Phys., 4:1467 (1963).
66. R. Penrose, J. Math. Phys., 8:345 (1967).
67. R. Penrose and M. A. H. MacCallum, Phys. Rep., 6C, No. 4 (1973).
68. W. E. Couch and E. T. Newman, J. Math. Phys., 13:929 (1972).
69. L. Mariot, C. R. Acad. Sci. Paris, 238:2055 (1954).
70. I. Robinson, J. Math. Phys., 2:290 (1961).
71. J. N. Goldberg and R. P. Kerr, J. Math. Phys., 5:172 (1964).
72. H. Urbantke, Acta Phys. Aust., 41:1 (1975).
73. H. Flanders, Differential Forms and Applications, Academic Press, New York (1963).
74. G. Debney, J. Math. Phys., 15:992 (1974).
75. E. Cartan, Bull. Soc. Math. Fr., 41:53 (1913).
76. E. Cartan, Theory of Spinors, MIT Press, Cambridge, Massachusetts (1966).
77. I. M. Gel'fand, R. A. Minlos, and Z. Ya. Shapiro, Representations of the Rotation and Lorentz Groups [in Russian], Fizmatgiz, Moscow (1962).
78. L. Infeld and B. L. van der Waerden, Sitz. Preuss. Akad. Wiss. Phys.-Math. Kl., 380 (1933).
79. V. Fock and D. Ivanenko, Z. Phys., 59:718 (1929).
80. W. L. Bade and H. Jehle, Rev. Mod. Phys., 25:714 (1953).
81. D. Brill and J. Wheeler, Rev. Mod. Phys., 29:465 (1957).
82. J. Weber, General Relativity and Gravitational Waves, Interscience, New York (1961).
83. S. Sternberg, Lectures on Differential Geometry, Prentice-Hall, New Jersey (1965).
84. M. M. Postnikov, The Variational Theory of Geodesics [in Russian], Nauka, Moscow (1965).
85. R. Debever, C. R. Acad. Sci. Paris, 249:1324 (1959).
86. A. Schild, "Lectures on general relativity," in: Lectures in Applied Mathematics., Vol. 8, J. Ehlers, ed. (1967), p. 1.
87. R. J. Adler and C. Sheffield, J. Math. Phys., 14:465 (1973).
88. J. Plebanski, Acta Phys. Pol., 26:963 (1964).
89. R. Penrose, in: Gravitation: Problems and Perspectives [Russian translation], Naukova Dumka, Kiev (1972), p. 203.
90. D. Kramer, G. Neugebauer, and H. Stephani, Fortschr. Phys., 20:1 (1972).
91. W. Kinnersley, J. Math. Phys., 10:1195 (1969).
92. D. G. Bobrow, Symbol Manipulation Languages and Techniques, North-Holland, Amsterdam (1968).
93. M. R. Rosenthal, Numerical Methods in Computer Programming, Homewood (Ill.) (1966).
94. A. D. Payne, Comput. Phys. Commun., 4:100 (1972).
95. R. Geroch, A. Held, and R. Penrose, J. Math. Phys., 14:874 (1973).
96. A. Held, Commun. Math. Phys., 37:311 (1974).
97. J. N. Goldberg and R. Sachs, Acta Phys. Pol., Suppl., 22:13 (1962).
98. I. Robinson and A. Schild, J. Math. Phys., 4:484 (1962).
99. W. Kundt and A. H. Thompson, C. R. Acad. Sci. Paris, 254:4257 (1962).
100. I. Robinson and A. Trautman, Proc. Theory of Gravitation, PWN-Polish Sci. Publ., Warsaw (1964), p. 107.
101. I. Robinson, J. R. Robinson, and J. D. Zund, J. Math. Mech., 18:881 (1969).
102. W. Kundt, Z. Phys., 163:77 (1961).
103. N. Rosen, Phys. Z. Sowjetunion, 12:366 (1937).
104. H. Bondi, F. Pirani, and I. Robinson, Proc. Roy. Soc., A251:519 (1959).

105. J. Ehlers and W. Kundt, "Exact solutions of the gravitational field equations," in: An Introduction to Current Research, L. Witten, ed. (1962).
106. R. Penrose, Rev. Mod. Phys., 37:215 (1965).
107. J. Klekowska and M. E. Osinovsky, Lett. Nuovo Cimento, 7:633 (1973).
108. B. Carter, Phys. Lett., 26A:399 (1968).
109. P. Jordan, J. Ehlers, and W. Kundt, Akad. Wiss. Mainz, Abh. Math.-Naturwiss. Kl., No. 2 (1960).
110. W. van Stockum, Proc. Roy. Soc. Edinburgh, A57:135 (1937).
111. R. Tiwari and M. Misra, Proc. Nat. Inst. Sci. India, A28:771 (1962).
112. I. Robinson and A. Trautman, Proc. Roy. Soc., A265:463 (1962).
113. E. T. Newman and L. Tamburino, J. Math. Phys., 3:902 (1962).
114. J. Foster and E. T. Newman, J. Math. Phys., 8:189 (1967).
115. C. Brans, J. Math. Phys., 12:1616 (1971).
116. M. Cahen and J. Spelkens, Bull. Acad. R. Belge, 53:817 (1967).
117. W. Kinnersley and M. Walker, Phys. Rev., D2:1359 (1970).
118. I. M. Dozmorov, Izv. Vyssh. Uchebn. Zaved. Fiz., No. 11, 68 (1971); VNIIOFI Preprint 70-26, Moscow (1970).
119. L. Derry, R. Isaacson, and J. Winicour, Phys. Rev., 185:1647 (1969).
120. E. T. Newman and R. Posadas, Phys. Rev. Lett., 22:1196 (1969).
121. E. T. Newman and R. Posadas, Phys. Rev., 187:1784 (1969).
122. I. Hauser, Phys. Rev. Lett., 33:1112 (1974).
123. D. E. Novoseller, Phys. Rev., D11:2330 (1975).
124. I. Robinson and J. Robinson, Int. J. Theor. Phys., 2:231 (1969).
125. I. Robinson, GRG, 6:423 (1975).
126. A. Held, Lett. Nuovo Cimento, 11:545 (1974).
127. R. Debever, Bull. Soc. Math. Belg. 23:360 (1971).
128. J. F. Plebanski and M. Demianski, Ann. Phys., 98:98 (1976); Preprint OAP-401, Pasadena (1975).
129. J. F. Plebanski, Ann. Phys., 90:196 (1975).
130. B. Carter, Commun. Math. Phys., 10:280 (1968).
131. M. Demianski and E. Newmann, Bull Acad. Pol. Sci., Ser. Math. Phys., 14:653 (1966).
132. V. P. Frolov, Teor. Mat. Fiz., 21:213 (1974).
133. M. Demianski, Acta Astron., 23:197 (1973).
134. M. Demianski, Phys. Lett., 42A:157 (1972).
135. T. Levi-Civita, Atti Accel. Lincei, Rend., 27:343 (1918).
136. E. T. Newman and L. Tamburino, J. Math. Phys., 2:667 (1961).
137. E. T. Newman, E. Couch, K. Chinnapared, A. Exton, A. Prakash, and R. Torrence, J. Math. Phys., 6:918 (1965).
138. Z. Perjes, Commun. Math. Phys., 12:275 (1969).
139. F. J. Ernst, Phys. Rev., 168:1415 (1968).
140. D. Kramer and G. Neugebauer, Commun. Math. Phys., 10:132 (1968).
141. D. Brill, Phys. Rev., 133:845 (1964).
142. F. Kottler, Ann. Phys., 4:56 (1918).
143. R. Kerr, Phys. Rev. Lett., 11:237 (1963).
144. E. T. Newman, L. Tamburino, and T. Unti, J. Math. Phys., 4:915 (1963).
145. A. Taub, Ann. Math., 53:472 (1951).
146. B. Bertotti, Phys. Rev., 116:1331 (1959).
147. I. Robinson, Bull. Acad. Pol. Sci., Ser. Math. Phys., 4:915 (1963).
148. W. de Sitter, Mon. Not. Roy. Astron. Soc., 78:3 (1917).
149. K. Schwarzschild, Sitz. Akad. Wiss. Berlin, 7:189 (1916).
150. H. Reissner, Ann. Phys., 50:106 (1916).
151. L. Norström, Proc. Amsterdam Acad., 20:1238 (1918).

152. L. Hughston and P. Sommers, Commun. Math. Phys., 32:147 (1973).
153. C. J. Talbot, Commun. Math. Phys., 13:45 (1969).
154. R. Kerr and G. Debney, J. Math. Phys., 11:2807 (1970).
155. J. Zund, Lett. Nuovo Cimento, 4:879 (1972).
156. J. Zund, Lett. Nuovo Cimento, 7:233 (1973).
157. R. W. Lind, GRG, 5:25 (1974).
158. D. Trim and J. Wainwright, J. Math. Phys., 12:2494 (1971).
159. D. W. Trim and J. Wainwright, J. Math. Phys., 15:535 (1974).
160. T. W. J. Unti and R. J. Torrence, J. Math. Phys., 7:535 (1966).
161. R. W. Lind, J. Math. Phys., 16:34 (1975).
162. R. W. Lind, J. Math. Phys., 16:39 (1975).
163. V. I. Khlebnikov, Izv. Vyssh. Uchebn. Zaved., Fiz., No. 3, 117 (1976).
164. V. I. Khlebnikov, Izv. Vyssh. Uchebn. Zaved., Fiz., No. 3, 122 (1976).
165. H. Urbantke, Acta Phys. Aust., 41:1 (1975).
166. G. C. Debney, Lett. Nuovo Cimento, 8:337 (1973).
167. P. C. Vaidya, Curr. Sci., 12:183 (1943).
168. P. C. Vaidya, Proc. Indian Acad. Sci., A33:264 (1951).
169. P. C. Vaidya, Nature, 171:260 (1953).
170. W. Kinnersley, Phys. Rev., 186:1335 (1969).
171. V. P. Frolov and V. I. Hlebnikov, Preprint FIAN, No. 27 (1975).
172. V. P. Frolov, Teor. Mat. Fiz., 27:337 (1976).
173. D. Kramer, in: Abstracts of Reports of the Third Soviet Conference on Gravitation [in Russian], Erevan (1972), p. 321.
174. L. P. Hughston, Int. J. Theor. Phys., 4:267 (1971).
175. J. K. Rao, J. Phys., A4:17 (1971).
176. R. K. Sachs, Proc. Roy. Soc., A270:103 (1962).
177. E. T. Newman and T. Unti, J. Math. Phys., 3:891 (1962).
178. R. Geroch, J. Math. Phys., 9:450 (1968).
179. S. W. Hawking and G. F. R. Ellis, The Large-Scale Structure of Space-time, Cambridge University Press (1973).
180. H. J. Seifert, GRG, 1:247 (1971).
181. B. G. Schmidt, GRG, 1:269 (1971).
182. R. Geroch, E. H. Kronheimer, and R. Penrose, Proc. Roy. Soc., A327:545 (1972).
183. B. G. Schmidt, Commun. Math. Phys., 29:49 (1973).
184. R. Geroch and E. T. Newman, J. Math. Phys., 12:314 (1971).
185. P. J. McCarthy, J. Math. Phys., 13:1837 (1972).
186. P. J. McCarthy, Proc. Roy. Soc., A330:517 (1972).
187. P. J. McCarthy, Proc. Roy. Soc., A333:317 (1973).
188. P. J. McCarthy and M. Crampin, Proc. Roy. Soc., A335:301 (1973).
189. M. Crampin and P. J. McCarthy, Phys. Rev. Lett., 33:547 (1974).
190. V. Cantoni, J. Math. Phys., 8:1700 (1967).
191. R. Penrose, Proc. Roy. Soc., A284:159 (1965).
192. M. Walker, "Asymptotically flat space-times," Preprint Max Planck Inst. Phys. Astrophys. Munich (1972).
193. A. Komar, Phys. Rev., 134B:1430 (1964).
194. D. Klarfeld and A. Komar, Phys. Rev., D4:987 (1971).
195. S. W. Hawking, Nature, 248:30 (1974).
196. V. P. Frolov, Usp. Fiz. Nauk, 118:473 (1976).
197. V. P. Frolov, I. V. Volovich, and V. A. Zagrebnov, Preprint FIAN, No. 60 (1976); Teor. Mat. Fiz., 29:191 (1976).
198. M. Murenbeeld and J. R. Trollope, Phys. Rev., D1:3220 (1970).

199. V. A. Ruban, Dokl. Akad. Nauk SSSR, 204:1086 (1972).
200. V. A. Ruban, in: Gravitation and the Theory of Relativity, No. 9 [in Russian], Izd. Kazan. Univ. (1973), p. 38.
201. L. A. Tamburino and J. H. Winicour, Phys. Rev., 150:1039 (1966).
202. E. Köhler and M. Walker, GRG, 6:507 (1975).
203. S. Schmidt, M. Walker, and P. Sommers, GRG, 6:489 (1975).